DESIGN OF BUILDINGS AND BRIDGES FOR WIND

DESIGN OF BUILDINGS AND BRIDGES FOR WIND

A Practical Guide for ASCE-7 Standard Users and Designers of Special Structures

EMIL SIMIU, PH.D., P.E.
TOSHIO MIYATA, PH.D.

WILEY

JOHN WILEY & SONS, INC.

This book is printed on acid-free paper. ♾

Published by John Wiley & Sons, Inc., Hoboken, New Jersey
Published simultaneously in Canada

Library of Congress Cataloging-in-Publication Data:

Simiu, Emil
Design of buildings and bridges for wind: a practical guide for ASCE-7 standard users and designers of special structures/Emil Simiu, Toshio Miyata.
p. cm.
Includes bibliographical references and index.
ISBN-13: 978-0471-65743-9
ISBN-10: 0-471-65743-3 (cloth)
1. Wind-pressure. 2. Buildings—Aerodynamics. 3. Bridges—Aerodynamics. 4. Wind resistant design. I. Miyata, Toshio, 1938– II. Title.
TA654.5.S54 2006
624.1′75—dc22

2005021300

Printed in the United States of America

10 9 8 7 6 5 4 3 2

CONTENTS

PREFACE

Wind loads considered in the design of structures and their envelopes must be consistent with the requirements that structural integrity and serviceability are assured during the structures' expected life. For most common types of structures, standard provisions on wind loads are in principle adequate for design purposes. The ASCE 7 Standard, among other standards, has incorporated a great deal of wind engineering knowledge accumulated within the last half-century. However, this has resulted in provisions that some practitioners view as overly complicated and unwieldy. In addition, owing to the Standard's present structure, even the simplified procedure incorporated in the ASCE 7-02 and 7-05 versions at the urgent request of numerous professionals cannot be used without resorting to provisions belonging to the Standard's original, more complicated "analytical" procedure.

One of the main objectives of this book is to help the reader better understand the ASCE 7 Standard provisions and apply them with confidence. Following an introductory chapter on wind engineering fundamentals, the book includes a guide to the Standard that "takes the user by the hand" through most of the ASCE 7 provisions, provides explanatory material, and illustrates their use with numerous numerical examples.

Wind loads and effects on special structures cannot in general be estimated simply by using standard provisions. Rather, they require wind tunnel model tests and information on the extreme wind speeds at the site. In addition, if the structure is flexible (as is the case for tall buildings), dynamic analyses are required as well. Finally, the wind tunnel data, the extreme wind speed data, and if it is significant, the dynamic response are functions of wind direction. Estimates of wind effects with any specified mean recurrence interval must therefore incorporate this dependence in a conceptually and statistically correct manner. These requirements are attended to with varying degrees of success by wind engineering consultants.

Some design offices have a policy of requiring wind effects estimates from more than one consultant. This is prudent and is justified by the fact that estimates can differ from each other significantly. For example, recent estimates of the New York World Trade Center towers response to wind performed by two well-known North American consultants differ from each other by about 40%. It is therefore clear that consultant reports need to be scrutinized carefully. This book enables structural engineers to do so in an effective manner. In addition, structural engineers are enabled by this book to calculate wind effects on their own, just as structural engineers can calculate seismic effects on their own. To calculate wind effects the structural engineer must make use of aerodynamic data supplied by wind tunnel operators in electronic form and of extreme wind speed data obtained from wind climatological consultants or available from public sources, many of which are indicated in the book. The wind effects of interest include internal forces and quantities associated with interaction equations for specified mean recurrence intervals (e.g., 50 years or 500 years), as well as accelerations and deflections needed to check serviceability requirements.

Wind effects on individual members are functions of influence coefficients that differ from member to member. In the past, the lack of sufficiently powerful computer resources has not allowed this dependence to be taken into account in a rigorous manner. The capability to do so is now available routinely. It has created a bridge between the wind engineer and the structural engineer that makes it possible to combine the wind and structural phases of structural design for wind in a far more accurate manner than was possible heretofore. Software implementation of this approach is included on the website http://www.nist.gov/wind.

Since this book is addressed to practicing engineers, it contains essentially no mathematical developments, except for two applications. The first application is tall building dynamics, which is governed by Newton's second law and therefore involves simple differential equations. Until about two decades ago the solution of large systems of differential equations posed insurmountable computational problems, and tall building dynamics calculations were performed by using the spectral (frequency-domain) approach, which transforms differential equations into algebraic equations. This approach is not always transparent and intuitive, and suppresses phase information that is useful for adding correctly wind effects from various sources (e.g., from two perpendicular lateral motions). The software referenced in the book uses the time-domain approach, which is clear and effective. A second application in which mathematical developments are included is the flutter and buffeting of long-span bridges, for which a routine, effective time-domain approach has not yet been developed.

The book provides information on two relatively new areas that despite their importance have been relatively neglected in wind engineering. A first area is structural reliability, which is useful for the development of wind load factors. The role of errors and uncertainties in the development of the importance factor or of wind load factors is not accounted for adequately in the ASCE 7 Standard. For example, the same wind load factor is specified in the Standard for ordinary buildings and for tall buildings whose response depends on damping. A reliability-based approach that takes into account the considerable uncertainties associated with damping could yield, for at least some tall buildings, larger load factors than the load factor specified in the ASCE Standard. Estimates of wind effects that fail to account for the dependence of the load factor upon uncertainties typical of tall buildings are therefore incomplete, and may be unsafe.

A second area in which substantial progress is currently being made is the estimation of ultimate capacities under fluctuating wind loads. Efforts in this direction will make it possible to achieve more rational designs than current linear methods of analysis allow, resulting in safer and more sustainable structures.

A main chapter of the book is devoted to wind effects on bridges. The outlook of the chapter is broad, and its sources are ubiquitous, but since Japan is one of the countries with the richest experience in the areas of long-span bridge research, development, design, and construction, much of the chapter is informed by that experience. The material on suspension bridges covers fundamental aerodynamic and aeroelastic considerations; analysis; design; wind tunnel testing; vibration mitigation for decks, towers, and cables; aerodynamic interactions, and specification issues. To our knowledge this is the most extensive material on the aeroelasticity of long-span bridges available to date.

We wish to express our warm appreciation to Dr. W. P. Fritz and Dr. J. A. Main of the National Institute of Standards and Technology for development of material used in Chapters 2 and 4 and of software on the dynamic analysis of tall buildings and rigid buildings, respectively; Professor S. Diniz of the University of Minas Gerais for developing material on structural reliability used in Chapter 2; Dr. A. Jain of Hart-Weidlinger, Santa Monica, for permission to reproduce material from his doctoral thesis; Dr. A. Grazini of the University of Perugia for creating the fluctuating pressures animation referenced in the book; Professor L. Caracoglia of Northeastern University, Dr. Dat Duthinh of the National Institute of Standards and Technology, Professor M. Grigoriu of Cornell University, Professor K. Gurley of the University of Florida, Dr. M. Iancovici of the Technical University of Civil Engineering, Bucharest, Professor T. Kijeswski-Correa, University of Notre Dame, Professor H. Yamada and Dr. H. Katsuchi of Yokohama National University, and Professor Emeritus A. Okajima of Kanazawa University, for useful comments and criticism; the Honshu–Shikoku Bridge Authority for kind permission to use valuable data and materials; Ms. K. Beall of the National Institute of Standards and Technology, Dr. F. Păulet-Crăiniceanu of the Technical University of Iaşi, Dr. E. Dragomirescu of Yokohama National University, and

Mr. T. Yamazaki of Chodai Co., Ltd. for their capable advice and assistance on the presentation of graphics. We also wish to thank J. Harper, Editor, and M. Torres, Senior Production Editor, for their unfailing courtesy and capable help.

This book is not a U.S. government publication, and the views expressed do not necessarily represent those of the U.S. government or any of its agencies.

We have been motivated by the hope that the book will serve the profession in its efforts to design structures for wind better and more effectively. We will welcome with appreciation our readers' comments and criticisms.

We dedicate this book to the memory of Dr. Richard D. Marshall, Professor Robert H. Scanlan, and Professor Hatsuo Ishizaki, in affectionate tribute to their integrity and lasting contributions.

EMIL SIMIU
Rockville, Maryland
TOSHIO MIYATA
Yokohama, Japan
June 2005

CHAPTER 1

INTRODUCTION

The purpose of this book is to provide structural engineers with the practical knowledge and tools required for the proficient design of buildings and bridges for wind loads. The book is concerned with both ordinary and special structures.

Ordinary structures can and should be designed by using standard provisions for wind loads. However, owing in part to the ASCE 7 Standard's development by successive and more or less disorderly accretions, the wind loading provisions of the Standard have become increasingly difficult to apply. One main objective of this book is to provide clear and detailed guidance to the use of the Standard.

Special structures require the direct use of extreme wind speed, wind tunnel, and structural data. Design information based on such data is commonly obtained from wind engineering consultant reports. However, because such reports do not—or do not yet—have to conform to uniform standards of practice, response estimates for the same building can differ by as much as 50% or even more, depending on the wind engineering laboratory providing them (see, e.g., [2.1, 2.26][1]). Figure 1.1.1 shows a façade of the New Orleans Hyatt Regency hotel after hurricane Katrina's passage (2005). The wind speeds that caused the damage do not appear to have exceeded the basic design wind speed specified in the ASCE 7 Standard. The disastrous performance of the cladding demonstrates the importance of selecting appropriate design procedures. These examples show that structural engineers need to scrutinize and evaluate consultant reports effectively. This book provides, in a clear, user-friendly, and comprehensive manner, the wind engineering

[1] Note that references listed at the end of Chapter 2 are referred to in Chapters 1, 3, and 4 as well as in Chapter 2.

Figure 1.1.1. A facade of the New Orleans Hyatt Regency hotel after the passage of hurricane Katrina (2005). Permission by Mark Gong.

knowledge and tools needed for this purpose by practicing structural engineers. For a vast majority of buildings, including super-tall buildings, the book also enables structural engineers to perform independently detailed estimates of the response to wind for strength and serviceability design purposes, much as structural engineers perform, independently, estimates of building response to earthquakes. For such independent estimates of the response to wind to be possible, it is necessary to make use of extreme wind speed data available in the public domain or provided by wind climatological consultants; aerodynamic data obtained in the wind tunnel on rigid models and supplied to the structural engineer in electronic form; and structural information obtained, on the basis of preliminary designs, by standard structural analyses. Response calculations are based on clear algorithms and can be performed routinely and efficiently by using software provided on the site http://www.nist.gov/wind.

For exceptionally flexible buildings and for long-span suspension bridges, it is necessary to perform aeroelastic wind tunnel tests. For buildings, aeroelastic tests are not performed routinely. For long-span bridges, aeroelastic tests are indispensable.

Once the response is calculated on the basis of extreme wind climate and aerodynamic considerations, and if the structure being designed is flexible, on the basis of dynamic considerations as well, it is necessary to apply appropriate safety margins (e.g., wind load factors) that make allowance for uncertainties in the parameters that govern the response. For ordinary structures, wind load factors are specified in standards and codes. For special structures, however, estimating the response with superior accuracy is pointless if the wind load factor does not adequately reflect the uncertainties inherent in the situation at hand. For example, the ASCE 7-05 Standard specifies a wind load factor of 1.6, and an additional safety factor, called the *importance factor,* which for structures whose failure would entail significant loss of life is 1.15. However, the specification of the wind load and importance factors does not differentiate between rigid and flexible structures, even though the response of flexible structures depends on damping ratios and natural frequencies whose imperfect knowledge increases the overall uncertainty. Simple, user-friendly structural reliability-based procedures are provided that make it possible to account in a systematic and comprehensive manner for the uncertainties associated with response to wind, thus allowing the designer, code writer, or building official to develop safety margins commensurate with the uncertainties affecting the structure of interest.

Chapter 2 is devoted to basic wind engineering topics relevant to various aspects of structural design: meteorological wind storm descriptions, wind micrometeorology, extreme wind climatology, aerodynamics and wind tunnel testing, structural dynamics, aeroelasticity, and structural reliability. *Wind micrometeorology* issues include, among others, the variation of wind speeds with height for various types of surface, the dependence of wind speeds on averaging time, and atmospheric turbulence features. *Extreme wind climatol-*

ogy material enables designers to select and process extreme wind data needed for determining wind effects with specified mean recurrence intervals. Information is provided to various sources of data, and numerical examples are used to familiarize the engineer with estimating procedures. The presentation of topics in aerodynamics, wind tunnel testing, structural dynamics, and aeroelasticity is intended to help the structural engineer to understand the underlying rationale of standard provisions and to converse knowledgeably with the wind tunnel professional. Relevant basic notions of structural reliability are presented briefly so that the engineer can make allowance for the role of uncertainties and develop appropriate safety margins for special structures subjected to wind.

Chapter 3 is a detailed presentation of and guide to the wind loading provisions of the ASCE 7-05 Standard. A large number of numerical examples are worked out in detail. A critique of the ANSI 7 Standard provisions for wind loads is provided with a view to facilitating their use and contributing to their improvement.

Chapter 4 is concerned with rigid (low- and medium-height) and flexible (tall) buildings. Whereas ordinary rigid buildings should be designed by using standard provisions, reasons of safety and/or economy may dictate that special rigid buildings be designed in accordance with more elaborate wind loading estimates. The database-assisted design (DAD) approach is appropriate for this purpose. The approach is described in detail from an aerodynamics, extreme wind climatology, structural, and reliability point of view.

Flexible building response to wind cannot in general be estimated by using standard provisions. Rather, three main approaches are available for use in practice, all of which are based on wind tunnel testing and therefore require various degrees of interaction between the wind engineer and the structural engineer:

1. More effectively than any other approach, the DAD approach bridges the gap between wind engineering and structural engineering, allowing structural engineers to use the power of the computer to achieve safe, serviceable, and economical designs in a differentiated and realistic manner. The use of the DAD approach renders unnecessary additional tests for measuring local pressures.
2. The HFFB (high-frequency force balance) approach is useful for conducting relatively inexpensive estimates of aerodynamic performance and for foundation design but can yield inadequate estimates of wind effects for the design of structural members.
3. The aeroelastic testing approach is also utilized.

HFFB and aeroelastic testing both require additional testing for determining local pressures. The advantages and shortcomings of the testing approaches are discussed in some detail. The effect on the dynamic response of the distance between elastic and mass centers is also discussed. The important problem of estimating mean recurrence intervals of wind effects as functions of the directional features of the extreme wind climate, the building's aerodynamics, and the dynamic response is discussed with a view to ensuring that its solution is clearly understood. This is essential, since large differences among tall building estimates by various wind engineering consultants are due in large part to lack of clarity and rigor in dealing with the wind directionality problem. In Chapter 4 we also address load factors estimation, the estimation of ultimate structural capacity under fluctuating wind loads, and serviceability issues. Software for calculating the response of both rigid and flexible buildings is included in the site http://www.nist.gov/wind.

Chapter 5 deals comprehensively with estimation of the wind-induced response of bridges to wind, including vortex-induced response, galloping, torsional flutter, classical flutter, the performance of cables, the design of bridge towers, and the mitigation or suppression of wind-induced oscillations. Wind tunnel testing of bridges is also discussed in some detail. Japan is one of the countries that possess a rich experience in the areas of long-span bridge research, development, design, and construction. Specific aspects of that experience are used to complement the material presented in Chapter 5.

CHAPTER 2

WIND ENGINEERING FUNDAMENTALS

2.1 INTRODUCTION

In this chapter we review briefly basic issues relevant to structural design for wind in the following areas: meteorology (Sect. 2.2), wind micrometeorology (Sect. 2.3), extreme wind climatology (Sect. 2.4), aerodynamics and wind tunnel testing (Sect. 2.5), structural dynamics (Sect. 2.6), aeroelasticity (Sect. 2.7), and structural reliability (Sect. 2.8). The fundamental material provided in this chapter is aimed at helping the designer to use, interpret, and assess current standard provisions as well as procedures for determining wind effects in special structures.

2.2 WINDSTORMS

Wind is caused fundamentally by heat radiated by the sun. Radiation, thermodynamic, and mechanical phenomena transform the thermal energy imparted to the atmosphere into mechanical energy associated with air motion, giving rise to various types of wind. In this section we describe windstorms of interest from a structural engineering viewpoint.

2.2.1 Large-Scale Storms

Extratropical storms occur at and above midlatitudes. Because their vortex structure is less well defined than in tropical storms, their winds are loosely called *straight winds.*

Tropical cyclones, known as *typhoons* in the Far East and as *cyclones* in Australia and the Indian Ocean, generally originate between 5° and 20° latitudes. *Hurricanes* are defined as tropical cyclones with sustained surface wind speeds of 74 mph or more. Tropical cyclones are translating vortices with diameters of hundreds of miles and counterclockwise (clockwise) rotation in the northern (southern) hemisphere. Their translation speeds vary from about 3 to 30 mph. As in a stirred coffee cup, the column of fluid is lower at the center than at the edges. The difference between edge and center atmospheric pressures is called *pressure defect.* Rotational speeds increase as the pressure defect increases, and as the radius of maximum wind speeds, which varies from 5 to 60 miles, decreases.

The structure and flow pattern of a typical tropical cyclone is shown in Fig. 2.2.1. The *eye of the storm* (region I) is a roughly circular, relatively dry core of calm or light winds surrounded by the *eye wall.* Region II contains the storm's most powerful winds. Far enough from the eye, winds in region V, which decrease in intensity as the distance from the center increases, are parallel to the surface. Where regions V and II intersect, the wind speed has a strong updraft component that alters the mean wind speed profile and is currently not accounted for in structural engineering practice. The source of energy that drives the storm winds is the warm water at the ocean surface. As the storm makes landfall and continues its path over land, its energy is depleted and its wind speeds gradually decrease.

In the United States, hurricanes are classified in accordance with the Saffir–Simpson scale (Table 2.2.1).

2.2.2 Local Storms

Foehn winds (called *chinook winds* in the Rocky Mountains area) develop downwind of mountain ridges. Cooling of air as it is pushed upward on the windward side of a mountain ridge causes condensation and precipitation. The dry air flowing past the crest warms as it is forced to descend and is highly turbulent. A similar type of wind is the *bora,* which occurs downwind of a plateau separated by a steep slope from a warm plain.

Jet-effect winds occur where features such as gorges cause wind speeds to increase.

Thunderstorms occur as heavy rain drops, formed by condensation of water vapor contained in ascending warm, moist air, drag down the air through which they fall, causing a downdraft that spreads on the Earth's surface (Fig. 2.2.2).

Tornadoes are small vortexlike storms that can contain winds in excess of 250 mph. A U.S. map of tornado wind speeds with probability of exceedance of 10^{-5} per year is included in the Commentary to the ASCE 7-05 Standard [2.2].

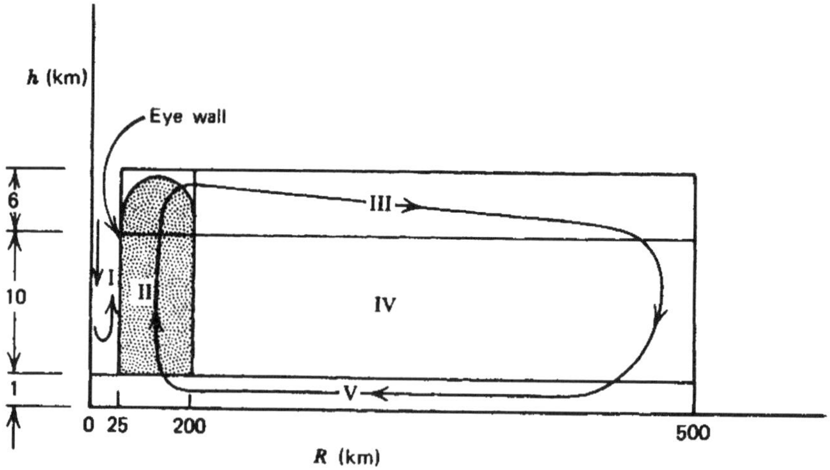

Figure 2.2.1. Structure of a hurricane. (From [2.5].)

2.3 WIND MICROMETEOROLOGY

Micrometeorological features of the wind flow at scales of interest to engineers include:

- The dependence of wind speeds on averaging time (Sect. 2.3.1)
- The variation of wind speed with height (Sect. 2.3.2), which depends on surface roughness and fetch (i.e., distance over which that surface roughness extends upwind of the structure), topography, and storm type and intensity

TABLE 2.2.1. Saffir–Simpson Scale and Corresponding Wind Speeds[a]

Hurricane Category	Damage Potential	1-min Speed at 10 m Above Open Water[c] [mph (m/s)][b]	North Atlantic Examples
1	Minimal	74–95 (33.1–42.5)	Agnes, 1972
2	Moderate	96–110 (42.6–49.2)	Cleo, 1974
3	Extensive	111–130 (49.3–58.1)	Betsy, 1965
4	Extreme	131–155 (58.2–69.3)	David, 1979
5	Catastrophic	>155 (>69.3)	Camille, 1969

Source: Courtesy of H. Saffir, who provided the official speed values for the Saffir–Simpson scale.

[a] For the definition of 1-min wind speeds, see Sect. 2.3.1.1. 1-min wind speeds at 10 m above open terrain can be shown to be lower by approximately 15% than the 1-min speeds at 10 m over open water. For details, see Numerical Example 2.3.8.

[b] Official speeds are in mph; speeds in m/s are approximate equivalents.

[c] See Commentary, ASCE 7-05 Standard, Table C6-1.

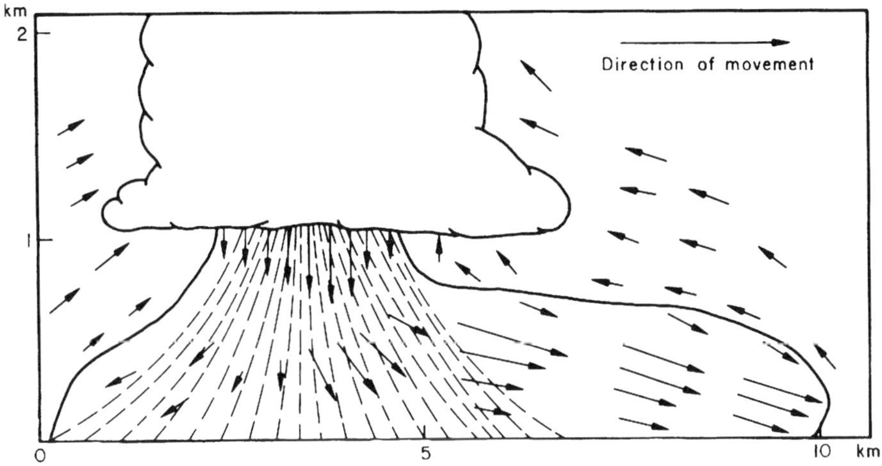

Figure 2.2.2. Section through a thunderstorm. (From [2.5].)

- The atmospheric turbulence (Sect. 2.3.3), which affects the definitions of wind speeds (e.g., 3-s peak gusts versus mean hourly speeds), and can strongly influence the aerodynamic loading

The models of Sects. 2.3.1, 2.3.2, and 2.3.3 apply to strong straight winds but are commonly also used for other types of wind, including hurricanes. Moderate wind speeds have features, including lower turbulence intensities, that differ from those of strong winds in ways affecting the design of structures susceptible to vortex-shedding effects (Sect. 2.3.4).

2.3.1 Wind Speeds and Averaging Times

Wind speeds over horizontal terrain with uniform roughness over a sufficiently long fetch (see Sect. 2.3.2.5) depend on averaging time in a well-established way. We first list definitions of wind speeds as functions of various averaging times (Sect. 2.3.1.1). We then consider the dependence of wind speeds on averaging time for the reference case of speeds at 10 m elevation above open terrain (Sect. 2.3.1.2). The current conventional way in which thunderstorm wind speeds averaged over various time intervals are modeled in practice is noted in Sect. 2.3.1.3. The case of wind speeds above surfaces other than open terrain is considered in Sect. 2.3.2.4.

2.3.1.1 Definitions of Wind Speeds as Functions of Averaging Time. If the flow were laminar, wind speeds would be the same for all averaging times. However, owing to turbulent fluctuations, such as those recorded in Fig. 2.3.1, the definition of wind speeds depends on averaging time.

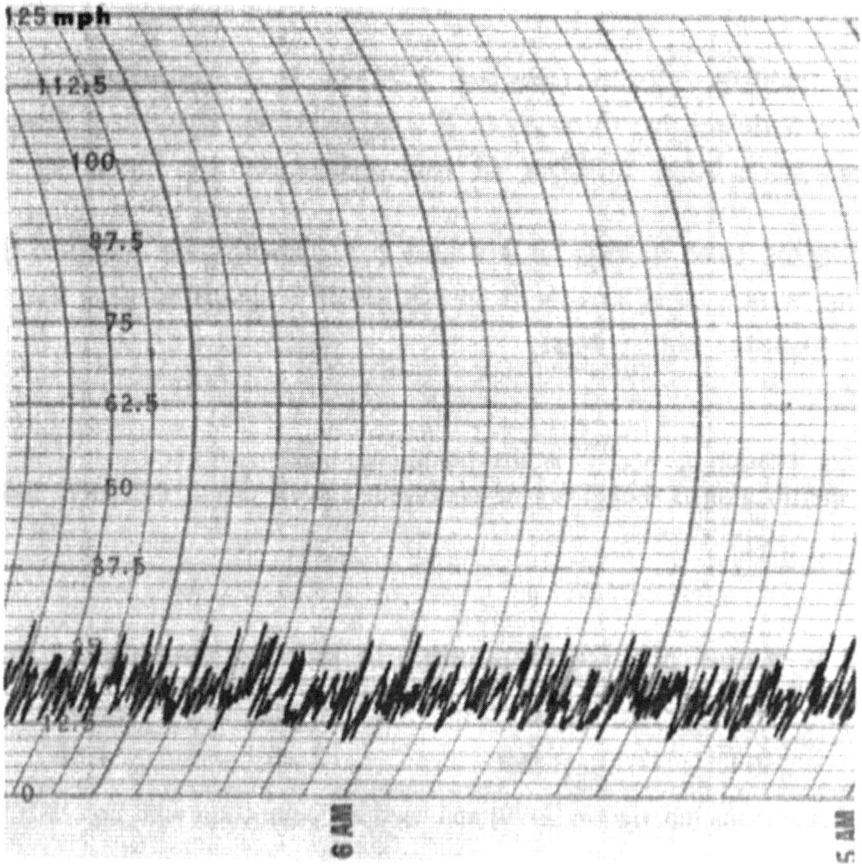

Figure 2.3.1. Extratropical wind speed record. (From [2.5].)

- The *peak 3-s gust speed* is a storm's largest speed averaged over 3 s. In 1995 it was adopted in the ASCE Standard as a measure of wind speeds. Similarly, the peak 5-s gust speed is the largest speed averaged over 5 s. The 5-s speed is reported by the National Weather Service ASOS (Automated Surface Observing System) and is about 2% less than the 3-s speed. The 28-mph peak of Fig. 2.3.1 is a 3-s speed.
- The *hourly wind speed* is the speed averaged over 1 h. It is commonly used as a reference wind speed in wind tunnel simulations: hence the need to convert 3-s speeds recorded at weather stations to 1-h wind speeds. The hourly wind speed is used to define extreme speeds by the National Building Code of Canada [2.3]. In Fig. 2.3.1 the statistical features of the record do not vary significantly (i.e., the record may be viewed as *statistically stationary*) over an interval of several hours; the hourly wind speed is about 18.5 mph, about 1/1.52 times the peak 3-s gust. The hurricane winds of Fig. 2.3.2 can be viewed as statistically

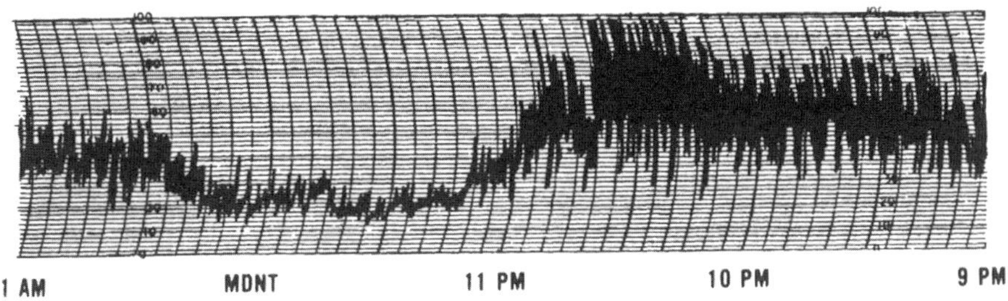

Figure 2.3.2. Hurricane wind speed record (mph). (From [2.5].)

stationary for about 20 min (10:20 to 10:40 h), and its largest *20-min speed* (i.e., wind speed averaged over 20 min) is about 64 mph. The thunderstorm wind record (taken at a 45-m elevation) of Fig. 2.3.3 can be viewed as statistically stationary for about 4 min (around 11:20 h).

- *Sustained wind speeds,* defined as wind speeds averaged over intervals of the order of 1 min, are used in both engineering and meteorological practice. The *fastest 1-min wind speed* is the storm's largest 1-min average wind speed. Before 1995 the ASCE 7 Standard used the *fastest-mile wind speed,* defined as a storm's largest wind speed averaged over

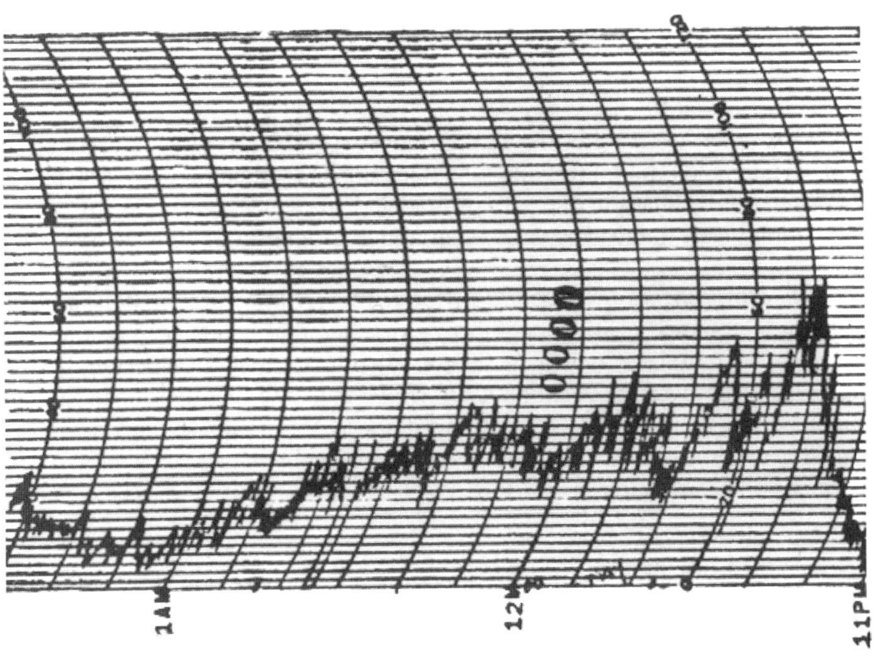

Figure 2.3.3. Thunderstorm wind speed record (mph). (Courtesy of the National Severe Storms Laboratory, National Oceanic and Atmospheric Administration, Washington, DC.)

the time required for the passage at a point of a volume of air with a horizontal length of 1 mile. (This definition is based on a wind speed recording device used until the 1990s.) For fastest-mile speeds, V_f, in mph, the averaging time in seconds is $t = 3600/V_f$.

- *10-min wind speeds* are wind speeds averaged over 10 min and are used for design purposes by the Eurocode [2.4], in conformity with World Meteorological Organization (WMO) practice.

2.3.1.2 Relation Between Wind Speeds with Different Averaging Times for Open Terrain. The approximate *mean ratio* r of the t-s speed to the hourly (3600-s) speed at 10 m above ground in open terrain is listed for selected values of t as follows:

t (s)	3	5	40	60	600	3600
r	1.52	1.49	1.29	1.25	1.1	1.0

(see [2.5, p. 70] or Fig. C6-2, ASCE 7-05 Commentary [2.2]). For winds over other than open terrain, see Sect. 2.3.2.4.

Numerical Example 2.3.1. *Conversion of fastest-mile wind speed to mean hourly speed and to peak 3-s gust for open terrain.* For a fastest-mile wind speed at 10 m over open terrain of 90 mph, the averaging time is 3600/90 = 40 s, and the corresponding hourly speed and peak 3-s gust are 90/1.29 = 69.8 mph and 69.8 × 1.52 = 106 mph, respectively.

Numerical Example 2.3.2. *Conversion of peak 3-s gust speed to mean hourly speed for open terrain.* Let the peak 3-s gust speed at 10 m above ground in open terrain be 64 mph. For wind tunnel testing and structural purposes, winds characterized by that gust speed are modeled by winds with a 64 mph/1.52 = 42 mph mean hourly speed at 10 m above ground in open terrain.

2.3.1.3 Conventional Simplified Model for Thunderstorm Winds. It is commonly assumed in wind engineering practice that wind speed records are statistically stationary over a period of 10 min to 1 h. This assumption is currently made even for *thunderstorm winds,* meaning that over open terrain, thunderstorm winds with a peak 3-s gust speed V_{3s} are conservatively represented for aerodynamic and structural purposes by winds in a large-scale storm with mean hourly speed $\overline{V} = V_{3s}/1.52$.

2.3.2 Wind Profiles

Strong winds increase with height above ground. Wind speeds are lower near the surface, owing to flow retardation by friction between the surface and the moving air. The retardation becomes increasingly weaker at higher elevations. For high speeds averaged over 10 min or longer above horizontal surfaces with uniform roughness over a sufficiently long fetch, the *logarithmic law* describes mean wind profiles up to elevations that increase as the wind speed increases and can reach several hundred meters (Sect. 2.3.2.1). An alternative description of wind profiles is the *power law* (Sect. 2.3.2.2).

The logarithmic law has long superseded the power law in meteorological practice. It is used in the Eurocode and in the Commentary to the ASCE 7-05 Standard. However, the power law remains in use in the Canadian Building Code (for mean hourly wind speeds) and in the ASCE 7 Standard (for fastest-mile speeds before 1995 and peak 3-s gust speeds as of 1995). The use of the logarithmic law is fully compatible with the specification of 3-s peak gust speeds in ASCE Standard wind maps and simply requires that the latter be converted to mean hourly or 10-min speeds, as shown in Sect. 2.3.1.

Wind speeds over surfaces other than open terrain can be calculated from wind speeds over open terrain by using the *relation between wind speeds in different roughness regimes* (Sect. 2.3.2.3). The wind profile is affected by the length of the *fetch* (Sect. 2.3.2.5) and by *topographical features* such as hills (Sect. 2.3.2.6).

2.3.2.1 Wind Profiles over Horizontal Terrain with Long Fetch: Logarithmic Law. The logarithmic (or log) law describes the variation with height and surface roughness of strong mean speeds with averaging times of 10 min to 1 h in straight winds. Its expression is

$$\overline{V}(z) = \overline{V}(z_{\text{ref}}) \frac{\ln(z/z_0)}{\ln(z_{\text{ref}}/z_0)} \tag{2.3.1}$$

where $\overline{V}(z)$ and $\overline{V}(z_{\text{ref}})$ are the mean wind speeds at elevation z and z_{ref}, respectively, z_{ref} is a reference elevation, and z_0 is an empirical measure of the surface roughness called the *roughness length.* Approximate ranges for z_0, and z_0 values that according to Section C.6.5.6 of the ASCE 7-05 Commentary are approximately consistent with ASCE 7-05 Standard terrain roughness specifications, are shown in Table 2.3.1.

The values of Table 2.3.1 differ from those specified in the Eurocode. In the latter, $z_0 = 0.003$ m for sea or coastal areas exposed to the open sea; $z_0 = 0.01$ m for lakes or flat and horizontal areas with negligible vegetation and no obstacles; $z_0 = 0.05$ m for areas with low vegetation and isolated obstacles such as trees or buildings with separations of maximum 20 obstacle heights (e.g., villages, suburban terrain, permanent forest); $z_0 = 0.3$ m for areas with regular cover of vegetation or buildings or with isolated obstacles with separations of maximum 20 obstacle heights (villages, subruban terrain, forests); $z_0 = 1$ m for areas in which at least 15% of the surface is covered with buildings whose average height exceeds 15 m.

The following relation was proposed by Lettau [2.7]:

$$z_0 = 0.5H_{\text{ob}} \frac{S_{\text{ob}}}{A_{\text{ob}}} \tag{2.3.2}$$

TABLE 2.3.1. Roughness Lengths z_0 Proposed in the ASCE 7-05 Commentary

Type of Surface	Roughness Length [ft (m)]
Water[a]	0.016–0.03 (0.005–0.01)
Open terrain[b]	0.05–0.5 (0.015–0.15)
Urban and suburban terrain, wooded areas[c]	0.5–2.3 (0.15–0.7)

[a] The larger values apply over shallow waters (e.g., near shorelines). Approximate typical value corresponding to ASCE 7-05 Exposure D: 0.016 ft (0.005 m) (ASCE Commentary C6.5.6). According to [2.6], for strong hurricanes $z_0 \approx 0.001$ to 0.003 m.

[b] Approximate typical value corresponding to ASCE 7-05 Exposure C: 0.066 ft (0.02 m) (ASCE Commentary C6.5.6).

[c] Value corresponding approximately to ASCE 7-05 Exposure B: 0.5 ft (0.15 m); this value is smaller than the typical value for ASCE 7-05 Exposure B [1 ft (0.3 m)] to account, conservatively, for the possible presence of open spaces such as parking lots (ASCE Commentary C6.5.6).

where H_{ob} is the average height of the roughness elements in the upwind terrain, S_{ob} is the average vertical frontal area presented by the obstacle to the wind, and A_{ob} is the average area of ground occupied by each obstruction, including the open area surrounding it.

Numerical Example 2.3.3. *Application of the Lettau formula.* Check Eq. 2.3.2 against the Eurocode value $z_0 = 1$ m indicated earlier, assuming that the average building height is $H_{\text{ob}} = 15$ m, the average dimensions in plan of the buildings are 16 m × 16 m, and $A_{\text{ob}} = 1600\ \text{m}^2$. We have $S_{\text{ob}} = 15 \times 16 = 240\ \text{m}^2$. Equation 2.3.2 therefore yields $z_0 = 0.5 \times 15 \times 240/1600 = 1.125$ m.

2.3.2.1.1 Surface Layer. The logarithmic law is not valid throughout the boundary layer but only over its lowest part, called the *surface layer*. According to results of micrometeorological research (see [2.5, Sect. 2.2.3 and references quoted therein]), the depth z_s of the surface layer *increases* with wind speed and terrain roughness and depends on angle of latitude φ, as follows:

$$z_s \approx \begin{cases} 100\,\dfrac{\overline{V}(z_{\text{ref}})}{\sin\varphi\,\ln(z_{\text{ref}}/z_0)} & \text{feet}\quad [\overline{V}(z_{\text{ref}})\text{ in mph}] \qquad (2.3.3a) \\ 68.3\,\dfrac{\overline{V}(z_{\text{ref}})}{\sin\varphi\,\ln(z_{\text{ref}}/z_0)} & \text{meters}\quad [\overline{V}(z_{\text{ref}})\text{ in m/s}] \qquad (2.3.3b) \end{cases}$$

Numerical Example 2.3.4. *Application of the log law: surface layer depth.* For a mean hourly speed $\overline{V}$ (32.8 ft) = 56 mph and a roughness length $z_0 = 0.1$ ft, Eq. 2.3.1 yields a mean hourly speed at 65.6 ft above ground $\overline{V}$ (65.6 ft) = 62.8 mph. For an angle of latitude $\varphi = 30°$ Eq. 2.3.3*a* yields a theoretical surface layer depth $z_s \approx 1900$ ft (590 m).

Before the development of atmospheric boundary layer theory it was widely believed—erroneously—that the surface layer depth is independent of wind speed. On the basis of measurements in low wind speeds it was further believed that regardless of wind speed, $z_s \approx 200$ ft (± 100 ft) [2.8].

2.3.2.1.2 Geostrophic Height. The geostrophic height is defined as the elevation z_{bl} above which surface friction effects are negligible. Its approximate expression is

$$z_{bl} \approx 10 z_s \tag{2.3.4}$$

[2.5]. The wind speed at the geostrophic height is termed the *geostrophic speed.* The ASCE 7 Standard uses the term *gradient* instead of *geostrophic.* To avoid confusion we will do so also.[1]

2.3.2.1.3 Veering Angle. The veering angle is the angle between the surface wind direction and the wind direction at a higher elevation. For buildings with heights of 300 to 500 m, *preliminary* measurement results suggest that, on average, the direction of the mean wind speed at the top of the building is obtained from the direction of the mean speed at 10 m above ground by adding to the latter an angle of about 7 to 10°, respectively, clockwise in the northern hemisphere and counterclockwise in the southern hemisphere. For buildings with heights of, say 100 to 300 m, the difference is about 3 to 7°, respectively. The angles should be increased by a factor of, say, 1.5 for buildings in urban terrain. The veering angle increases with height above the surface and reaches its maximum for winds at the gradient height. For elevations below 100 m, the veering angle may be assumed to be zero [2.9].

2.3.2.2 Wind Profiles over Horizontal Terrain with Long Fetch: Power Law. The expression for the power law is

$$V(z) = V(z_{ref}) \left(\frac{z}{z_{ref}} \right)^{1/\alpha} \tag{2.3.5}$$

In Eq. 2.3.5 the exponent $1/\alpha$ depends on surface roughness and on the wind speed averaging time, the profiles being flatter as the averaging time decreases. The power law applied to 3-s peak gust wind profiles has the same form as Eq. 2.3.5; however, its power law exponent is denoted by $\hat{\alpha}$ rather than by α. Table 2.3.2 shows power law exponents and gradient heights adopted by the ASCE 7-93 Standard for sustained wind speeds (including fastest-mile wind speeds), the National Building Code of Canada (NBC) for

[1] Strictly speaking this is not correct in meteorological terms: *geostrophic* pertains to straight winds and *gradient* pertains to winds belonging to vortexlike storms.

TABLE 2.3.2. Power Law Exponents and Gradient Heights Used in the 1993 to 2005 Versions of the ASCE Standard, and in the National Building Code of Canada [2.3]

Exposure		A[a]	B[b]	C[c]	D[d]
ASCE 7-93[e]	$1/\alpha$	1/3	1/4.5	1/7	1/10
	z_g ft (m)	1500 (457)	1200 (366)	900 (274)	700 (213)
NBC[f]	$1/\alpha$	0.4	0.28	0.16	—
	z_g ft (m)	1700 (520)	1300 (400)	900 (274)	—
ASCE 7[g]	$1/\hat{\alpha}$	—	1/7	1/9.5	1/11.5
(1995–2005)	z_g ft (m)	—	1200 (366)	900 (274)	700 (213)

[a] Centers of large cities.
[b] Suburban terrain, towns.
[c] Open terrain (e.g., airports).
[d] Water surfaces.
[e] Sustained speeds.
[f] Mean hourly speeds.
[g] Peak 3-s gust speeds.

mean hourly speeds, and the 1995 and subsequent versions of the ASCE 7 Standard for peak 3-s gusts [2.10]. Differences between exponents corresponding to 3- and 5-s peak gusts may be neglected for practical purposes.

Equation 2.3.5 is valid up to a height z_g, purported to represent the gradient height. The height z_g differs from the height z_{bl} as estimated from theory (Eq. 2.3.4).

Numerical Example 2.3.5. *Application of the power law.* Let $z_{ref} = 32.8$ ft (10 m), $V_{3s}(z_{ref}) = 55$ mph, and $\hat{\alpha} = 1/9.5$ (open terrain). From Eq. 2.3.5, at 100 ft above ground $V_{3s}(100 \text{ ft}) = 55\ (100/32.8)^{1/9.5} = 62$ mph.

2.3.2.3 Relation Between Wind Speeds in Different Roughness Regimes. Information on design wind speeds is usually based on wind speed measurements over open terrain (airports). Owing to the stronger effects of ground friction in rougher terrain, wind speeds at any given elevation within the atmospheric boundary layer are lower over built-up than over open terrain. At the top of the boundary layers, where ground friction effects are negligible, wind speeds are the same over surfaces with any type of roughness. The relation between wind speeds at various elevations in different roughness regimes is based on this fact. This relation is, strictly speaking, valid only for straight winds, but it is also used in current practice for hurricane and thunderstorm winds.

2.3.2.3.1 Logarithmic Law Description. For a storm with mean hourly or 10-min speed $\overline{V}(z_{open})$ at elevation z_{open} over open terrain with roughness z_{0open}, the corresponding speed at elevation z upwind of a site with roughness z_0 may be written as

$$\overline{V}(z) = \overline{V}(z_{0\text{open}}) \left(\frac{z_0}{z_{0\text{open}}}\right)^{0.07} \frac{\ln(z/z_0)}{\ln(z_{\text{open}}/z_{0\text{open}})} \tag{2.3.6}$$

Equation 2.3.6, proposed by Biétry in 1976, is used in the Eurocode.

Numerical Example 2.3.6. *Relation between wind speeds in different roughness regimes: log law description.* The mean hourly speed is $\overline{V}(z_{\text{open}}) = 56$ mph (25 m/s) at $z_{\text{open}} = 32.8$ ft (10 m) above open terrain with $z_{0\text{open}} = 0.066$ ft (0.02 m). Calculate the mean hourly wind speed $\overline{V}(z)$ at 150 ft (45.7 m) above *suburban terrain* with $z_0 = 2.3$ ft (0.7 m). Equation 2.3.6 yields $\overline{V}$ (150 ft) = 48.3 mph (21.6 m/s). Above *open terrain,* $\overline{V}$ (150 ft) = 69.7 mph (31 m/s) (Eq. 2.3.1). Note the significant retardation of the flow at 150 ft elevation over suburban terrain, where the wind speed is 48.3 mph, with respect to the flow at the same elevation above open terrain, where the speed is 69.7 mph.

2.3.2.3.2 Power Law Description. For strong winds, given the mean hourly speed $\overline{V}(z_{\text{open}})$ at the reference height z_{open} above open terrain with power law exponent $1/\alpha_{\text{open}}$, the mean hourly wind speed at height z above built-up terrain with power law exponent $1/\alpha$ is

$$\overline{V}(z) = \overline{V}(z_{\text{open}}) \left(\frac{z_{g,\text{open}}}{z_{\text{open}}}\right)^{1/\alpha_{\text{open}}} \left(\frac{z}{z_g}\right)^{1/\alpha} \tag{2.3.7}$$

where the product of the first two terms on the right-hand side is the gradient speed above open terrain, $\overline{V}(z_{g,\text{open}})$. Since gradient speeds are not affected by surface roughness, the gradient speed over built-up terrain, $\overline{V}(z_g)$, is equal to $\overline{V}(z_{g,\text{open}})$. The last term in Eq. 2.3.7 transforms $\overline{V}(z_g)$ into $\overline{V}(z)$ at height z above built-up terrain. An equation similar to Eq. 2.3.7 is also used (with the appropriate values of the parameters z_g and $\hat{\alpha}$ from Table 2.3.2), for 3-s peak gust speeds, denoted by V or V_{3s}, instead of $\overline{V}$, and for sustained wind speeds such as fastest-mile speeds or 1-min speeds. In the ASCE 7 Standard, $V(z_{\text{open}} = 32.8$ ft), denoted in the Standard by V, is the 3-s basic wind speed, and the product of the last two terms in Eq. 2.3.7 is denoted by $\sqrt{K_z}$.

Numerical Example 2.3.7. *Relation between wind speeds in different roughness regimes: power law description.* Denote the 3-s peak gust speed by V. Let V(32.8 ft) = 86 mph above open terrain ($\hat{\alpha} = 9.5$, $z_g = 900$ ft, Table 2.3.2). Equation 2.3.7 yields V(148 ft) = 101 mph (*open terrain*). Using Table 2.3.2 and Eq. 2.3.7, above *suburban terrain* ($\hat{\alpha} = 7.0$ and $z_g = 1200$ ft), V(32.8 ft) = 73 mph and V(148 ft) = 90 mph.

2.3.2.4 Relation Between Wind Speeds Averaged over Different Time Intervals for Winds over Any Type of Surface. In Sect. 2.3.2.1.2 we considered the relation between wind speeds averaged over various time intervals for the case of open terrain. In this section we consider this relation for the case of winds over any type of surface. The following approximate relation may be used:

$$V_t(z) = \overline{V}(z)\left[1 + \frac{\eta c(t)}{2.5\ln(z/z_0)}\right] \tag{2.3.8}$$

where $V_t(z)$ is the speed averaged over t seconds and $\overline{V}(z)$ is the speed averaged over 1 h for terrain with surface roughness z_0. The coefficients η and $c(t)$ may be assumed to have the approximate values given in Table 2.3.3. For an application, see Numerical Example 2.3.8.

Numerical Example 2.3.8. *Conversion of Saffir–Simpson scale winds to wind speeds above open terrain.* Calculate the 1-min wind speeds at 10 m above open terrain for category 4 hurricanes as defined in the Saffir–Simpson scale. From Table 2.2.1, the 1-min speeds at 10 m above open water that define weakest and strongest category 4 hurricanes are 131 and 155 mph, respectively. In this example we perform the conversion by using the power law description. We then perform the conversion by using the log law description, and comment on the conversions included in the Commentary to the ASCE 7-05 Standard [2.34].

Power law description. For a 131-mph speed at 10 m above water, the 1-min speed at 10 m above open terrain is $131 \times (700/33)^{1/10} \times (33/900)^{1/7} = 0.846 \times 131 = 111$ mph (Table 2.3.2, ASCE 7-93 power law description). For a 155-mph 1-min speed at 10 m above open water, the 1-min speed at 10 m above open terrain is $0.846 \times 155 = 131$ mph (see footnote *a*, Table 2.2.1), and the peak 3-s gust speed at 10 m above ground in open terrain is about $(1.52/1.25) \times 0.846 \times 155 = 1.03 \times 155 =$ *159 mph,* where 1.52 and 1.25 are, respectively, the mean ratio between the peak 3-s gust speed and the mean hourly speed, and the ratio between the 1-min speed and the mean hourly speed (Sect. 2.3.1.2). In contrast, the value listed in the to ASCE 7-05 Commentary is *191 mph.* Assuming that the ASCE 7-93 power law profiles

TABLE 2.3.3. Factors $\eta(z_0)$ and $c(t)$

(*a*)	z_0 (m)	0.005	0.03	0.30	1.00
	$\eta(z_0)$	2.55	2.45	2.30	2.20

(*b*)												
	t	1	10	20	30	50	100	200	300	600	1000	3600
	$c(t)$	3.00	2.32	2.00	1.73	1.35	1.02	0.70	0.54	0.36	0.16	0.00

are reasonably correct—an assumption that is confirmed by the subsequent calculations—this is tantamount to an overestimation of the wind pressures inherent in the ASCE 7-05 Commentary by a factor of 1.44.

Logarithmic law description. The conversion depends on the assumed values of the surface roughness lengths z_0 for open water and open terrain. Relative large values of z_0 are applicable to wind flow over water near shorelines where the water is shallow, as opposed to flow over open water. According to results in [2.6], for hurricane winds over open water it is reasonable to assume that $z_0 = 0.003$ m or even less. Equation 2.3.8 yields, with $\eta(z_0 \approx 0.003\text{ m}) \approx 2.55$ and $c(60\text{ s}) \approx 1.29$ (Table 2.3.3):

$$V^w_{60\text{s}}(10\text{ m}) = \overline{V}^w(10\text{ m})\left[1 + \frac{2.55 \times 1.29}{2.5\ln(10/0.003)}\right]$$

that is,

$$\overline{V}^w(10\text{ m}) = 0.86\ V^w_{60\text{s}}(10\text{ m})$$

where the superscript w signifies "over open water." Assuming conservatively that over open terrain $z_0 = 0.04$ m, Eq. 2.3.6 yields

$$\overline{V}^w(10\text{ m}) = \overline{V}(10\text{ m})\left[\frac{0.003}{0.04}\right]^{0.07}\frac{\ln(10/0.003)}{\ln(10/0.04)}$$

that is,

$$\overline{V}(10\text{ m}) = 0.816\overline{V}^w(10\text{ m})$$

where the unsubscripted quantity is the mean hourly wind speed over open terrain. It follows that

$$\begin{aligned}\overline{V}(10\text{ m}) &= 0.86 \times 0.816 V^w_{60\text{s}}(10\text{ m})\\ &= 0.7 V^w_{60\text{s}}(10\text{ m})\end{aligned}$$

Therefore, the peak 3-s gust over open terrain is

$$\begin{aligned}V_{3s}(10\text{ m}) &= 1.52 \times 0.7 \times V^w_{60\text{s}}(10\text{ m})\\ &= 1.06\ V^w_{60\text{s}}(10\text{ m})\end{aligned}$$

This result is close to the result obtained by using the ASCE 7-93 power law model. To the speed $V^w_{60\text{s}}(10\text{ m}) = 155$ mph there corresponds, then, a peak 3-s gust at 10 m over open terrain of about *164 mph* rather than the *191 mph* speed listed in the ASCE 7-05 Commentary.

The excessively large peak 3-s gust speeds over open terrain listed in the ASCE 7-05 Commentary as equivalent to the Saffir–Simpson scale 1-min speeds over open water are due to the following inconsistency. The surface roughness length on the basis of which those peak 3-s gusts were calculated in ASCE 7-05 corresponds to flow over water near the shoreline rather than over open water as would have been required for consistency with the Commentary's definition of Saffir–Simpson scale speeds. Even though the consequences of this inconsistency were brought to the attention of the ASCE 7 Committee many months before the printing of the ASCE 7-05 Standard, the difficulty of submitting the requisite modifications to public balloting without delaying the printing of the Standard apparently prevented the corrections from being made. It was decided, instead, that the Commentary would qualify those conversions from 1-min speeds over open water to 3-s peak gust speeds over open terrain by referring to them as "approximate."

2.3.2.5 Wind Profiles near a Change of Surface Roughness, and Terrain Exposure. The surface roughness is not the sole factor that determines the wind profile at a site. The profile also depends on the distance (the *fetch*) over which that surface roughness prevails upwind of the site. The terminology used in the ASCE 7 Standard therefore distinguishes between *surface roughness* and *exposure*. For example, a site is defined as having exposure B if it has surface roughness B *and* if surface roughness B prevails over a sufficiently long fetch. For design purposes the wind profile at a site with exposure B may then be described by the power law with parameters corresponding to surface roughness B. In Sects. 2.3.2.1 to 2.3.2.3 we specifically considered only the case of long fetch. ASCE 7 criteria on the requisite fetch are reproduced in Sect. 3.2.3.2; see also ASCE 7-05 Commentary, Sect. C6.5.6.4.

2.3.2.6 Wind Profiles over Nonhorizontal Terrain. Topographic features alter the local wind environment and create wind speed increases (speed-up effects). A procedure specified in the ASCE 7 Standard for the calculation of speed-up effects on two- or three-dimensional isolated hills and two-dimensional ridges and escarpments is presented in Sect. 3.2.4.

2.3.3 Atmospheric Turbulence

Except for winds with relatively low speeds under special temperature conditions, the wind flow is not laminar (smooth). Rather, it is *turbulent*—it fluctuates in time and space; that is, at any one point in space, the wind speed is a random function of time (Fig. 2.3.1), and at any one moment in time the wind speed is a random function of position in space.

Atmospheric flow turbulence is of interest in structural engineering applications for three reasons. First, it affects the definition of the wind speed specified in engineering calculations (Sects. 2.3.1.2 and 2.3.2.4). Second, the turbulence can influence significantly the wind flow around a structure and therefore the wind-induced forces (Sects. 2.5.5.2 and 2.5.6.2). Third, the flow fluctuations produce dynamic effects in flexible structures such as tall buildings or long-span bridges. For both aerodynamic and dynamic reasons, wind tunnel simulations attempt to reproduce basic features of turbulent atmospheric flows (Sect. 2.5.6 and 2.7.3.2). We introduce several useful descriptors of atmospheric turbulence: turbulence intensities, integral turbulence lengths, and turbulence spectra and co-spectra.

2.3.3.1 Descriptors of Atmospheric Turbulence. Descriptors of atmospheric turbulence are needed for engineering calculations of buffeting effects on flexible buildings, towers, and bridges and to check that wind tunnel simulations conform roughly to models of atmospheric flow.

2.3.3.1.1 Turbulence Intensities. The longitudinal turbulence intensity, $I_u(z) = \overline{u^2(z,t)}^{1/2}/\overline{V}(z)$, is the ratio, at a point with elevation z, of the root mean square of the longitudinal wind speed fluctuations $u(z, t)$ to the mean speed $\overline{V}(z)$, $u(z, t)$ being parallel to $\overline{V}(z)$. Vertical and lateral turbulence intensities are defined similarly. The following relation holds [see Table 2.3.3*a* for approximate values of $\eta(z_0)$]:

$$I_u(z) \approx \frac{\eta(z_0)}{2.5 \ln(z/z_0)} \tag{2.3.9}$$

Numerical Example 2.3.9. *Calculation of longitudinal turbulence intensity.* For $z_0 = 0.03$ m, $z = 20$ m, $I_u(z) \approx 0.15$.

2.3.3.1.2 Integral Turbulence Scales. Integral turbulence scales are measures of the spatial extent of the fluctuating wind speed at a fixed time. The integral turbulence scale L_u^x is an indicator of the extent to which a longitudinal wind speed fluctuation u will engulf a structure in the along-wind direction and will thus affect its windward and leeward sides simultaneously. The scales L_u^y and L_u^z are measures of the transverse and vertical spatial extent of the longitudinal fluctuation u. The scale L_w^x is a measure of the longitudinal spatial extent of the vertical wind speed fluctuation w. If the mean wind is normal to a bridge span and L_w^x is large, a vertical speed gust will act on the entire width of the bridge deck.

Information on turbulence lengths is quite uncertain. Measurements suggest that L_u^x increases with height above ground and as the terrain roughness decreases. Roughly, at 15 m above ground, in suburban terrain $L_u^x \approx 50$ m, and in open terrain $L_u^x \approx 75$ m; in open terrain at 30 m above ground, 60 m $< L_u^x <$ 450 m (average 200 m), and at 150 m above ground, 120 m $< L_u^x <$ 630 m (average 400 m) [2.5]. Also, very roughly,

$$L_u^y \approx 0.2L_u^x \qquad L_u^z \approx 6z^{0.5}\ (L_u^z \text{ and } z \text{ in meters}) \qquad L_w^x \approx 0.4z \qquad (2.3.10a,b,c)$$

2.3.3.1.3 Spectra of Turbulent Wind Speed Fluctuations. A randomly varying signal (e.g., the record of Fig. 2.3.1) can be represented as a sum of many small sinusoidal components with frequencies n ($0 \leq n < n_{max}$, where n_{max} is a sufficiently high cutoff frequency). The excitation's sinusoidal components with frequencies equal or close to the natural frequencies of a flexible structure can cause that structure to experience resonant response (Sect. 2.6.3). For a signal $g(t)$ that is statistically stationary (see Sect. 2.3.1.1), the *spectral density function* $S_g(n)$, or *spectrum,* is a convenient measure of the relative contribution to the signal of each sinusoidal component. The area under the spectral density curve is by definition equal to the mean square value of the signal:

$$\int_0^{n_{max}} S_g(n)\, dn = \overline{g^2(t)} \qquad (2.3.11)$$

The spectral ordinate $S_g(n)$ is a measure of the contribution to $\overline{g^2(t)}$ of the signal's sinusoidal component with frequency n. Expressions for the spectra of the longitudinal, lateral, and vertical velocity fluctuations in atmospheric flow are provided as needed in specific applications.

2.3.3.1.4 Cross-Spectra of Turbulent Fluctuations. The cross-spectra indicate the extent to which sinusoidal components of fluctuations with frequencies n at different points in space are in tune with each other or evolve at cross-purposes (i.e., are or are not mutually *coherent*). For components with high frequencies, the distance in space over which wind speed fluctuations are mutually coherent is small. For low-frequency components that distance is relatively large, on the order of integral turbulence scales. The sinusoidal components of the flow may be associated with conceptual, intuitively appealing *eddies,* the turbulence being viewed as a superposition of eddies of various sizes. Large (small) eddies are associated with sinusoidal components with low (high) frequencies. An eddy corresponding to the frequency n is said to envelop a structure if the distance over which the fluctuations with frequency n are relatively coherent is comparable to the relevant dimension of the structure. Expressions for cross-spectra of turbulent fluctuations are provided as needed in subsequent applications (see Chapter 5, Appendix 5C).

2.3.4 Wind Flows in Various Types of Storms and Flows for Moderate Wind Speeds

The models presented in Sections 2.3.1 to 2.3.3 were developed for strong straight winds. The corresponding models may be different for other types of storms. Up to elevations of about 500 m or so, and at higher elevations over built-up terrain, *hurricane* wind flows at sufficiently large distances from the eye wall appear to be similar to straight wind flows. Above about 500 m or so, wind speeds appear to *decrease* with height, at least over the ocean [2.6]. For hurricane wind profile measurements, see also [2.5, pp. 77, 78].

A simplified model of hurricane flow, which accounts for the accelerations associated with the flow curvature, suggests that the effects of those accelerations on the wind profile are negligible [2.11]. No studies appear to exist on the extent to which the mean wind profile is affected by upflow near the eye wall, which causes the flow not to be parallel to the Earth's surface, as is assumed in the derivation of the logarithmic law. For this reason, in current structural engineering practice no micrometeorological distinction is made between straight and hurricane mean wind speeds.

In current practice it is also assumed conservatively that the properties of *thunderstorm* flows are the same as those of straight winds provided that the 3-s peak gusts at 10 m above ground are the same for both flows (Sect. 2.3.1.3). *Chinook* wind flows (Sect. 2.2.2), which occur, for example, in Boulder, Colorado, may also differ significantly from straight wind flows, but no specific studies on their micrometeorology appear to be available.

Vortex-induced oscillations can occur at *moderate* wind speeds, at which the wind flow may exhibit features that differ from those typical of strong winds; in particular, the flows may be less turbulent and have flatter wind profiles. Although few relevant results are available, the designer should make conservative allowance for possible differences between moderate and strong winds. This is true, in particular, for vortex-induced vibrations of long bridges, which should be tested in smooth flow (see Chapter 5).

2.4 EXTREME WIND SPEEDS

Structures must be designed for wind-induced effects with specified probabilities of being exceeded in any one year. For example, allowable stresses in a structural member are commonly designed for a nominal probability of nonexceedance of 0.02 per year. The probabilities of wind-induced effects depend on the probability distributions of the extreme wind speeds.

In Sect. 2.4.1 we discuss nondirectional and directional speed wind speed data in nonhurricane and hurricane-prone regions. In Sect. 2.4.2 we provide simple, intuitive definitions of exceedance probabilities and mean recurrence intervals. In Sects. 2.4.3 and 2.4.4 we consider approaches and methods used to estimate extreme winds with specified mean recurrence intervals, and associated estimation errors.

2.4.1 Wind Speed Data

2.4.1.1 Micrometeorological Homogeneity of the Data. Extreme wind speed analyses must be conducted on sets of data that are micrometeorologically homogeneous, meaning that all the data in a set must correspond to the same height above surfaces with the same exposure (e.g., ocean surface, open terrain, suburban built-up terrain, prevailing over a sufficient distance upwind), and to essentially the same averaging time (e.g., 3 s for peak wind gust speeds, 1 min, 10 min, or 1 h). Where data do not satisfy the micrometeorological homogeneity requirement, they have to be transformed so that the requirement is satisfied (see Sect. 2.3).

2.4.1.2 Directional and Nondirectional Wind Speeds. Standard provisions for wind loads are based primarily on the use of *nondirectional* extreme wind speeds, that is, extreme speeds considered independent of direction. *Directional* extreme wind speeds, that is, extreme wind speeds as functions of direction, are used for estimates of wind effects on structures for which wind tunnel data are available for a sufficient number of wind directions.

Numerical Example 2.4.1. *Directional and nondirectional wind speeds.* To illustrate the definitions of directional and nondirectional wind speeds, we consider for two consecutive years the following largest peak 3-s gusts in mph:

	Directional Speed								Nondirectional Speed (Largest Speed from Any Direction)
Year	NE	E	SE	S	SW	W	NW	N	
1	90	**100**	81	96	87	89	93	78	100
2	78	94	83	**107**	80	83	72	76	107

The largest *nondirectional* speeds are also shown (in bold type) in the list of directional speeds.

2.4.1.3 Wind Speed Data Sets. We mention some data sets available from various sources.

2.4.1.3.1 Data in the Public Domain (Available Free of Charge). See Internet site http://www.nist.gov/wind; click on the data sets link.

- *Largest yearly directional 3-s peak gust speeds* in m/s at 10 m above ground in open terrain at about 35 U.S. weather stations for up to about 30 years of record, listed for each of eight octants (at 45° intervals).
- *Nondirectional annual 3-s peak gust speeds* in m/s at 10 m above ground in open terrain for about 15- to 50-year periods of record at about 100 U.S. weather stations.

- *Simulated directional tropical storm/hurricane 1-min speeds in knots* at 10 m elevation in open terrain for 55 locations along the Gulf and Atlantic coasts (Fig. 2.4.1). The speeds were obtained by Monte Carlo simulation from roughly 100 years of recorded hurricane climatological data, such as pressure defects, radii of maximum wind speeds, and translation speed and direction (see Sect. 2.2.1 for the definitions of these terms).

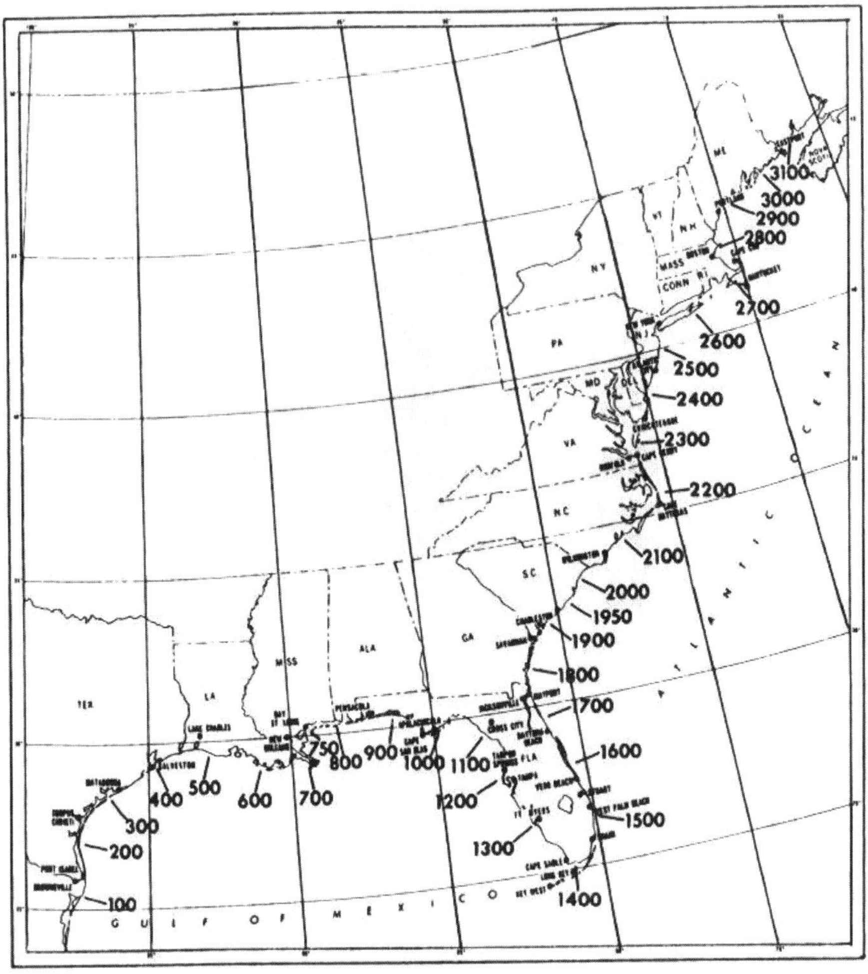

Figure 2.4.1. Locator map with coastal distance intervals (mileposts) marked (knots). (From F. P. Ho, R. W. Schwerdt, and H. W. Goodyear, *Some Climatological Characteristics of Hurricanes and Tropical Storms, Gulf and East Coast of the United States,* NOAA Technical Report NWS 15, National Oceanic and Atmospheric Administration, Washington, DC, 1975.)

> For each location the data consist of estimated hurricane arrival rates and sets of 999 1-min speeds in knots for 16 directions at 22.5° intervals (1 knot ≈ 1.15 mph; ratio between 1-min and 3-s speeds: 0.82; ratio between 1-min and 1-h speeds: 1.25; 1 mph = 0.447 m/s).

At any given site, as many of 20 to 40% of the total number of simulated hurricane wind speeds are negligibly small. Such small or vanishing speeds occur, for example, where the hurricane's translation velocity counteracts the rotational velocity.

Hurricane wind data sets that also cover interior areas, in addition to the coastline, are available commercially [2.12]. For coastal stations, estimated extreme wind speeds based on [2.12] and on the publicly available data sets are substantially similar.

2.4.1.3.2 Data Available for a Fee from the National Weather Service. *Directional peak gust speeds* above open terrain for each of 36 directions (at 10° intervals) are recorded during every hour for periods of as many as about 20 years within the framework of Atmospheric Surface Observing System (ASOS). Speeds associated with thunderstorms can be identified specifically, allowing the extraction of separate sets of nonthunderstorm and thunderstorm wind speeds. Before being used in statistical analyses the ASOS data must be reduced to a common elevation by using information on anemometer elevation history. If peak 5-s gust speeds are listed, they can be transformed to 3-s peak gust speeds through multiplication by the factor 1.02.

2.4.1.4 Superstations. Extreme wind speed estimates improve as the sizes of the samples on which they are based increase (Sect. 2.4.4). Some analysts have suggested basing statistical estimates on larger data sets obtained by pooling data from several stations (i.e., by creating a *superstation* from several stations). A uniform extreme wind climate based on those estimates is then assumed to prevail over the entire superstation zone. This approach is warranted in principle, but it has pitfalls that can lead to incorrect estimates. A superstation must satisfy the following criteria: (1) its component stations must be comparable in meteorological and micrometeorological terms (e.g., a superstation consisting of two stations, one of which is in a hurricane-prone region while the other is not, would yield incorrect wind speed estimates for both stations); (2) the data of the component stations should be mutually independent (e.g., creating a superstation by pooling the data measured at two immediately neighboring stations would not necessarily increase the total amount of useful information and would not result in significantly more precise estimates, although it might decrease observation errors); (3) a station cannot be used as a component station in more than one superstation, since this would create artificially large zones of climatological uniformity. The development of the ASCE 7 peak gust map was shown to violate criteria (1) and (3), with serious consequences with respect to the quality of the estimates of winds speeds specified in the map for nonhurricane regions [2.13].

2.4.2 Cumulative Distributions, Exceedance Probabilities, and Mean Recurrence Intervals

In Sect. 2.4.2.1 we introduce various notions intuitively by using the example of a fair die and show how they also apply to extreme speeds. In Sect. 2.4.2.2 we consider the important case of mixed distributions: for example, distributions in extreme wind climates subjected to significant nonhurricane winds in addition to hurricane winds, or to nonthunderstorm and thunderstorm winds.

2.4.2.1 Estimation of Event Probabilities for a Fair Die: Extension to Extreme Wind Speeds. We denote the outcome of throwing a die once by O. The probability, denoted by $P(O \leq n)$ ($n = 1, 2, \ldots, 6$), that the outcome O is less than or equal to n is called the *cumulative distribution function* (CDF) of the event O. The probabilities $P(O \leq n)$ for $n = 1, 2, \ldots, 5, 6$ are 1/6, 2/6, . . . , 5/6, 6/6, respectively. The probability that the outcome is larger than n, $P(O > n) = 1 - P(O \leq n)$, is called the *probability of exceedance* of the outcome n. The *mean recurrence interval* (MRI) of the event $O > n$, defined as the inverse of the probability of exceedance of that event, is the average number of trials (throws) required for $O > n$. The MRI is also called the *mean return period.*

Numerical Example 2.4.2. *Mean recurrence interval of an outcome of throwing a die.* For a fair die the probability of exceedance $P(O > 5) = 1 - P(O \leq 5) = 1 - 5/6 = 1/6$. The MRI of the event $O > 5$ is $1/(1/6) = 6$ trials, that is, the outcome "six" occurs, on average, once in six trials.

Conceptually, there is little difference between the statement "the outcome of throwing a die once exceeds n" and the statement "the largest wind speed occurring in any one year exceeds v," except that the CDF of the largest speed in a year, $P(V \leq v)$, is continuous, whereas $P(O \leq n)$ is discrete (this requires a small adjustment in the counting of probabilities for the continuous distribution case; see p. 33, footnote 3). For any given n, $P(O \leq n)$ is the same for any one trial (throw) and is independent of the outcomes of other trials. Similarly, except for, say, possible global warming effects, which are not accounted for in codes, $P(V \leq v)$ is the same for any one trial (year) and is independent of speeds occurring in other years.

The speed v with an $\bar{N}$-year MRI is called the *$\bar{N}$-year speed.* The MRI, in years, is

$$\bar{N}(v) = \frac{1}{1 - P(V \leq v)} \tag{2.4.1}$$

Numerical Example 2.4.3. *Mean recurrence interval of a wind speed v.* Assume that for $v = 90$ mph, $P(V \leq 90 \text{ mph}) = 0.98$. The probability of exceedance of the speed $v = 90$ mph in any one year is $1 - 0.98 = 0.02$. The MRI of the event $V > 90$ mph is $\bar{N} = 1/0.02 = 50$ years.

2.4.2.2 Mixed Extreme Wind Climates. We consider wind speeds in regions exposed to both nonhurricane and hurricane winds. We are interested in the probability that in any one year, wind speeds are less than or equal to a specified speed v regardless of their origin. Let the random variables V_H and V_{NH} denote, respectively, the largest hurricane wind speed in any one year and the largest nonhurricane wind speed in any one year. Further, let the probability that $V_H \leq v$ and the probability that $V_{NH} \leq v$ be denoted, respectively, by $P(V_H \leq v)$ and $P(V_{NH} \leq v)$. The random variable of interest is the maximum yearly speed regardless of whether it is a hurricane or a nonhurricane wind speed, denoted by $\max(V_H, V_{NH})$. The statement "$\max(V_H, V_{NH}) \leq v$" and the statement "$V_H \leq v$ and $V_{NH} \leq v$" are equivalent. Therefore, $P[\max(V_H, V_{NH}) \leq v] = P(V_H \leq v$ and $V_{NH} \leq v)$. Since it is reasonable to assume that V_H and V_{NH} are independent random variables, it follows that

$$P[\max(v_H, v_{NH}) \leq v] = P(v_H \leq v)P(v_{NH} \leq v) \tag{2.4.2}$$

The probability distributions $P(v_{NH} \leq v)$ and $P(v_H \leq v)$ can be obtained as shown in Sect. 2.4.3. With an appropriate change of notation Eq. 2.4.2 is also applicable to non-thunderstorm and thunderstorm wind speeds.

Numerical Example 2.4.4. *MRI of specified wind speeds in a mixed wind climate.* Assume that at a given site, the MRI of the 100-mph peak 3-s gust speed due to nonhurricane winds is $\bar{N}_{NH} = 120$ years, and that the MRI of the 100-mph peak 3-s gust speed due to hurricanes is $\bar{N}_H = 50$ years. The respective CDF values are $P(V_{NH} \leq 100$ mph$) = 1 - 1/\bar{N}_{NH} = 0.99167$ and $P(V_H \leq 100$ mph$) = 1 - 1/\bar{N}_H = 0.98$. By Eq. 2.4.2, the CDF of the 100-mph wind speed due to either nonhurricane or hurricane winds is $P[\max(v_H, v_{NH}) \leq 100$ mph$] = P(v_H \leq 100$ mph$)P(v_{NH} \leq 100$ mph$) = 0.99166 \times 0.98 = 0.9718$. By Eq. 2.4.1, the MRI of the 100-mph wind speed at the site is $1/(1 - 0.9718) = 35.5$ years.

2.4.3 Estimates of $\bar{N}$-Year Wind Speeds

In Sect. 2.4.3.1 we discuss two approaches to the estimation of the extreme wind speeds: the epochal approach and the peaks-over-threshold approach. Section 2.4.3.2 presents methods for estimating extreme speeds corresponding to any specified MRI $\bar{N}$. For nonhurricane wind speeds, estimates may be obtained by fitting epochal data or peaks-over-threshold data to extreme value distributions. Although the reverse Weibull distribution appears to be a more appropriate model than the extreme value type I distribution, we provide estimators for both. For hurricane wind speeds we review parametric and nonparametric estimation methods.

2.4.3.1 Epochal and Peaks-over-Threshold Approaches to Estimation of Extremes. In the *epochal* approach, a cumulative probability distribution function is fitted to a set of wind speeds which consists of the largest speed recorded at the site of interest in each of a number of consecutive fixed epochs. The epoch being chosen most commonly is one year. The data set then consists of the largest annual wind speed for each year of the period of record.

In the *peaks-over-threshold* (POT) approach, a generalized Pareto cumulative distribution function is fitted to the excess of the speeds v over a sufficiently high threshold u. Speeds must be independent, meaning that they do not belong to the same storm system, that is, they are separated from each other in time by, say, at least 5 days. The advantage of the POT approach is that it allows the use of larger data samples, since speeds other than the largest annual speeds can also be included in the data sample.

Numerical Example 2.4.5. *Sample sizes in epochal and POT approaches.* Assume that in year 1 the largest speed is 69 mph, whereas in year 2 the largest speed is 85 mph, and the second and third largest speeds are 70 and 67 mph. If a threshold of 68 mph is chosen, the speeds during years 1 and 2 included in the sample are 85, 70, and 69 mph (three speeds). In the epochal approach only two speeds are included in the sample: 69 mph (year 1) and 85 mph (year 2).

If the threshold is too high, the advantage of a larger sample size is lost. For example, if in Numerical Example 2.4.5 the threshold were 80 mph, only one speed—85 mph—would be included in the 2-year sample. If the threshold is too low, the sample will include wind speeds that may not to be representative of the extreme wind climate and will result in biased estimates.

Figure 2.4.2 shows, as functions of threshold speed (in mph) and sample size (i.e., number of data above the threshold speed), estimates of 100-, 1000- and 100,000-year fastest-mile wind speeds at 6.1 m above ground in open terrain at Green Bay, Wisconsin. The data consisted of the maximum wind speed for each of the successive 8-day intervals within a 15-year record. For thresholds between about 38 and 32 mph (sample sizes of about 35 to 127), the estimated 100-year speeds are stable around 60 mph. This is indicative of a reliable estimate. The reliability of the estimate is poorer as the MRI increases (this is clearly seen for the 100,000-year estimates, which are not useful for practical purposes). For higher thresholds the estimates are less stable; that is, they vary fairly strongly as a function of threshold. For lower thresholds the estimates are increasingly biased with respect to the 60-mph estimate, owing to the presence in the data sample of low speeds unrepresentative of the extremes. *Including low speeds in a sample used for inferences on extreme speeds results in biased estimates,* as would be the case if the heights of children were included in a sample used to estimate the height of adults. Nevertheless, some practitioners use speeds recorded every hour, the vast majority of which are low, for making inferences on extreme wind speeds (see Sect. 4.2.2.3). Modern extreme value statistics recognizes, however, that

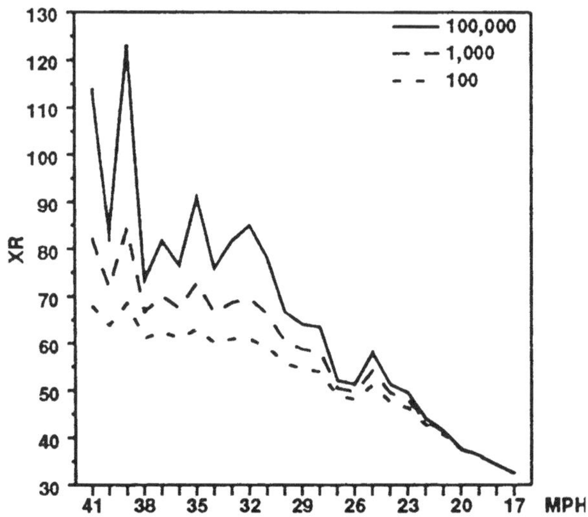

Figure 2.4.2. Estimated wind speeds with 100-, 1000-, and 100,000-year mean recurrence intervals at Green Bay, Wisconsin, as functions of threshold (mph). Note that for low wind speed thresholds (below 32 mph, say), the estimates are increasingly biased; that is, the wind speeds are increasingly underestimated. (From [2.14].)

to obtain dependable estimates of extreme values, it is necessary to "let the tails speak for themselves" instead of allowing estimates to be biased by data with small values, as is the case in Fig. 2.4.2 for wind speeds below about 32 mph.

2.4.3.2 Methods for the Estimation of Extreme Wind Speeds Numerous methods—some more precise than others—are available for estimating the distribution of extreme speeds (see software provided on the site http://www.nist.gov/wind). The methods covered in this section are adequate for engineering purposes. More precise methods need to be used for very small samples (e.g., 10 years, say, or less). Sections 2.4.3.2.1 and 2.4.3.2.2 present estimators for nonhurricane and hurricane extreme wind speeds, respectively.

2.4.3.2.1 Nonhurricane Extreme Wind Speeds: Gumbel and Reverse Weibull Distributions. A reasonably persuasive theoretical and empirical basis exists for the assumption that extreme value distributions of the largest values are adequate for describing extreme wind speeds probabilistically. Of these distributions two have found acceptance among extreme wind climatologists: the *Gumbel distribution* [also known as the *extreme value type I* (EV I) *distribution*] and following recent progress in the field of extreme value statistics, the *reverse Weibull* (RW) *distribution* [also known as the *extreme value type III* (EV III) *distribution* of the largest values]. The evidence suggesting that the RW distribution is a more appropriate model than the EV I distribution is statistical [2.14–2.16] as well as practical. The practical evidence is the following. According to results of calculations based on the EV I distribution of the extreme speeds, the safety index characterizing the reliability of structures under wind loads is far lower than for structures under gravity loads [2.17]. It would follow from this result that the probability of failure of structures subjected to wind loads is much higher than failure statistics appear to indicate. In contrast, the use in calculations of the RW distribution, which yields lower extreme wind speeds than the Gumbel distribution, does not result in similar discrepancies. The RW distribution has been adopted in the Australian/New Zealand Standard [2.30, Commentary Section C3.2]. See also [2.31, 2.32].

It has been argued that because the RW distribution (or its counterpart, the generalized Pareto distribution with negative tail length parameter) is valid only in an asymptotic sense, it is a less appropriate model of the extreme winds than is the EV I distribution. This argument has no merit insofar as the EV I distribution is itself valid only in an asymptotic sense. The RW distribution differs from the EV I distribution only in that it contains a tail parameter $c < 0$; in the limit of vanishing c the RW approaches the EV I distribution. There is merit, however, to the argument that owing to the asymptotic assumption, unless the size of the sample is extremely large, extreme value distributions are approximate. However, there are indications that the inaccuracy inherent in the asymptotic assumption is acceptably small and that for the RW distribution, it can be accounted for by slightly increasing the estimated tail length parameter, as indicated in the comment following Eqs. 2.4.5. Independently of the approximation inherent in the asymptotic assumption, it remains true that low thresholds lead to bias errors, as shown in Fig. 2.4.2.

EPOCHAL APPROACH, METHOD OF MOMENTS. We describe an approximate method, the *method of moments,* for estimating the $\overline{N}$-year speed from a sample of the n maximum yearly nonhurricane wind speeds. The method of moments is derived in [2.5, Sect. A1.6] and is shown by calculations to be adequate for samples of size, say, 20 or larger [2.5, Sect. 3.2.2]. The method relies on calculated sample means $\overline{V}$ and standard deviations $s(v)$ of the sample of n wind speeds.[2] Here we apply the method of moments to both the EV I and RW distributions.

If applied to the Gumbel (EV I) distribution, the method of moments yields the following estimator of the $\overline{N}$-year speed:

$$v_{\text{EV I}}(\overline{N}) \approx \overline{V} + 0.78(\ln \overline{N} - 0.577)s(v) \tag{2.4.3}$$

[2]Note that $\overline{V}$ here denotes the mean of a sample of wind speeds rather than the mean hourly speed during a storm.

If applied to the reverse Weibull (RW) distribution, the method of moments yields the following estimator of the $\overline{N}$-year speed:

$$v_{\mathrm{RW}}(\overline{N}) = \overline{V} + s(v)A(c)\left\{B(c) - \left[-\ln\left(1 - \frac{1}{\overline{N}}\right)\right]^{-c}\right\} \tag{2.4.4}$$

$$A(c) = \frac{1}{\{\Gamma(1 - 2c) - [\Gamma(1 - c)]^2\}^{1/2}} \qquad B(c) = \Gamma(1 - c) \tag{2.4.5a,b}$$

where the tail parameter $c < 0$ and Γ denotes the gamma function. For nonhurricane speeds it is prudent to assume that in Eq. 2.4.4 $c = -0.1$, even if the estimated c is smaller, and therefore $A(c) = 8.726$, $B(c) = 0.95135$. As is suggested by Monte Carlo simulations [2.18], such conservatism accommodates the fact that the asymptotic assumption inherent in the use of extreme value distributions is only approximately correct for analyses based on sample sizes available in practice.

Software for performing estimates by using the epochal and the POT approaches is available on the site http://www.nist.gov/wind. If the POT software is used, and if the estimated tail length parameter is $c < -0.1$, it is prudent to reanalyze the data by assuming $c = -0.1$. In conducting POT analyses it is important to keep in mind that each individual storm is represented in a data sample by no more than one wind speed—the largest speed occurring within that storm (provided that the speed exceeds the threshold being considered). Various techniques for achieving this requirement are available. They consist essentially of assuring that peak wind speeds are separated by a sufficiently large time interval (e.g., the typical duration of a storm, i.e., five to eight days, say). An additional consideration is the need to assure that there are not too many identical speeds in a sample. If there are identical speeds, they can be made unequal by adding to them a number larger than -0.5 and smaller than 0.5, created by a random number generator. This assures that ordering of the wind speeds is achieved for data analysis purposes.

Numerical Example 2.4.6. *EV I and RW estimates by the epochal approach.* Assume that for an $n = 17$-year record at a site, the yearly peak 3-s gust speeds (in mph) are: 80, 76, 80, 80, 74, 80, 90, 82, 91, 81, 74, 74, 63, 76, 66, 69, 64. The epochal approach makes use of the estimated mean $\overline{V} = 76.65$ mph and standard deviation $s = 7.83$ mph of the n largest annual speeds.

From Eqs. 2.4.3 and 2.4.4 (in which $c = -0.1$) we obtain, respectively, the extreme value type I-based and reverse Weibull-based estimates:

- $\overline{N} = 50$ years: $v_{\mathrm{EV\ I}} = 97.0$ mph; $\overline{N} = 500$ years: $v_{\mathrm{EV\ I}} = 111.1$ mph
- $\overline{N} = 50$ years: $v_{\mathrm{RW}} = 95.4$ mph; $\overline{N} = 500$ years: $v_{\mathrm{RW}} = 105.0$ mph

For $\overline{N} = 50$ years the estimated extreme speeds based on the reverse Weibull are only slightly lower than those based on the Gumbel distribution, but the differences increase for higher MRIs.

In this section we have dealt only with estimates of nondirectional extreme wind speeds. For directional wind speeds it appears that more dependable estimates can be obtained by using POT software, listed on the site http://www.nist.gov/wind.

In certain applications large sets of directional wind speeds are needed. Such sets can in principle be created by simulation from smaller sets of recorded data.

2.4.3.2.2 Hurricane Wind Speeds

PARAMETRIC ESTIMATION METHODS. Figure 2.4.3 shows the tail length parameter c (see Eq. 2.4.4) and the nondirectional hurricane hourly mean speeds at an elevation of 10 m in open terrain, estimated from (1) the 1-min speeds for coastal station (milepost) 2550 (near New York City; see Fig. 2.4.1), and (2) the hurricane arrival rate (i.e., mean rate of occurrence of hurricane winds per year), listed for that station on the site http://www.nist.gov/wind (click on the data sets link). The estimates were obtained by using the POT method in conjunction with software listed on the same site. The software is based on the generalized Pareto distribution, which for the tail length parameters $c < 0$ and $c \to 0$ is equivalent to the RW and EV I distributions, respectively. Note that in Fig. 2.4.3*a*, $c < 0$. Estimates are seen to be poor for $m < 35$ or so. Estimated speeds based on results for thresholds corresponding to sample sizes $35 < m < 200$ are about 32, 43, and 46 m/s for 50-, 500-, and 2000-year speeds, respectively. If low wind speeds are included in the sample (i.e., for lower thresholds, corresponding to sample sizes of $m > 200$ or so), the estimates gradually deteriorate.

NONPARAMETRIC ESTIMATION METHODS. In some applications *nonparametric* estimates of hurricane wind speeds can be used. This is possible because of the large size of the simulated data samples. Consider a set of n nondirectional hurricane wind speed data at a location (milepost) where the hurricane arrival rate is u per year. If the rate were $u = 1$ hurricane per year, the estimated probability that the highest speed in the set would be exceeded is $1/(n + 1)$, and the corresponding estimated MRI would be $\bar{N} = n + 1$ years (on average, $n + 1$ "trials" would be required to get a hurricane exceeding that highest speed).[3] The estimated probability that the qth highest speed in the set is exceeded is $q/(n + 1)$, and the estimated MRI in years is $\bar{N} = (n + 1)/q$.[4]

[3] Using $n + 1$ instead of n is required by the continuous nature of the wind speed variate, which implies the possibility of occurrence of wind speeds higher than the highest wind speed in the set; a similar possibility does not exist in the case of the die.

[4] Note the similarity to Numerical Example 2.4.2.

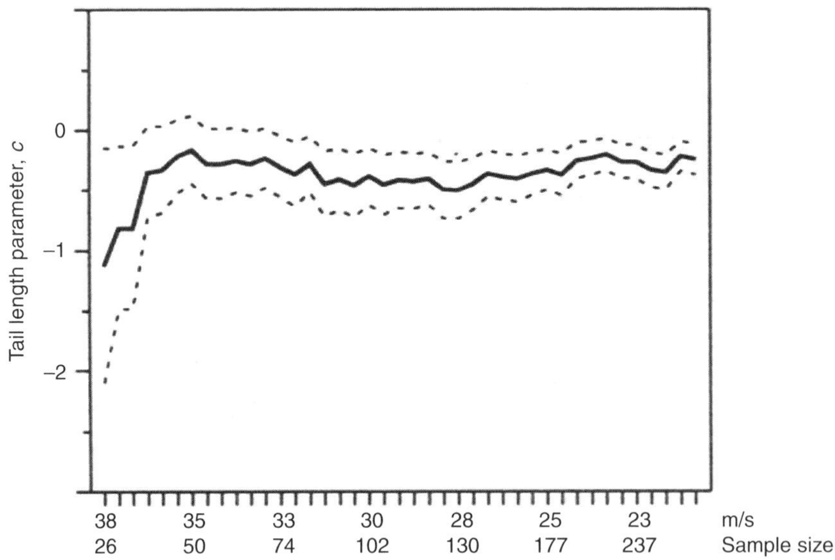

(a)

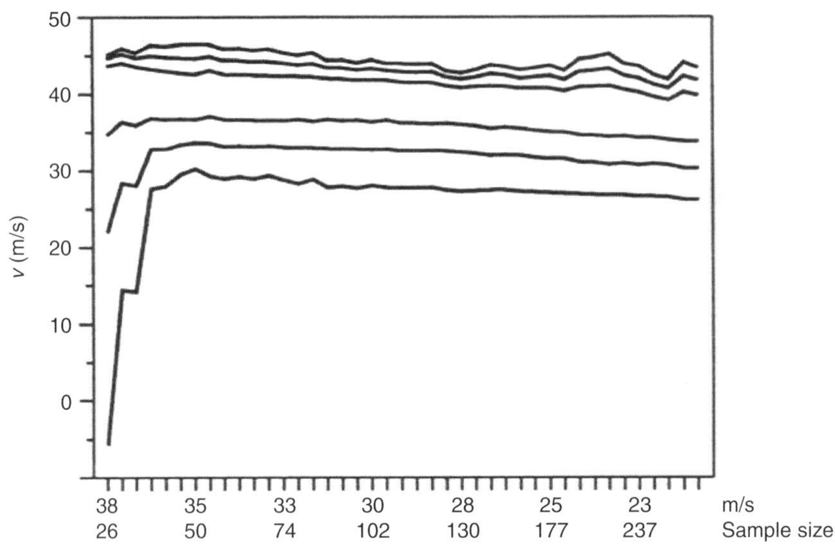

(b)

Figure 2.4.3. Peaks-over-threshold estimates of (*a*) tail length parameter *c* of a reverse Weibull distribution and (*b*) wind speeds with 25- 50-, 100-, 500-, 1000-, and 2000-year mean recurrence intervals, milepost 2550 (1-min speeds at 10 m above ground in open terrain), as functions of wind speed threshold and sample size. Note that estimates tend to be most stable for thresholds between about 32 m/s (sample size 85) and 34 m/s (sample size 56). (From [2.19].)

In general, $u \neq 1$, and the estimated MRI is therefore $\overline{N} = (n + 1)/qu$ years. For example, if $n = 999$ hurricane wind speed data, and $u = 0.5$ per year, the estimated MRI of the event that the highest wind speed in the sample will occur $\overline{N}$ is $= (n + 1)/u = 1000/0.5 = 2000$ years, the estimated MRI of the second-highest speed is 1000 years, and so on.[5] The rank of the speed with a specified MRI of $\overline{N}$ is $q = (n + 1)/\overline{N}u$, and the estimated CDF of that speed is $1 - 1/\overline{N}$.[6]

Numerical Example 2.4.7. *Nonparametric MRI estimates.*[7] We estimate the MRIs of speeds blowing from the southwest (i.e., from the 22.5° sector centered on the 225° direction) at milestone 2250 (near New York City), where $u = 0.305$ per year, and the wind speed data are rank-ordered in Table 2.4.1.

TABLE 2.4.1. Rank-Ordered Peak 3-s Gust Speeds (m/s) from the Southwest at 10m Above Open Terrain for the 22.5° Sector at Milepost 2550[a]

Rank, q	SW 225°	Rank, q	SW 225°	Rank, q	SW 225°
1	54	19	19	39	14
2	39	20	19	40	14
3	33	21	18	41	14
4	30	22	18	42	13
5	27	23	18	43	13
6	26	24	17	44	13
7	26	25	17	45	13
8	23	26	17	46	13
9	23	27	17	47	13
10	22	28	17	48	12
11	22	29	17	49	12
12	21	30	17	50	12
13	20	31	17	51	11
14	20	32	17	52	10
15	20	33	17	53	10
16	19	34	16	54	9
17	19	35	16	55	0
18	19	36	16		

[a](1-min speed in knots = 0.625 × 3-s speed in m/s). 1 knot = 1.15 mph; 1 mph = 0.447 m/s; 3-s speed = 1.216 × 1-min speed. (*Note:* 1.216 = 1.52/1.25, where 1.52 = ratio of 3-s speed to 1-h speed, and 1.25 = ratio of 1-min speed to 1-h speed; see Sect. 2.3.1.2.)

[5] $1/u$ may be interpreted as an average epoch in years.

[6] A formula based on the theory of Poisson processes, which takes into account the possibility that two or more hurricanes may occur at a site in any one year and is more exact for short MRIs (e.g., 5 years), is $\overline{N} = 1/\{1 - \exp\{-u[q/(n + 1)]\}$. For example, for $n = 999$, $u = 0.5$, and $q = 2$, $\overline{N} = 1000.5$ years.

[7] This example was contributed by W. P. Fritz.

It is sufficient to consider the first 55 rank-ordered data, since higher-rank data are negligibly small.

The qth largest speed in the set of 999 speeds corresponds to a mean recurrence interval $\bar{N} = (n + 1)/qu = 1000/0.305q$. For the highest and second-highest speeds listed in Table 2.4.1, $\bar{N} = 1000/0.305 = 3279$ years and $\bar{N} = 1000/(0.305 \times 2) = 1639$ years. The peak 3-s gust speed with a 100-year mean recurrence interval has rank $q = 1000/(0.305 \times 100) = 32.78$ (i.e., 33) and is seen from Table 2.4.1 to be 17 m/s. Note that the precision of the estimates is poorer for higher-ranking speeds, owing to the relatively large differences between successive higher-ranking speeds in Table 2.4.1 (e.g., 54 m/s vs. 39 m/s for the highest vs. the second-highest speed. This problem is less acute for winds blowing from any direction (i.e., for nondirectional wind speeds) and for data samples of size considerably larger than 999.

2.4.4 Error Estimates

Inherent in such estimates as those obtained in Sect. 2.4.3 or by other estimation methods are errors of three types:

- *Observation errors,* which affect measurements of the wind speeds or of the parameters used in wind speed simulation models (e.g., the radius of maximum wind speeds).
- *Physical modeling errors,* which affect the transformation of wind speeds at a given elevation in a given type of terrain exposure averaged over a given time interval (e.g., 1 min) into wind speeds at a different elevation and/or in a different type of exposure and/or averaged over a different time interval (e.g., 3 s). Physical modeling errors also pertain to the physical models by which the parameters used in hurricane wind speed simulations yield simulated wind speeds.
- *Sampling errors,* due to the limited size of the data sample on which estimates of extreme wind speeds are based.[8] Sampling errors affect inferences on wind speeds based on samples of both measured and simulated wind speeds. It has been argued that because the size of simulated samples of hurricane wind speeds can be arbitrarily large (e.g., $m_{max} = 1000$ or even 100,000), the corresponding sampling errors are in practice negligible. However, this view disregards the fact that sampling errors in the estimation of hurricane speeds are due predominantly to the relatively small size (about 100 years or less) of the samples of parameters used in hurricane wind speed simulations (Sect. 2.4.4.2).

[8] Assume that a coin is thrown twice and that the outcomes are "heads" after both throws. An inference that the probability of getting "heads" is unity based on a sample of two outcomes would be incorrect; owing to the small size of the sample of outcomes, it would be affected by sampling errors in the estimation of the frequency of occurrence of "heads."

In this section we consider only sampling errors. For other types of error, see Sect. 2.8.

2.4.4.1 *Sampling Errors in the Estimation of Nonhurricane Wind Speeds.* For the epochal approach the standard deviation of the sampling errors in the estimation of $\bar{N}$-year speeds with a Gumbel (EV I) CDF is, with the notations of Eq. 2.4.3,

$$\mathrm{SD}[v(\bar{N})] = 0.78\Big[1.64 + 1.46(\ln \bar{N} - 0.577) + 1.1(\ln \bar{N} - 0.577)^2\Big]^{1/2} \frac{s}{\sqrt{n}} \qquad (2.4.6)$$

The use of Eq. 2.4.7 for speeds with a reverse Weibull CDF (Eq. 2.4.4) is conservative.

Numerical Example 2.4.8. *Sampling errors estimates for nonhurricane wind speeds.* We refer to the data of Numerical Example 2.4.6. For a sample of size $n = 17$ of largest annual peak 3-s gust speeds, $s = 7.83$ mph. Equation 2.4.6 yields $\mathrm{SD}[v(\bar{N} = 50 \text{ years})] = 6.41$ mph.

2.4.4.2 *Sampling Errors in the Estimation of Hurricane Wind Speeds.* Sampling errors based on a hurricane recording period of about 100 years have coefficients of variation[9] of about 8, 10, 15, and 20% for $\bar{N} = 50$, 100, 1000, and 10,000 years, respectively [2.20]. Samples of hurricane wind speed samples obtained by numerical simulation can be very large (e.g., 10,000 speeds or more). However, for such large samples, the sampling errors are due predominantly to the relatively short period of record (i.e., about 100 years) from which the physical information used in the simulations is obtained. Increasing the size of the simulated sample does not change this fact.

2.5 AERODYNAMICS AND WIND TUNNEL TESTING

A body immersed in a turbulent flow experiences flow-induced fluctuating pressures due to (1) flow fluctuations caused by the presence of the body, and (2) turbulent wind speed fluctuations in the oncoming flow. The purpose of this section is to present basic aerodynamics principles that will help to interpret procedures for determining wind-induced effects on structures. The principles presented in this section are also used to discuss some features and limitations of wind tunnel tests of engineering structures.

[9] The estimated *coefficient of variation* of the sampling error is defined as the ratio between the estimated standard deviation of the error and the estimated wind speed.

Computational fluid dynamics (CFD) has recently made enormous strides. However, techniques for obtaining time-dependent pressures induced by turbulent flows do not allow the routine and confident use of CFD as a substitute for wind tunnel testing of structures (see Sect. 5.3.6), although CFD can be a useful complement to such testing. The role of CFD in structural engineering applications may be expected to become more important in the future.

2.5.1 Flow Velocities, Pressures, Shear Stresses, and the Reynolds Number

2.5.1.1 Bernoulli Equation. For a steady flow with no viscous effects, the Bernoulli equation relates the velocity V and the pressure p along a streamline as follows:

$$p_0 + \tfrac{1}{2}\rho V_0^2 = p + \tfrac{1}{2}\rho V^2 = \text{const} \tag{2.5.1}$$

where ρ is the fluid's specific mass and V_0 is the velocity far upwind on the streamline. In the atmosphere the pressure p_0 is the atmospheric pressure. At the point where a streamline intersects the upstream side of a fixed body, the velocity is $V = 0$ and the pressure on the body, called *stagnation pressure,* is $p = p_0 + \frac{1}{2}\rho V_0^2$. At any point along a streamline where $V > V_0$ ($V < V_0$), the pressure $p - p_0$ is negative (positive); that is, the pressure p is lower (higher) than the atmospheric pressure.

To show that Eq. 2.5.1 holds, consider a fluid element (Fig. 2.5.1) in the direction of the streamline subjected to the force $p\ dy\ dz$, the force $-(p + dp)\ dy\ dz$, and the *inertial force*

$$\rho\ dx\ dy\ dz\ \frac{dV}{dt} \equiv \rho\ dy\ dz\ V\ dV \tag{2.5.2}$$

where $V = dx/dt$. (The inertial force is equal to the mass of the element $\rho\ dx\ dy\ dz$ times the acceleration dV/dt.) The equation of equilibrium among these three forces yields $-dp = \rho V\ dV$ or, upon integration, Eq. 2.5.1. The term $\frac{1}{2}\rho V^2$ is called the *dynamic pressure.* The derivative dp/dx is called the *pressure gradient* in the x-direction.

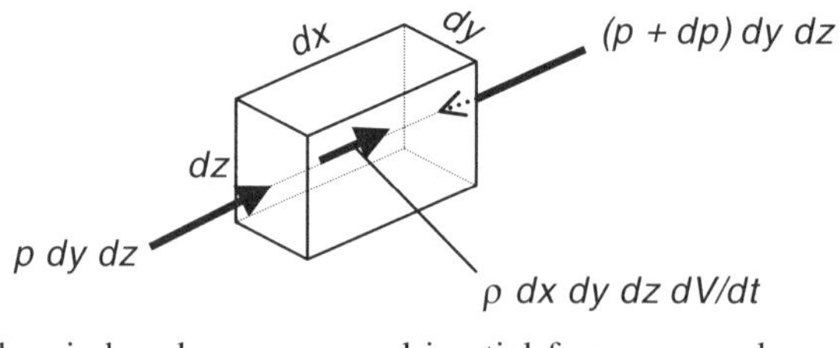

Figure 2.5.1. Flow-induced pressures and inertial force on an elemental volume of a fluid in motion.

2.5.1.2 Reynolds Numbers. In addition to pressures, real flows experience *shear stresses.* For example, if the flow velocity changes across the coordinate axis z normal to a streamline (Fig. 2.5.2), the shear stress is

$$\sigma = \mu \frac{dV}{dz} \tag{2.5.3}$$

The constant μ is the *fluid viscosity*. Viscosity causes the fluid to adhere to the boundaries, where the flow velocity vanishes.

Bernoulli's equation shows that in a steady flow, the pressure $p - p_0$—which is on the order of $\frac{1}{2}\rho V^2$ and has the dimensions of force per unit area—creates on a volume of fluid with linear dimensions L inertial forces on the order of $\rho V^2 L^2$. The viscous stresses are on the order of $\mu V/L$ and create viscous (friction) forces on the order of $(\mu V/L)L^2 = \mu VL$. The ratio between the inertial forces and the friction forces for the volume of fluid is therefore on the order of $VL/(\mu/\rho)$. This ratio is nondimensional and is called the *Reynolds number* of the flow. The ratio $\upsilon = \mu/\rho$ is called the fluid's *kinematic viscosity.* The Reynolds number can then be written

$$\text{Re} = \frac{VL}{\upsilon} \tag{2.5.4a}$$

For airflow at the usual temperature and atmospheric pressure conditions, Eq. 2.5.4*a* is

$$\text{Re} = 67{,}000\ VL \tag{2.5.4b}$$

where V and L are given in m/s and m, respectively.

Numerical Example 2.5.1. *Reynolds numbers in the model and the prototype.* For a circular cylinder with diameter $D = 30$ m in flow with the characteristic velocity $V = 40$ m/s, $\text{Re} = 67{,}000 \times 40 \times 30 = 8 \times 10^7$ (Eq. 2.5.4*b*). For a 1: 400 model of the cylinder in flow with a characteristic velocity of the

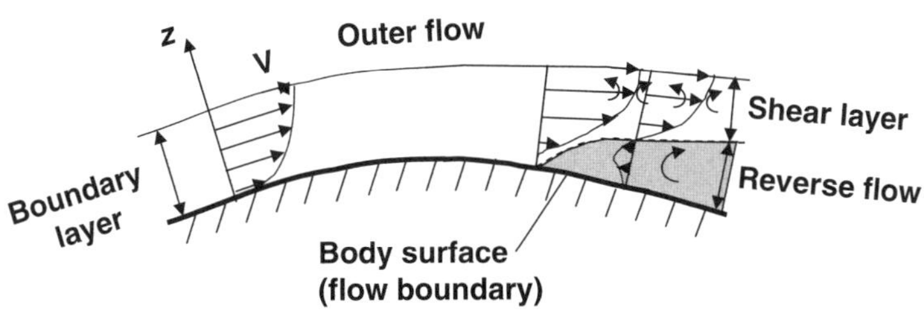

Figure 2.5.2. Velocity profile in the boundary layer and in the separation zone of a flow near a curved body surface. (After *Aérodynamique,* Centre Scientifique et Technique du Bâtiment, Nantes, France, 1980.)

order of $V_{\mathrm{model}} = 10$ m/s (typical of commercial wind tunnels), Re $= 5 \times 10^4$, that is, about three orders of magnitude less than in the full-scale case.

The flow is controlled by its boundary conditions (i.e., the geometry of the body) and the Reynolds number. The choice of the characteristic velocity V and length L is arbitrary, provided that it is indicated explicitly and that experimental or theoretical results are reported in a manner consistent with that choice. For example, for a cylinder with diameter D immersed in a flow with mean oncoming velocity V, Re may be based on a characteristic velocity V and a characteristic length L equal to the cylinder diameter D (this is the standard choice). Alternatively, Re could be based on a characteristic velocity V and, say, a characteristic length $L = D/2$. The significance of the Reynolds number for the model testing of structures is discussed in Sect. 2.5.6.

2.5.2 Boundary Layers, Flow Separation and Reattachment, Negative Pressures, and Drag

Friction, which causes the velocity to vanish at the flow boundary, retards the flow within a *boundary layer* (Fig. 2.5.2), in which shear stresses are present (Eq. 2.5.3). The retardation effect (and the shears) are weaker as the distance from the boundary increases. Outside the boundary layer the effects of friction become negligible, and in the absence of turbulence in the oncoming flow, an ideal, frictionless flow prevails.

In a flow through a nozzle, the velocity through the nozzle's constricted area is higher than the velocity through the larger area upstream of the nozzle, since the same amount of fluid must flow per unit of time through both areas. A similar explanation applies to *hills,* which, like a nozzle, tend to constrict the area available for the passage of the fluid and cause the occurrence of speed-up effects (Sects. 2.3.2.6 and 3.2.4). Conversely, if the area through which the flow occurs increases downwind, the flow decelerates. In addition, friction at a boundary slows down the flow, an effect that can lead to *flow reversal* and *flow separation,* typically associated with the formation of a turbulent *shear layer* (Figs. 2.5.2 and 2.5.3).

A visualization of flow separation for a boxlike shape and of the turbulent flow in the separation zone is shown in Fig. 2.5.4*a*. If the shape of the deck is streamlined (Fig. 2.5.4*b*), the separation zone decreases substantially, and the turbulent flow above the upper face of the deck almost disappears.

The increase in the downwind area through which the flow occurs is a function of the curvature of the boundary surface. For given Reynolds number Re the flow is more prone to separation if the radius of curvature is small. At sharp edges, where the radius of curvature is very small or zero, separation will occur except in flows with very low Re (e.g., honeylike flows). Depending on the flow and the boundary shape, a separated flow may experience *reattachment* (Fig. 2.5.3).

In zones between a shear layer and a flow boundary, or between shear layers in the wake of a body, pressures are equal to the negative pressures

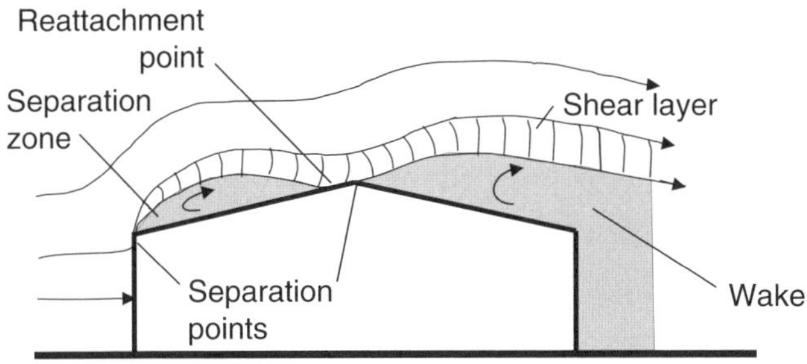

Figure 2.5.3. Flow about a building with sharp edges. (After *Aérodynamique,* Centre Scientifique et Technique du Bâtiment, Nantes, France, 1980.)

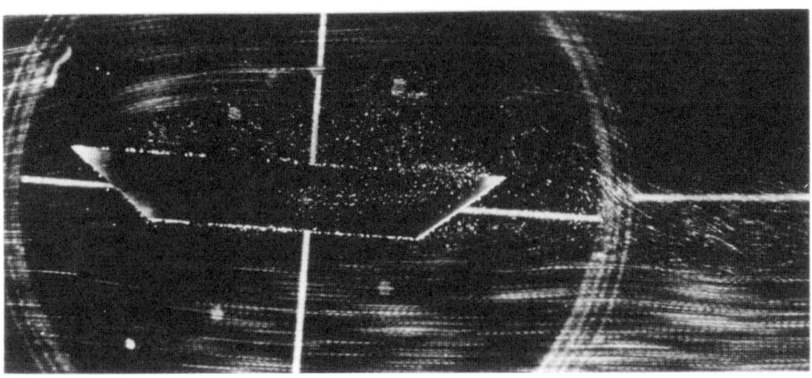

(*a*)

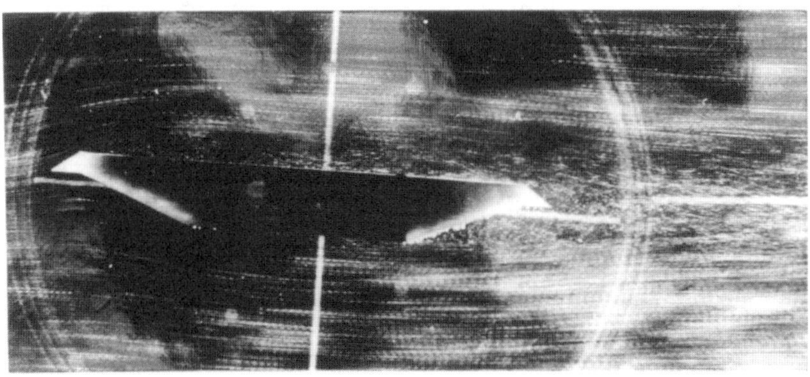

(*b*)

Figure 2.5.4. Visualization of water flow over (*a*) a model bridge deck section and (*b*) a partially streamlined model bridge deck section. (Courtesy of the National Aeronautical Establishment, National Research Council of Canada.)

just outside the shear layers. In separation zones and on its leeward side, a body is therefore typically subjected to negative pressures (Fig. 2.5.3). Like the flow in the separation zone, those pressures fluctuate in time. The resultant of the forces due to positive and negative pressures acting on a body in the direction of the mean flow is called the drag force, or *drag*.

2.5.3 Vortex Shedding, the Strouhal Number, and Vortex-Induced Fluctuating Lift

2.5.3.1 Vortex Shedding and the Strouhal Number. Cylindrical or prismatic bodies immersed in uniform, smooth flow shed in their wake alternating vortices at a dominant frequency n_s given by the relation

$$n_s = \mathrm{St}\,\frac{V}{D} \tag{2.5.5}$$

where the Strouhal number St depends on the cross section of the body and the Reynolds number, V is the constant flow velocity, and D is a characteristic dimension of the cross section. Vortex shedding also occurs under three-dimensional conditions, that is, in nonuniform and turbulent flows and for tapered and/or relatively short bodies. Examples of vortex shedding are shown in Figs. 2.5.5 and 2.5.6. In both cases the vorticity is regular.

2.5.3.2 Vortex-Induced Fluctuating Lift. At any given instant the fluctuating flow in a prism's or cylinder's wake is *asymmetrical* about a line parallel

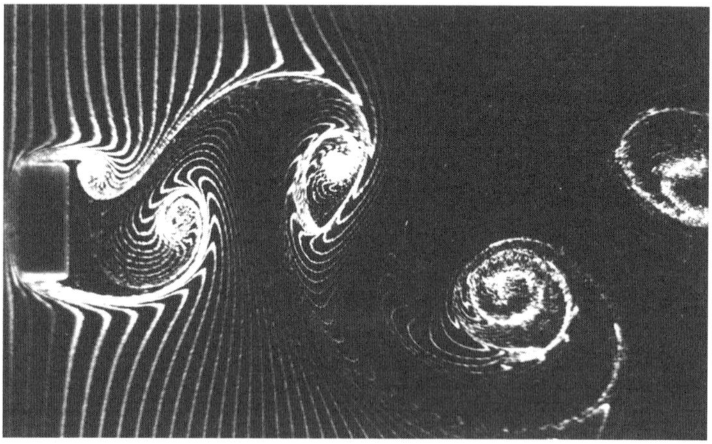

Figure 2.5.5. Flow around a rectangular cylinder (Re = 200). (From Y. Nakamura, "Bluff-body aerodynamics and turbulence," *J. Wind Eng. Ind. Aerodynam.*, **49,** 65–68, 1993. Copyright 1993, with permission from Elsevier.)

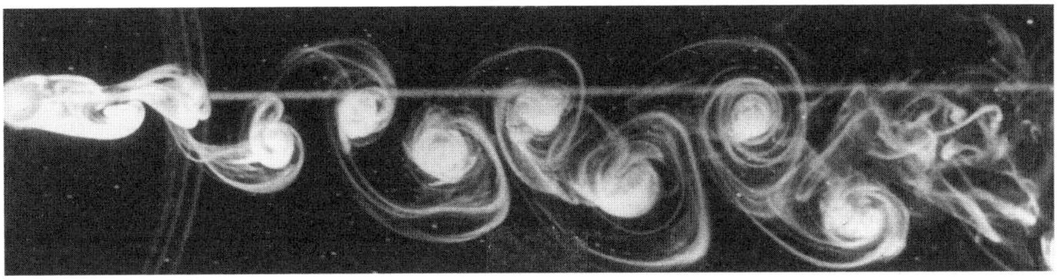

Figure 2.5.6 Vortices shed in the wake of a circular cylinder in a water tunnel. (From the National Aeronautical Establishment, National Research Council of Canada.)

to the oncoming flow (Fig. 2.5.5) and therefore so are the pressures the flow induces on the cylinder. The effect of the asymmetrical pressures is to induce on the body a transverse fluctuating load *perpendicular* to the oncoming flow, called the lift force or *lift* (see also Sect. 2.7.1).

2.5.4 Pressure and Force Coefficients and Internal Pressures

2.5.4.1 Nondimensional Pressure and Force Coefficients. Since the quantity $\frac{1}{2}\rho V^2$ has the dimensions of pressure, the ratio

$$C_p = \frac{p - p_0}{\frac{1}{2}\rho V^2} \tag{2.5.6}$$

where V is a reference wind speed (e.g., the wind speed at the elevation of a building's eave or top) and is a nondimensional quantity termed the *pressure coefficient.* Similarly, drag or lift force coefficients may be defined as ratios of drag forces or lift forces to the reference force $\frac{1}{2}\rho V^2 A$, where A is a reference area. Moment coefficients are defined as ratios between aerodynamic moments and reference moments $\frac{1}{2}\rho V^2 AB$, where B is a reference linear dimension.

Pressures generally vary randomly in time and in space, as can be seen clearly in the animation of pressures induced by turbulent flow included in the main page of the site http://www.nist.gov/wind. The random variation of pressures in space means that pressures at different points do not act in tune (i.e., have imperfect spatial coherence), so peak forces acting on an area tend to decrease as the area increases. This fact is reflected in the definition of *effective wind areas* used in the ASCE Standard (see Sect. 3.3.4).

2.5.4.2 Internal Pressures. Loads on cladding or roof components are determined by both external and internal pressures. If the building has an opening on the windward (leeward) side and is otherwise sealed, the wind flow will create a positive (Fig. 2.5.7*a*) or negative (Fig. 2.5.7*b*) internal pressure. In most cases the opening or porosity distribution over the building envelope

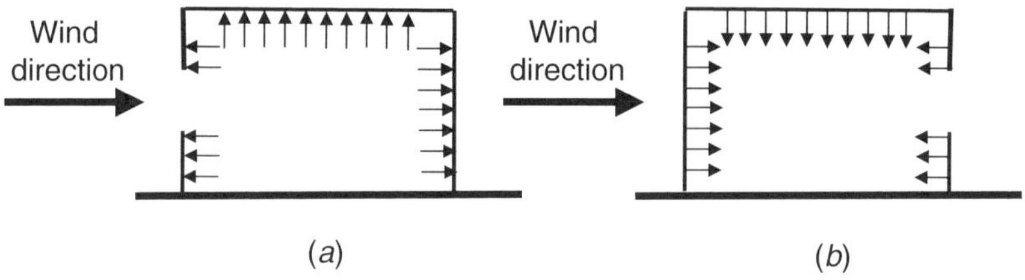

Figure 2.5.7. Mean internal pressures in buildings with openings.

is not well known, and internal pressures could be positive or negative. The ASCE Standard provisions specify internal pressures that depend on the size and distribution of openings in the building envelope (Sect. 3.5.2, item 8). For additional material and references, see [2.33].

2.5.5 Reynolds Number and Drag: Drag Reduction for Bodies with Rounded Shapes

2.5.5.1 Reynolds Number Effects on Drag for Circular Cylinders. We consider briefly the case of the circular cylinder in uniform laminar flow. Figure 2.5.6 shows vorticity shed periodically in a cylinder's wake for $30 \le \text{Re} < 5000$ or so. Note the coherence of the vortex formations. For higher Re the flow undergoes various changes, until for $\text{Re} > 10^5$ or so the wake becomes turbulent, no longer exhibiting strong coherence; the vortex shedding is irregular, but it has a dominant frequency corresponding, for rough cylinders, to a Strouhal number $\text{St} \approx 0.2$ up to $\text{Re} = 10^7$ [2.27].[10] For $10^5 < \text{Re} < 2 \times 10^5$, $C_D \approx 1.2$. At $\text{Re} \approx 2 \times 10^5$ the drag coefficient drops dramatically, from $C_D \approx 1.2$ to $C_D \approx 0.25$, then increases gradually with Reynolds number until it reaches a value $C_D \approx 0.6$ for $\text{Re} = 10^7$.

The drop in the drag coefficient can be explained as follows. At relatively low velocities V, the flow in the boundary layer that forms near the cylinder surface is laminar (smooth). Higher velocities V cause the flow in the boundary layer upwind of the separation points to become unstable, that is, to change from laminar to turbulent. The turbulence causes an exchange of fluid particles between the boundary layer and the zone just outside it. The outside fluid that, owing to this exchange, penetrates the boundary layer, has on average higher velocities than the velocities in the boundary layer flow within which the flow is retarded. The effect of these higher velocities is that flow reversal and separation occur farther downwind than would have been the case in the absence of the turbulent transport, and the wake becomes narrower. This reduces the cylinder's area subjected to negative pressures and therefore

[10] The dependence of St on Re appears to be more complicated for smooth cylinders (see Fig. 5.2.5).

reduces the drag. The cylinder with a narrower wake behaves as if it were aerodynamically more streamlined. Turbulence present in the oncoming flow can also cause a narrowing of the wake and a consequent drag reduction.

2.5.5.2 Drag Reduction Due to Turbulence near Boundaries. Turbulence promoting exchange of fluid between the boundary layer and the higher-velocities zone outside it can be created artificially by roughening the surface of the body in various ways. For example, a shift of the flow separation downstream, with consequent narrowing of the wake and decrease in drag, is achieved in golf balls through the provision of striations that cause turbulence. Drag reduction for hyperbolic cooling towers [2.5, p. 404] is achieved by narrowing the wake through the provision on the towers' rib surfaces,[11] whose effect is similar to the effect of the striations of a golf ball.

2.5.6 Wind Tunnel Modeling of Full-Scale Flows

2.5.6.1 Reynolds Number Similarity and Its Violation in Conventional Wind Tunnels. In principle, for similarity between prototype (i.e., full-scale) and wind tunnel flows to be achieved, the respective Reynolds numbers must be the same. This requirement is referred to as *Reynolds number similarity.* In conventional wind tunnels, the fluid used for testing models of structures is air at atmospheric pressure, and Reynolds number similarity is unavoidably violated (for an illustration, see Numerical Example 2.5.1).

2.5.6.2 Reynolds Numbers and Flows Around Bodies with Rounded Shapes and with Sharp Edges. The reproduction in the wind tunnel of the aerodynamic behavior of rounded bodies is difficult, owing to the violation of Reynolds number similarity. Separation occurs farther downwind in the prototype (i.e., at full scale), where Re is larger (i.e., where inertial forces are larger relative to viscous forces) than in the wind tunnel. However, rendering the model surface rougher, or adding to the body surface wires or other devices that "trip" the flow, reduces the discrepancies between prototype and model-scale flows.

Unlike bodies with rounded shapes, bodies with sharp edges have fixed separation points (Fig. 2.5.3) whose position at the edges is independent of Reynolds number. It has therefore been hypothesized that flows around such bodies are similar at full scale and in the wind tunnel even if Reynolds number similarity is violated.

This is only partially true. In a wind tunnel the friction forces are larger in relation to the inertial forces than at full scale, so the counterparts of full-scale high-frequency flow fluctuations (i.e., the smaller eddies) are damped

[11] This was not known to designers of hyperbolic towers in the People's Republic of China during the years of the Cultural Revolution, a period of isolation from the outside world [2.28].

out. This affects the local vorticity at edges and corners in the wind tunnel, resulting in local pressures typically weaker than at full scale. Also, the damping out in the wind tunnel of higher-frequency turbulence components, which are effective in bringing about the exchange of fluid particles between the boundary layer and the outer flow, results in weaker promoton of reattachment in the model than in the prototype. The first tests of building models in boundary-layer flows were performed in the early 1930s in Goettingen by Flachsbart,[12] under Prandtl's supervision.

2.5.6.3 Variation of Test Results from Laboratory to Laboratory. Independent tests conducted by six prominent wind tunnel laboratories on models of two industrial buildings with 20- and 32-ft eave height [2.25], and by three prominent North American laboratories on models of a smaller building [2.26], showed that tests results can vary significantly from laboratory to laboratory. For this reason, as a matter of policy, some structural engineering firms engaged in the design of important structures commission wind tunnel tests from more than one laboratory.

For an industrial building with a 20-ft eave height, the ratio of the larger to the smallest reported estimate of the wind-induced bending moments at a typical portal frame knee varied from 1.4 to 2.4, with coefficients of variation of 0.15 to 0.35. For a similar building with a 32-ft eave height, the ratio varied from 1.15 to 1.55 (coefficients of variation of 0.05 to 0.15) for open terrain and from 1.3 to 2.15 (coefficients of variation of 0.12 to 0.30) for suburban terrain [2.25]. The variability was greater for the lower building and for the building in suburban terrain. The discrepancies among results reported in [2.26] were similar. Such discrepancies may be due to the difficulty of defining the reference mean flow at relatively low eave levels, especially in suburban terrain, where the turbulence is relatively high. An issue that may warrant debate is whether an increase in model scale would improve this state of affairs. Such an increase would worsen the ratio between integral turbulence scale and characteristic building dimension, but the consequent errors may be less severe than those due to poor modeling of the mean flow at the eave height.

It is likely that for taller buildings the variability of results obtained in different wind tunnels would be lower, particularly in open terrain. In Sect. 2.7 we discuss briefly a rational way to accommodate such variability through the application of simple reliability concepts.

2.6 STRUCTURAL DYNAMICS

Under wind loads, flexible structures can experience dynamic effects, that is, effects involving resonance. A well-known example of resonance is the effect

[12] Because he refused to divorce his Jewish wife, Nazi authorities prohibited Flachsbart from publishing after 1934 [2.29]. His work was forgotten until 1986, when it was rediscovered in the United States and partially reproduced in [2.5, p. 173].

on a bridge of a military formation marching in lock step at a frequency equal to the bridge's fundamental frequency of vibration. The effects of successive steps are additive: A first step causes a deflection whose maximum is reached when the second step strikes. The second step causes an additional deflection, and subsequent steps keep adding to the response.

The mathematical model for the dynamic response of a simple one-degree-of-freedom oscillator is Newton's second law (Sect. 2.6.1). Newton's second law is also applied to continuous structures such as buildings, towers, or bridges (Sect. 2.6.2).

2.6.1 Dynamics of the Single-Degree-of-Freedom Oscillator

The system of Fig. 2.6.1 consists of a particle of mass m concentrated at point B and of a member AB with linear elastic behavior and negligible mass. The particle is subjected to the force $F(t)$. Its displacement $x(t)$ is opposed by (1) a restoring force, kx supplied by the member AB; and (2) a damping force, $c\ dx/dt \equiv -c\dot{x}$,[13] where the stiffness k and the damping coefficient c are constant. Newton's second law states that the particle's mass times its acceleration, $M\ddot{x}$, is equal to the total force applied to the particle. The equation of motion of the system is

Figure 2.6.1. Single-degree-of-freedom system.

[13]Here and elsewhere in the book the dot denotes differentiation with respect to time, that is, $\dot{x} \equiv dx/dt$.

$$M\ddot{x} = -c\dot{x} - kx + F(t) \tag{2.6.1}$$

With the notations $n_1 = \sqrt{k/M}/2\pi$ and $\zeta_1 = c/2\sqrt{kM}$, where n_1 denotes the frequency of vibration of the oscillator and ζ_1 is called the damping ratio, Eq. 2.6.1 becomes

$$\ddot{x} + 2\zeta_1(2\pi n_1)\dot{x} + (2\pi n_1)^2 x = \frac{F(t)}{M} \tag{2.6.2}$$

For structures ζ_1 is typically small (on the order of 1%). In the particular case of a harmonic load, $F(t) = F_0 \cos 2\pi nt$, the response can be written as

$$x(t) = H(n)F_0 \cos(2\pi nt - \theta) \tag{2.6.3}$$

$$H(n) = \frac{1}{4\pi^2 n_1^2 M\{[1 - (n/n_1)^2]^2 + 4\zeta_1^2(n/n_1)]^2\}^{1/2}} \tag{2.6.4}$$

In Eq. 2.6.3, θ is a phase angle dependent on n, n_1, and ζ_1. The function $H(n)$ is called the *mechanical admittance function.* Note that for the case $n = n_1$, the amplitude of the response is largest and is inversely proportional to the damping ratio ζ_1. In this case the motion exhibits *resonance* (or *resonant amplification*). Instead of the damping ratio ζ_1, some analysts use as a measure of damping the *logarithmic decrement,* defined as

$$\delta_1 = \frac{2\pi\zeta_1}{\sqrt{1 - \zeta_1^2}} \tag{2.6.5}$$

in which it may typically be assumed for practical purposes that the denominator is unity.

2.6.2 Dynamics of Continuous Systems

Typically, a structure is not well modeled by a concentrated mass. A linearly elastic structure with continuously distributed mass per unit length, $m(z)$, and low damping can be shown to vibrate in resonance with the exciting force if the latter has certain sharply defined frequencies called the structure's *natural frequencies of vibration.* Associated with each natural frequency is a *mode,* or *modal shape,* of the vibrating structure. The first four normal modes $x_i(z)$ ($i = 1, 2, 3, 4$) of a vertical cantilever beam with running coordinate z are shown in Fig. 2.6.2. The natural modes and frequencies are structural properties independent of the loads (see Sect. 4.3.1.2.1).

A deflection $x(z, t)$ due to forcing of a continuous system can in general be written in the form

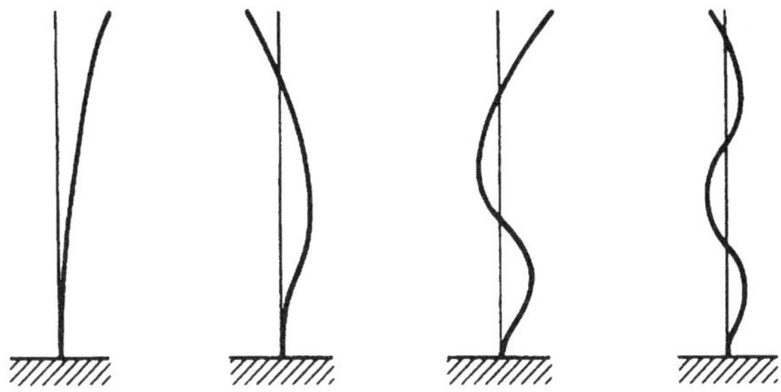

Figure 2.6.2. First four normal modal shapes of a cantilever beam.

$$x(z, t) = \sum_i x_i(z)\xi_i(t) \tag{2.6.6}$$

where the functions $\xi_i(t)$ are called the *generalized coordinates* of the system and $x_i(z)$ are modal shapes in the ith mode of vibration. For a building, similar expressions hold for deflections $y(z, t)$ in a horizontal direction normal to the direction x and for horizontal torsional angles $\varphi(z, t)$. For buildings whose centers of mass and elastic centers do not coincide, the motions in the x, y, and φ modes are coupled (Sect. 4.3.1.2.1). Equations of motion that make use of Eq. 2.6.6 and their y-direction and torsional counterparts are developed in Sect. 4.3.1.2.2.

2.6.3 Wind Loading and Dynamic Response in the Time and Frequency Domains

For tall buildings, current computer capabilities allow the routine solution of the equations of motion in the time domain (Sect. 4.3). This more direct and transparent approach was not possible in the 1950s and subsequent decades, when the spectral approach, which obviates the need for solving large systems of ordinary differential equations, was widely used.

Nevertheless, in the present state of the art, the spectral approach continues to be used for long-span bridge response analyses. The spectrum of a statistically stationary fluctuating signal (see Sect. 2.3.1.1) is a function of frequency that provides a measure of the contributions of each of its sinusoidal components to the signal's mean square value (Section 2.3.3.1.3). The cross-spectrum of two statistically stationary signals (Sect. 2.3.3.1.4) is a function of frequency n that provides a measure of the extent to which the signals' components with frequencies n are or are not in tune with each other (have imperfect mutual coherence).

If the signals consist of fluctuating speeds at two points in a turbulent flow, or of fluctuating pressures at two points of a body surface, their cross-spectrum is not only a function of frequency n, but for any fixed n the cross-spectrum is also a decreasing function of the distance between the two points, meaning that the smaller the distance between the two points, the larger is the degree to which the components with frequency n of the two signals are in tune with each other. Consider, for example, the aerodynamic pressures generated at two points of a vertical building's upwind face by oncoming fluctuating flow. If the lateral and vertical integral turbulence scales (Sect. 2.3.3.1.2) of the incoming flow are larger than the distance between the two points, lower-frequency components of the two pressures act to a large degree in tune (additively), meaning that the cross-spectrum at low frequencies is relatively large. In contrast, the high-frequency pressure components tend to act independent of each other; that is, their cross-spectrum is small or zero. For mathematical details on the cross-spectral approach, see [2.5]. For an application to long-span bridges, see Appendix 5C.

2.7 AEROELASTICITY

Flow-induced structural motions can modify the flow in a manner that, in turn, affects the structural motions. Structural motions that enhance the aerodynamic action of the flow on the structure are called *self-excited,* and the behavior associated with them is termed *aeroelastic.* The flutter of the Brighton Chain Pier Bridge (called "undulation" in the nineteenth century) (Fig. 2.7.1), and more than a century later the flutter of the original Tacoma–Narrows Bridge (Fig. 2.7.2) are notorious examples of aeroelastic behavior. Tall chimneys and buildings can also respond aeroelastically. Boston's John Hancock building, which has a relatively flat shape in plan, has experienced severe transverse and torsional motions that warranted the installation of a large tuned-mass damper system at its top. These motions may have been due to vortex-induced aeroelastic effects.[14]

To describe the interaction between aeodynamic forces and the structural motions they induce, it is in principle necessary to solve the full equations of motion describing the flow, with time-dependent boundary conditions imposed by the moving structure. For bluff bodies in turbulent flow, this problem defies analytical capabilities, and despite continual progress, it remains difficult to solve the problem dependably by computational fluid dynamics methods. Therefore, the aeroelastic characterization of civil engineering structures relies largely on laboratory testing and empirical modeling. Applicability to the prototype of laboratory test results and associated empirical models needs to be assessed carefully, owing to violation of the Reynolds number similarity criterion in commercial wind tunnel testing. However, for carefully modeled

[14] Recent research on tall buildings with a large depth/width ratio appears to suggest that this was indeed the case. To the authors' knowledge, for legal or other reasons, detailed technical reports on the wind-induced behavior of the John Hancock building are not available in the public domain.

Figure 2.7.1 Brighton chain pier failure, 1836. (From J. S. Russel, "On the vibration of suspension bridges and other structures; and the means of preventing injury from this cause." *Trans. R. Scott. Soc. Arts,* 1841.)

structures, aeroelastic test results are generally assumed to yield reasonably realistic results.

The flow-body interaction is destabilizing or stabilizing if an energy transfer occurs from the flow into the body or from the body into the flow, respectively. By analogy with viscous damping forces, which tend to reduce the energy of an oscillating body, aeroelastic forces are referred to as *negative aerodynamic damping* forces if they tend to *increase* the body's oscillatory energy, that is, if they are destabilizing. Stabilizing aeroelastic forces are referred to as *positive aerodynamic damping* forces.

The purpose of this section is to provide an introduction to basic aeroelastic phenomena of interest in the design of flexible structures. Additional details are available in Chapters 4 and 5. Section 2.7.1 discusses aeroelasticity phenomena associated with vortex lock-in. Section 2.7.2 is concerned with galloping. Because a galloping motion takes place in one dimension, its aeroelastic behavior is relatively simple and provides an excellent point of departure for understanding the basic flow–structure interactions that produce flutter. Fundamentally, galloping occurs in some types of body because, as the body moves, the angle of attack of the relative velocity of the flow with respect to the body changes. In this sense the aeroelastic phenomena that

Figure 2.7.2. Flutter of the Tacoma Narrows bridge, 1940. (From F. B. Fraquharson, ed., *Aerodynamic Stability of Suspension Bridges,* Part 1, Bulletin 116, University of Washington Engineering Experiment Station, Seattle, WA, 1949–1954.)

govern flutter motion are similar. They are more complicated, however, for two reasons. First, they involve effects due to vorticity, which are absent in the galloping case. Second, they are associated with motions in three degrees of freedom instead of just one degree of freedom. Section 2.7.3 considers flutter and buffeting in the presence of aeroelastic effects. The interesting relation between the nature of the separated flow near the body and some aspects of galloping and flutter is discussed in detail in Chapter 5. For the static aeroelastic phenomenon called *torsional divergence instability,* which rarely occurs in bridges, see [2.5, Sects. 6.4 and 13.1.2] and [2.21, Sect. 6.3].

2.7.1 Vortex-Induced Oscillations

2.7.1.1 Vortex-Induced Lock-in. The shedding of vortices in the wake of a body gives rise to fluctuating *lift forces*. If the body is flexible or if it has elastic supports, it will experience motions due to aerodynamic forces and, in particular, to the fluctuating lift force. As long as the motions are sufficiently small they do not affect the vortex shedding, and Eq. 2.5.5 remains valid. If the vortex-shedding frequency n_s, and therefore the frequency of the associated lift force, is equal to a natural frequency of vibration of the body n_1, resonant amplification can occur. Experiments show that this is the case not only at the flow speed $n_s D/\mathrm{St}$ (Eq. 2.5.5), but also at any speed V within an interval $n_s D/\mathrm{St} - \Delta V < V < n_s D/\mathrm{St} + \Delta V$, where $\Delta V/V$ depends on cross-sectional shape and mechanical damping and is on the order of a few percent. Within that interval the vortex shedding frequency no longer conforms to Eq. 2.5.5, but aligns itself to the body's frequency, n_s.

This is an aeroelastic effect: While the flow affects the body motion, the body motion in turn affects the flow insofar as it produces *lock-in,* that is, synchronization of the vortex-shedding frequency with the frequency of vibration of the body.

2.7.1.2 Vortex-Induced Oscillations and Increased Lift-Force Along-Span Correlations. A second aeroelastic effect due to a cylinder's vibration is of interest in practice. If the cylinder is infinitely rigid, vortex-induced lift forces per unit span at different stations along the cylinder are imperfectly correlated; that is, they are not perfectly in tune with each other. However, if the cylinder oscillates under excitation by vortex-induced lift forces, the oscillations bring about an increase in the correlation among those forces. This, in turn, results in increased amplitudes of the cylinder's oscillations.

2.7.1.3 Modeling of Vortex-Induced Oscillations as a Single-Degree-of-Freedom Motion. A third aeroelastic effect characterizing vortex-induced oscillations is the development of aeroelastic forces associated with flow modifications induced by the oscillations. A variety of vortex-induced oscillation models are available in which the aeroelastic forces depend on adjustable parameters fitted to match experimental results. By construction, the solutions of the model equations with parameters fitted to those results provide a reasonable description of the aeroelastic motions observed. However, an empirical model may not be valid as a motion predictor for conditions that differ significantly from the experimental conditions for which the fitted parameters were obtained.

For long elastically supported cylinders in uniform smooth flow, we review Scanlan's simple but useful single-degree-of-freedom model [2.5]:

$$m[\ddot{y} + 2\pi\zeta n_1 \dot{y} + (2\pi n_1)^2 y] = \frac{1}{2}\rho V^2 D \left[Y_1(K)\left(1 - \varepsilon \frac{y^2}{D^2}\right)\right] \frac{\dot{y}}{V} \quad (2.7.1)$$

where m is the body mass per unit length, ζ is the damping ratio, n_1 is the frequency of vibration of the body, D is the cylinder's diameter, V is the flow velocity, ρ is the specific mass of the fluid, $K = 2\pi Dn/V$, and the vortex-shedding frequency n satisfies the Strouhal relation $n = \mathrm{St}V/D$; Y_1 and ε are adjustable parameters that must be fitted to experimental results. This model allows for negative and positive aerodynamic damping at low and high body displacements, respectively, that is, for the aeroelastic transfer of energy from the flow to the body and from the body to the flow. Scanlan's model was used for the estimation of vortex-shedding effects on tall chimneys and on bridge decks [2.5].

2.7.2 Galloping

Galloping is a large-amplitude aeroelastic oscillation (one to 10 or more cross-sectional dimensions of the body) that can be experienced by cylinders or prisms with certain types of cross section (e.g., rectangular section, D-section, ice-laden power cables). The oscillations occur in a plane normal to the oncoming flow velocity, at frequencies that are much lower than the vortex-shedding frequencies for the same sections. The flow speeds that cause galloping motions are typically considerably larger than those that cause vortex lock-in. Flow reattachment, which is present at vortex lock-in and at flutter, does not occur in the galloping case; fully separated flows are thus a feature of galloping motions and result in the absence of vortex-induced effects on the body (Sects. 5.2.2 and 5.2.3).

Fundamental to the galloping phenomenon is the fact that the angle of attack of the relative flow velocity with respect to the body changes as the body experiences an incipient motion away from its position of equilibrium. This changed relative velocity creates in bodies with certain cross-sectional shapes asymmetrical pressure distributions that enhance that incipient motion, rather than suppress it, as would be the case if the body were aeroelastically stable.

We consider a square cylinder immersed in a flow with velocity V in the x-direction (Fig. 2.7.3). It is assumed that the body is elastically restrained and mechanically damped in the direction normal to the x-axis and that its natural frequency is much lower than the vortex-shedding frequency in the body's wake, so the effect of the vortices upon the lift is negligible. Let a small perturbation cause the cylinder to move *downward* from its position of

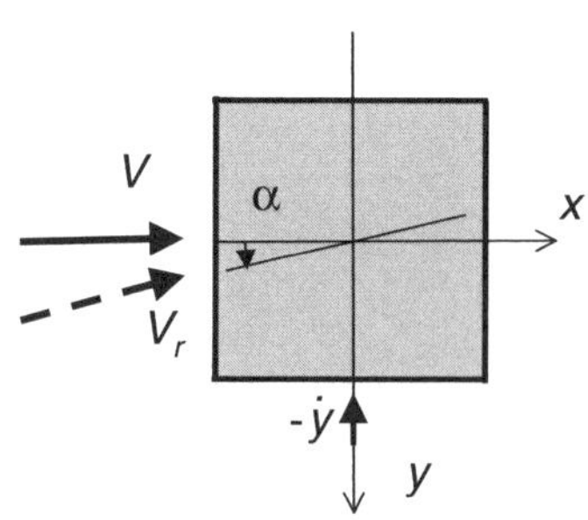

Figure 2.7.3 Flow velocity V and relative flow velocity V_r with respect to a square cylinder moving downward with velocity $\dot{y}$.

equilibrium. Associated with the motion is a downward velocity $\dot{y}$ of the body with respect to the flow. The relative vertical velocity of the flow with respect to the body is then $v = -\dot{y}$ (*upward*), and the total relative velocity, V_r, of the flow with respect to the body is the vector sum of the velocities V and $-\dot{y}$ (Fig. 2.7.3). The flow has an effective angle of attack with respect to the body $\alpha \approx -\dot{y}/V$; that is, *the relative velocity V_r makes an angle α with the axis x.* As indicated earlier, if the aerodynamic properties of the body are such that for $\alpha \approx 0$, the force induced on the body by the relative velocity V_r tends to push the downward-moving body farther downward rather than bringing it back to its original undeformed position (i.e., if V_r causes the average pressure on the upper side of the body to be larger than on the lower side), the body is unstable—it will experience galloping motion. We now obtain the necessary condition for this to be the case.

The drag and lift force along direction α may be written as

$$D(\alpha) = \tfrac{1}{2}\rho V_r^2 B C_D(\alpha) \tag{2.7.2a}$$

$$L(\alpha) = \tfrac{1}{2}\rho V_r^2 B C_L(\alpha) \tag{2.7.2b}$$

The projection of these components on the direction y is

$$F_y(\alpha) = -D(\alpha)\sin\alpha - L(\alpha)\cos\alpha \tag{2.7.2c}$$

We now write $F_y(\alpha)$ in the alternative form

$$F_y(\alpha) = \tfrac{1}{2}\rho V^2 B C_{F_y}(\alpha) \tag{2.7.2d}$$

Since $V = V_r \cos\alpha$, it follows that

$$C_{F_y}(\alpha) = \frac{-[C_L(\alpha) + C_D(\alpha)\tan\alpha]}{\cos\alpha} \tag{2.7.2e}$$

For $\alpha = 0$ it follows immediately from the differentiation of Eq. 2.7.2*e* with respect to α that

$$\left.\frac{dC_{F_y}}{d\alpha}\right|_{\alpha=0} = -\left(\frac{dC_L}{d\alpha} + C_D\right)_{\alpha=0} \tag{2.7.2f}$$

For small α, $\alpha = -\dot{y}/V$ and

$$F_y \approx \left.\frac{dF_y}{d\alpha}\right|_{\alpha=0} \left(-\frac{\dot{y}}{V}\right) \tag{2.7.2g}$$

$$= \frac{1}{2}\rho V^2 B \left(\frac{dC_L}{d\alpha} + C_D\right)_{\alpha=0} \frac{\dot{y}}{V} \tag{2.7.2h}$$

The sum of the mechanical and aerodynamic damping affecting the system is therefore

$$2m\omega\zeta + \frac{1}{2}\rho VB\left(\frac{dC_L}{d\alpha} + C_D\right)_{\alpha=0} = d \tag{2.7.2i}$$

The second term on the left-hand side of Eq. 2.7.2*i* can be interpreted as an aerodynamic damping term. For galloping to be possible, the aerodynamic damping must be negative, that is, the quantity between parentheses in Eq. 2.7.2*i* must be negative. This is known as the *Glauert–Den Hartog necessary condition for galloping*. This in turn requires that $(dC_L/d\alpha)_{\alpha=0} < 0$ (Fig. 2.7.4). Note that if the necessary condition for galloping is satisfied and the mechanical damping is vanishingly small, galloping can in theory occur for any flow velocity, however small.

Tests have shown that the derivatives $dC_{F_y}/d\alpha$ (Eq. 2.7.2*f*) do not depend on the frequency of the body motion and can be obtained from aerodynamic force measurements on the *fixed* body. Aerodynamic arguments developed in Chapter 5 explain this fact in some detail. $dC_{F_y}/d\alpha$ are called *steady-state aerodynamic lift coefficient derivatives* or simply, *steady-state aerodynamic derivatives.* We note in the next section that in the case of flutter, the aeroelastic behavior is characterized by quantities of a similar nature, called *flutter*

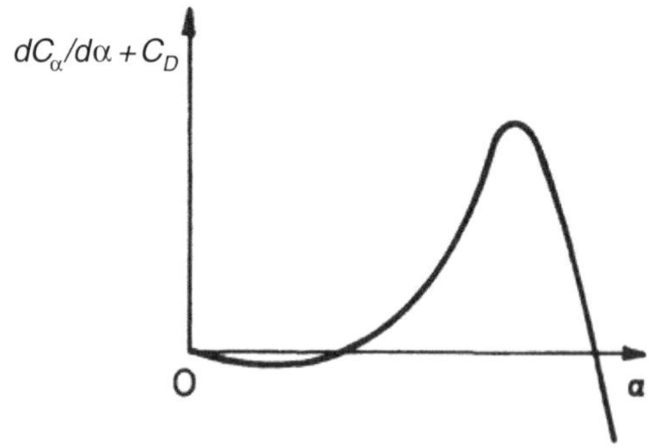

Figure 2.7.4. Dependence of $dC_\alpha/d\alpha + C_D$ on α.

aerodynamic derivatives, which unlike the steady-state derivatives that characterize galloping motion, depend on the oscillation frequency.

2.7.3 Flutter

Flutter is an aeroelastic phenomenon that occurs in flexible bodies with relatively flat shapes in plan. It involves oscillations with amplitudes that grow in time and can result in catastrophic structural failure. Like other aeroelastic phenomena, flutter entails the solution of equations of motion involving inertial, mechanical damping, elastic restraint, and aerodynamic forces (including self-excited forces) that depend on the ambient flow and the shape and motion of the body.

Assume that the mechanical damping is negligible. The motion of the body is aeroelastically *stable* if, following a small perturbation away from its position of equilibrium, the body will revert to that position, owing to stabilizing self-excited forces associated with the perturbation. As the flow velocity increases, the aerodynamic forces acting on the body change, and for a certain critical value of the flow velocity the self-excited forces cause the body to be neutrally stable. For velocities larger than the critical velocity the oscillations initiated by a small perturbation from the position of equilibrium will grow in time. The self-excited forces that cause these growing oscillations can be viewed as producing a *negative aerodynamic damping* effect.

The main difficulty in solving the flutter problem for bridges is the development of expressions for the self-excited forces. For thin airfoil flutter in incompressible flow, self-excited forces due to small oscillations have been derived from basic aerodynamic theory. However, for bridge sections the airfoil solutions are in general not applicable.

To date perhaps the most influential contribution to solving the flutter problem for bridges is a simple conceptual framework wherein the self-excited forces due to relatively small bridge deck oscillations can be characterized by fundamental functions called *flutter aerodynamic derivatives.* As pointed out earlier, in the galloping case the self-excited forces are fully characterized by steady-state derivatives, $dC_{F_y}/d\alpha$. As explained in Chapter 5, the latter may be obtained from measurements on the *fixed* body; in contrast, flutter derivatives depend on oscillation frequency and must be obtained from measurements on the *oscillating* body.

Although flutter is accompanied at all times by vortex shedding with frequency equal to the flutter frequency, it is a phenomenon distinct from vortex-induced oscillation. The latter entails aeroelastic flow–structure interactions only for flow velocities at which the frequency of the vortex shedding is equal or close to the structure's natural frequency. For velocities higher than those at which lock-in occurs, the oscillations are much weaker than at lock-in. In contrast, for velocities higher than those at which flutter sets in, the strength of the oscillations increases monotonically with velocity.

In Sect. 2.7.3.1 we consider two-dimensional bridge deck behavior in smooth flow. Section 2.7.3.2 is concerned, in a two-dimensional context, with bridge deck flutter in turbulent flow and the consequent buffeting of the bridge deck.

2.7.3.1 Formulation of the Two-Dimensional Bridge Flutter Problem in Smooth Flow. Bridge decks are typically symmetrical; that is, their elastic and mass centers coincide. The dependence of flutter derivatives on the oscillation frequency n of the fluttering body can be expressed in terms of the nondimensional *reduced frequency*

$$K = \frac{2\pi B n}{V} \tag{2.7.3}$$

where B is the width of the deck and V is the mean wind flow velocity. If the horizontal displacement p of the deck is also taken into account, the equations of motion of a two-dimensional section of a symmetrical bridge deck with linear viscous damping and elastic restoring forces in smooth flow can be written as

$$m\ddot{h} + c_h\dot{h} + C_h h = L_h \tag{2.7.4a}$$

$$I\ddot{\alpha} + c_\alpha\dot{\alpha} + C_\alpha \alpha = M_\alpha \tag{2.7.4b}$$

$$m\ddot{p} + c_p\dot{p} + C_p p = D_p \tag{2.7.4c}$$

where h, α, and p are the vertical displacement, torsional angle, and horizontal displacement, respectively (Fig. 2.7.5). A unit span is acted upon by the aerodynamic lift L_h, moment M_α, and drag D_p; and has mass m, mass moment of inertia I, vertical, torsional, and horizontal restoring forces with stiffness C_h, C_α, and C_p, respectively, and viscous damping coefficients c_h, c_α, and c_p.

We saw that a galloping body experiences a single-degree-of-freedom motion, and that for small displacements the aeroelastic force acting on the body is linear with respect to the time rate of change of the displacement, the proportionality factor being a function of aerodynamic origin (Eq. 2.7.2h). Flutter entails motions with three degrees of freedom (h, α, and p), and the

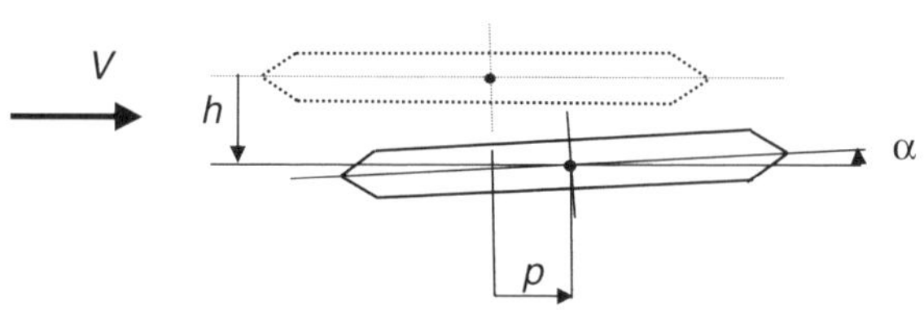

Figure 2.7.5. Notations.

expressions for the aeroelastic forces acting on the body are therefore lengthier than in the galloping case, although conceptually they are related. Indeed, for small displacements the aeroelastic forces can be written as sums of terms that like their galloping counterparts, are linear with respect to the rates of change of h, α, and p, the factors of proportionality also being functions of aerodynamic origin. However, unlike in the case of galloping, terms proportional to h, α, and p come into play as well. (For example, it is intuitively clear that the aerodynamic effects on the body of Fig. 2.7.5 depend on the angle of attack α.) In mathematical terms, the expressions for the aeroelastic forces can therefore be written as sums of terms as follows:

$$L_h = \frac{1}{2}\rho V^2 B\left[KH_1^*(K)\frac{\dot{h}}{V} + KH_2^*(K)\frac{B\dot{\alpha}}{V} + K^2H_3^*(K)\alpha + K^2H_4^*(K)\frac{h}{B} + KH_5^*(K)\frac{\dot{p}}{V} + K^2H_6^*(K)\frac{p}{B}\right] \tag{2.7.5a}$$

$$M_\alpha = \frac{1}{2}\rho V^2 B\left[KA_1^*(K)\frac{\dot{h}}{V} + KA_2^*(K)\frac{B\dot{\alpha}}{V} + K^2A_3^*(K)\alpha + K^2A_4^*(K)\frac{h}{B} + KA_5^*(K)\frac{\dot{p}}{V} + K^2A_6^*(K)\frac{p}{B}\right] \tag{2.7.5b}$$

$$D_p = \frac{1}{2}\rho V^2 B\left[KP_1^*(K)\frac{\dot{p}}{V} + KP_2^*(K)\frac{B\dot{\alpha}}{U} + K^2P_3^*(K)\alpha + K^2P_4^*(K)\frac{p}{B} + KP_5^*(K)\frac{\dot{h}}{V} + K^2P_6^*(K)\frac{h}{B}\right] \tag{2.7.5c}$$

Terms proportional to $\ddot{h}$, $\ddot{\alpha}$, and $\ddot{p}$ (i.e., *added mass terms,* reflecting the forces due to the body motion that result in fluid accelerations around the body) do not appear in Eqs. 2.7.5, as these terms are negligible in wind engineering applications. The role of the terms in h and p is to account for changes in the frequency of vibration of the body due to aeroelastic effects; the terms in α reflect the role of the angle of attack noted earlier. The quantities $\dot{h}/U$ and $B\dot{\alpha}/V$ are effective angles of attack (e.g., the ratio $\dot{h}/U$ has the same significance as in the case of galloping; i.e., it represents the angle of attack of the relative velocity of the flow with respect to the moving body). Those quantities are nondimensional. The coefficients H_i^*, A_i^*, and P_i^* are known as *Scanlan flutter derivatives*[15] and are also nondimensional.

[15] Equations 2.7.5 are formulated in terms of real variables, viewed by many practitioners to be best suited for structural engineering purposes. An alternative wherein the aeroelastic forces and the displacements they induce in the bridge are expressed in terms of complex variables is preferred by others, insofar as it may offer insights into phase relationships among various aeroelastic forces and displacements (see Sect. 5.2.5.3.1).

To summarize, each term in Eqs. 2.7.5 can be viewed as similar in form to terms of the type

$$L = \frac{1}{2}\,\rho V^2 B C_L = \frac{1}{2}\,\rho V^2 B\,\frac{dC_L}{d\alpha}\,\alpha \tag{2.7.6}$$

for small angles of attack α. Quantitites such as KH_i^* and $K^2A_i^*$ are therefore analogous to aerodynamic lift coefficient derivatives $dC_{Fy}/d\alpha$ that arise in the galloping case. Unlike in the galloping case, where owing to the absence of vortex-induced pressures on the body, the derivatives can be obtained experimentally from static tests (i.e., tests in which the body is at rest), for the flutter case the coefficients of the displacements and their time rate of change must be obtained experimentally from measurements on the oscillating body, which owing to its elongated shape is affected by vortex-induced pressures. For this reason those coefficients are called *motional* aerodynamic derivatives, which go over into *steady-state* aerodynamic derivatives such as $dC_L/d\alpha$ for $K \to 0$. In Sects. 5.2.2 and 5.2.3 we discuss in some detail the physical basis for the existence of steady-state derivatives in the galloping case and of motional derivatives in the flutter case. The solution of the flutter equations can be obtained if plots of the flutter derivatives H_i^*, A_i^*, and P_i^* are available from measurements as functions of K. It is assumed that the expressions for h, α, and p are proportional to $e^{i\omega t}$. These expressions are inserted into Eqs. 2.7.1. The determinant of the amplitudes of h, α, and p is set to zero.

For each value of K a complex equation in n is obtained. For a value of K denoted by K_c the corresponding equation yields a real solution n_c. The flutter velocity is

$$V_c = \frac{2\pi B n_c}{K_c} \tag{2.7.7}$$

(see Appendixes 5B and 5C for details). In Eqs. 2.7.5 the terms containing first derivatives of the displacements are measures of *aerodynamic damping*. If, among these terms, only those associated with the coefficients H_1^*, A_2^*, and P_1^* are significant, the total (structural plus aerodynamic) damping can be written as

$$c_h - \tfrac{1}{2}\rho U^2 BKH_1^* \qquad c_\alpha - \tfrac{1}{2}\rho U^2 BKA_2^* \qquad c_p - \tfrac{1}{2}\rho U^2 BKP_1^* \tag{2.7.8a,b,c}$$

for the vertical, torsional, and horizontal degree of freedom, respectively.

The original Tacoma Narrows bridge (Fig. 2.7.2) had negligible H_1^* values for all K, meaning that the total damping (Eq. 2.7.8*a*) for motion in the h direction was positive, thus precluding flutter in the vertical degree of freedom. However, A_2^* was positive for $K > 0.16$ or so. As the effect of horizontal deck motions appears to have been negligible, for sufficiently high flow velocity the total damping given Eq. 2.7.8*b* was negative, so flutter involving only the torsional degree of freedom occurred in wind with a mean velocity of about 20 m/s. The bridge's susceptibility to flutter was due to the use of an H section (the horizontal line in the H represents the deck, and the vertical lines represent the girders supporting it). Owing to their inherent instability, H bridge sections are no longer used.

2.7.3.2 Bridge Section Response to Turbulent Wind in the Presence of Aeroelastic Effects. The aerodynamic forces induced on a bridge by turbulent wind are due to (1) aeroelastic forces associated with flutter derivatives, (2) vortex-induced forces associated with the Strouhal number of the flow, and (3) buffeting forces induced by turbulent flow.

The expressions for the aeroelastic forces have the same form as for the smooth flow case (Eqs. 2.7.5). However, the aerodynamic coefficients H_i^*, A_i^*, P_i^* should be obtained from measurements in turbulent flow, since turbulence may affect the aerodynamics of the bridge deck by changing the configuration of the separation layers and the position of reattachment points. Through complex aerodynamic mechanisms, turbulence can affect the flutter derivatives and, therefore, the flutter velocity—in many instances favorably but possibly also unfavorably. The buffeting forces per unit span may be written as follows:

$$L_b = \frac{1}{2}\rho V^2 B\left[2C_L\frac{u(t)}{V} + \left(\frac{dC_L}{d\alpha} + C_D\right)\frac{w(t)}{V}\right] \tag{2.7.9a}$$

$$M_b = \frac{1}{2}\rho V^2 B\left[2C_M\frac{u(t)}{V} + \frac{dC_M}{d\alpha}\frac{w(t)}{V}\right] \tag{2.7.9b}$$

$$D_b = \frac{1}{2}\rho V^2 B\left[2C_D\frac{u(t)}{V}\right] \tag{2.7.9c}$$

For example, Eq. 2.7.9*c* is derived from the expression for the total (mean plus fluctuating) drag force D, where

$$D = \overline{D} + D_b = \tfrac{1}{2}\rho C_D B[V + u(t)]^2 \tag{2.7.10}$$

V is the mean flow velocity, $u(t)$ is the along-wind (longitudinal) component of the turbulent velocity fluctuation at time t, the mean drag force is defined as

$$\overline{D} = \tfrac{1}{2}\rho C_D B V^2 \tag{2.7.10a}$$

and the drag coefficient C_D is measured in turbulent flow. For the two-dimensional case the solution of the buffeting problem in the presence of aeroelastic effects is obtained from Eqs. 2.7.4, in which the right-hand sides consist of the sums $L_h + L_b$, $M_\alpha + M_b$, and $D_p + D$, respectively, as affected by the angle of attack at each time step. However, even though the two-dimensional case can provide useful insights into the behavior of a bridge, to be useful in applications to actual bridges the solution must be obtained for the three-dimensional case, in which the bridge displacement and the aerodynamic forces are functions of position along the span. Such a solution is presented in Appendix 5C.

2.8 STRUCTURAL RELIABILITY

Users of wind tunnel test results, code writers, and designers of special structures need assessments of uncertainties in the estimation of wind effects. Such assessments help to calculate safety margins that assure risk consistency and adequate safety levels. The uncertainties are due to the inherent randomness of variables (e.g., extreme wind speeds, pressure fluctuations) and to knowledge uncertainties pertaining to micrometeorological, climatological, aerodynamic, and dynamic parameters (e.g., surface roughness upwind of a structure, or damping ratio).

For design purposes we are interested in uncertainties affecting the peak wind effects $F_{pk}(\tau, \bar{N})$ (e.g., wind-induced bending moments) for a storm with duration τ (nominally taken to be 1 h) and mean recurrence interval $\bar{N}$, in years, of the wind speed (or the equivalent wind speed [2.22]; see Sect. 4.2.2.2.2) [2.24]. Wind load factors can be defined as functions of the uncertainties inherent in the estimation of $F_{pk}(\tau, \bar{N})$ (see Sect. 4.2.3).

One of the merits of a reliability-based framework for estimating uncertainties and safety margins (e.g., wind load factors) is that the uncertainties associated with the various parameters that determine wind effects are considered *collectively* rather than individually. If the uncertainty with respect to an individual factor has a modest relative contribution to the *overall* uncertainty of the wind effect of interest, a design decision based solely on the uncertainty inherent in that factor would be unwarranted.

For example, the issue of how much the size of aerodynamic time histories measured in the wind tunnel may be reduced without significant penalty from a structural safety point of view is best resolved within a structural reliability framework. Even though a shorter length of the time histories increases the uncertainty with respect to the magnitude of the peak aerodynamic coefficients, that increase can be small in relation to the *total* uncertainty associated with the overall wind effect of interest [2.24].

Similarly, a reliability framework can be useful for answering questions on the cost-effectiveness of equipment that may reduce uncertainties with respect to individual parameters (e.g., pressure coefficients, drag force) affecting the wind-induced response. Would such reductions have a significant contribution to the reduction in overall wind effect, or would the contribution be marginal?

Expenses for costly experimental facilities can be justified or avoided on the basis of reliability-based economic studies.

For example, the aerodynamic tests discussed in Sect. 2.5.6.3 entailed coefficients of variation across laboratories of roughly 10% for higher eaves and open terrain, and roughly 25% for lower eaves and suburban terrain. Taking these variabilities into account resulted in wind load factor increases, estimated as in Sect. 4.2.3 (see also [2.24]), of about 5 and 10% in the first and second cases, respectively. This suggests that design advantage inherent in the lower wind loads specified by the ASCE 7 Standard for buildings in suburban terrain may in fact be less significant than appears to be the case, since the wind load factors must be larger for such buildings than for their counterparts in open terrain. Improvements in wind tunnel simulation techniques, including possibly an increase in model scale, could reduce those variabilities. Whether larger wind tunnel facilities would reduce significantly the magnitude of the requisite wind load factors is not clear; careful assessments of simulation improvements due to larger wind tunnel size, and reliability-based economic studies accounting for the relative weight of aerodynamic uncertainties with respect to other types of uncertainties would be required to ascertain whether or not major expenditures on larger wind tunnels are warranted.

REFERENCES

2.1. *Federal Building and Fire Safety Investigation of the World Trade Center Disaster,* NIST NCSTAR 1-2A, Table 4-4, National Institute of Standards and Technology, Gaithersburg, MD, Sept. 2005, p. 80 (http://wtc.nist.gov/reports_october05.htm).

2.2. *Minimum Design Loads for Buildings and Other Structures,* ASCE Standard 7-05, American Society of Civil Engineers, Reston, VA, 2005.

2.3. *Canadian Structural Design Manual,* Supplement 4 to the National Building Code of Canada, National Research Council of Canada, Toronto, Ontario, 1971.

2.4. *Eurocode 1: Actions on Structures*—Part 1-4, *General Actions—Wind Actions,* CEN TC 250, Draft 2002-06.

2.5. E. Simiu and R. H. Scanlan, *Wind Effects on Structures,* 3rd ed., Wiley, New York, 1996.

2.6. M. D. Powell, T. A. Reinhold, and P. J. Vickery, "Reduced drag coefficient for high wind speeds in tropical cyclones," *Nature,* **422,** 279–283, 2003.

2.7. H. Lettau, "Note on aerodynamic roughness element description," *J. Appl. Meteorol.,* **8,** 828–832, 1969.

2.8. A. G. Davenport, "The relationship of wind structure to wind loading," in *Proceedings of the Symposium on Wind Effects on Buildings and Structures,* Vol. 1, National Physical Laboratory, Teddington, UK, Her Majesty's Stationery Office, London, 1965, pp. 53–102.

2.9. M. D. Powell, personal communication, 2005.

2.10. E. Simiu, F. Sadek, and M. A. Riley, "Discussion of 'Definition of wind profiles in ASCE 7' by Y. Zhou and A. Kareem," *J. Struct. Eng.,* **129,** 1564–1566, 2003.

2.11. E. Simiu, V. C. Patel, and J. F. Nash, "Mean wind profiles in hurricanes," *J. Eng. Mech. Div., ASCE,* **102,** 265–273, 1976.

2.12. P. J. Vickery and L. A. Twisdale, "Prediction of hurricane wind speeds in the United States," *J. Struct. Eng.,* **121,** 1691–1699, 1995.

2.13. E. Simiu, R. Wilcox, F. Sadek, and J. J. Filliben, "Wind speeds in ASCE 7 Standard peak-gust wind map: assessment," *J. Struct. Eng.,* **120,** 427–439, 2003.

2.14. E. Simiu and N. A. Heckert, *Extreme Wind Distribution Tails: A "Peaks over Threshold" Approach,* NIST Building Science Series 174, National Institute of Standards and Technology, Gaithersburg, MD, 1995; *J. Struct. Eng.,* **122,** 539–547, 1996.

2.15. J. D. Holmes and W. W. Moriarty, "Application of the generalized Pareto distribution to extreme value analysis in wind engineering," *J. Wind Eng. Ind. Aerodynam.,* **83,** 1–10, 1999.

2.16. E. Simiu and J. A. Lechner, "Discussion of 'Classical extreme value model and prediction of extreme winds' by J. Galambos and N. Macri," *J. Struct. Eng.,* **128,** 271–273, 2002.

2.17. B. R. Ellingwood, T. V. Galambos, J. G. MacGregor, and C. A. Cornell, *Development of a Probability Based Criterion for American National Standard A58,* NBS Special Publication 577, National Bureau of Standards, Washington, DC, 1980.

2.18. J. L. Gross, N. A. Heckert, J. A. Lechner, and E. Simiu, "Novel extreme value estimation procedures: application to extreme wind data," in *Extreme Value Theory and Applications,* J. Galambos, J. Lechner, and E. Simiu, eds., Kluwer Academic, Dordrecht, The Netherlands, 1994, pp. 139–158.

2.19. E. Simiu, N. A. Heckert, and T. A. Whalen, *Estimates of Hurricane Wind Speeds by the "Peaks over Threshold" Method,* NIST Technical Note 1416, National Institute of Standards and Technology, Gaithersburg, MD, 1996; *J. Struct. Eng.,* **124,** 445–449, 1998.

2.20. S. Coles and E. Simiu, "Establishing uncertainty in the extreme wind analysis of data generated by a hurricane simulation model," *J. Eng. Mech.,* **129,** 1288–1294, 2003.

2.21. R. H. Scanlan and E. Simiu, "Aeroelasticity in civil engineering," in *A Modern Course in Aeroelasticity,* E. H. Dowell, ed., Kluwer Academic, Dordrecht, The Netherlands, 2004.

2.22. A. Rigato, P. Chang, and E. Simiu, "Database-assisted design, standardization, and wind direction effects," *J. Struct. Eng.,* **127,** 855–860, 2001.

2.23. P. Hanzlik, S. Diniz, A. Grazini, M. Grigoriu, and E. Simiu, "Building orientation and wind effects estimation," *J. Eng. Mech.,* **131,** 254–258, 2005.

2.24. S. Diniz and E. Simiu, "Probabilistic description of wind effects and load factors for database-assisted design," *J. Struct. Eng.,* **131,** 507–516, 2005.

2.25. W. P. Fritz, B. Bienkewicz, O. Flamand, E. Ho, C. Letchford, and C. Bo, "International comparison of wind tunnel estimates of wind effects on an industrial building model: test-related uncertainties," *Proceedings of the 5th International Conference on Structural Safety and Reliability,* Rome, G. Augusti and G. Schuëller, eds., 2005.

2.26. F. Long, *Uncertainties in Pressure Coefficients Derived from Full and Model Scale Data,* Wind Science and Engineering Research Center, Department of Civil Engineering, Texas Tech University, Lubbock, TX, 2005.

2.27. W. C. L. Shih, C. Wang, D. Coles, and A. Roshko, "Experiments on flow past rough cylinders at large Reynolds numbers," *J. Wind Eng. Ind. Aerodynam.,* **49,** 351–368, 1993.

2.28. T. F. Sun, personal communication, 1983.

2.29. O. Mahrenholz, personal communication, 1998.

2.30. *Structural Design Actions—Wind Actions—Commentary AS/NZ 1170.2 Supplement 1:2002,* Standards Australia International, Sydney, and Standards New Zealand, Wellington, 2002.

2.31. E. Castillo, A. S. Hadi, N. Balakrishnan, and J. M. Sarabia, *Extreme Value and Related Models with Applications in Engineering and Science,* Wiley Series in Probability and Statistics, Wiley, New York, 1995.

2.32. R. D. Reiss and M. Thomas, *Statistical Analysis of Extreme Values,* Basel, Birkhäuser, 1997.

2.33. J. D. Holmes, *Wind Loading on Structures,* E&FN Spon, London, 2001.

2.34. E. Simiu, P. Vickery, and A. Kareem, "Relation between Saffir-Simpson hurricane scale wind speeds and peak 3-s gust speeds over open terrain," *J. Struct. Eng.* (in press).

CHAPTER 3

ASCE 7 STANDARD PROVISIONS

3.1 INTRODUCTION

The ASCE 7-02 Standard and the 7-05 Standard specify three procedures for determining design wind loads: the *simplified procedure,* the *analytical procedure,* and the *wind tunnel procedure,* also referred to in the standard as methods 1, 2, and 3, respectively. Sections 3.2 and 3.3 contain basic material required for the use of both the simplified and analytical procedures. The simplified and analytical procedures are presented in Sects. 3.4 and 3.5, respectively. The wind tunnel procedure is discussed in Chapters 2, 4, and 5.

Some of the material contained in ASCE 7-95, 7-98, and 7-02 versions of the Standard has been unmodified in subsequent versions. Where appropriate, figure captions and tables in this chapter identify the original version of the ASCE Standard in which the material shown first appeared and subsequent versions in which that material has remained unmodified. Appendix 3A is a list of the main differences between the ASCE 7-02 and the ASCE 7-05 Standards. Otherwise, throughout the chapter we refer solely to the provisions of the ASCE 7-05 Standard.

Henceforth, with few exceptions, we refer to the ASCE Standard 7-05 as ASCE, or the Standard and, for example, to ASCE 7-05 Standard Sect. 6.5.4 as ASCE 6.5.4. Section, figure, and table numbers not prefixed by the initials "ASCE" pertain to sections, figures, and tables in this book; for example, Sect. 3.2.2 and Fig. 2.5.7 designate a section and a figure in this book.

In some users' opinion, parts of the ASCE Standard and Commentary are not sufficiently clear and user-friendly to serve as design tools on their own. This is due to cluttering of the standard by repetitive and redundant material, the Standard's imperfect logical structure, and the lack of clarity and/or completeness of some of its provisions. Some of the Standard's shortcomings are noted in Appendix 3B. Examples of the application of the ASCE Standard provisons are helpful but in our opinion are not sufficient. What is needed, in addition, and is attempted in this chapter, is a clarification and, to the extent possible, a simplification of the standard's labyrinthine structure and presentation, with a view to easing the user's way through areas that some users find difficult to follow or outright confusing. We include, as needed, comments aimed at helping the user. We provide numerical examples that allow users to ascertain that their understanding and application of the provisions are correct. Our structuring and presentation of the material eliminates most of the shortcomings noted in Appendix 3B. We believe that the material as presented here can be useful not only to the designer, but also for future efforts to improve the structure of the standard, clarify its presentation, and reduce its unnecessary bulk.

This chapter is self-contained and enables the reader to become fully proficient in performing the requisite calculations. However, it is not intended to be a substitute for the Standard itself, of which the user will need a copy. Clear reference is made throughout the chapter to the relevant section, figure, or table in the Standard. The definition of each frequently used term is given when the term first appears in the text and can readily be located by using the book's index.

3.2 BASIC MATERIAL COMMON TO SIMPLIFIED AND ANALYTICAL PROCEDURES, I

The user of the simplified procedure (ASCE 6.4) must obtain material on *basic wind speeds, importance factors, exposure categories, and topographic factors* from the Standard's section on the analytical procedure (ASCE 6.5). It is helpful to streamline the use of the Standard by introducing that material in a preliminary section serving both the simplified and the analytical procedure. We review in this section the requisite material on basic wind speeds (Sect. 3.2.1), importance factors (Sect. 3.2.2), exposure categories (Sect. 3.2.3), topographic factors (Sect. 3.2.4), and enclosure classification (Sect. 3.2.5). Additional material is covered in Sect. 3.3.

3.2.1 Basic Wind Speeds (ASCE 6.5.4; see also Sect. 2.3.4)

The basic wind speed V is defined as the 3-s peak gust speed at 33 ft (10 m) above ground in open terrain with sufficiently long fetch in all directions (i.e., terrain with exposure C; see Sect. 3.2.3), with mean recurrence intervals (MRIs) defined as follows.

Outside hurricane-prone regions, V is defined as the speed with a nominal 50-year (MRI). The 50-year MRI is based on past common practice as applied to *allowable stress design.*

The wind speed used for *strength design* is obtained through multiplication of the basic wind speed *V* by $\sqrt{1.6}$, where 1.6 is the wind load factor specified by the Standard, and has a nominal MRI of about 700 years. (In earlier versions of the Standard the wind load factor was in effect 1.5, so the MRI of the wind speed used for strength design was only about 500 years—see ASCE Commentary C6.5.4.)

In hurricane-prone regions the basic wind speed *V* is no longer defined as the speed with an MRI of 50 years. Rather, it is defined as the hurricane wind speed with an MRI of 500 years, divided by $\sqrt{1.5}$ (ASCE 7-05 Commentary C6.5.4). Because the probability distributions of the wind speeds differ in hurricane-prone regions from those typical of non-hurricane regions, the nominal MRIs of the basic speeds so defined are longer than 50 years.

The basic wind speeds are specified in the *ASCE wind maps* (ASCE Fig. 6-1). They should be increased where records indicate that the wind speeds are higher than those indicated in the wind maps. Wind speeds in the special wind regions indicated in ASCE Fig. 6-1, and in mountainous terrain, gorges, and promontories, must be estimated on a case-by-case basis. Tornado winds are *not* accounted for in the ASCE wind maps (however, a tornado wind speed map is included in the ASCE Commentary Fig. C6-5). The wind speeds indicated in the wind maps are assumed to be the same for all horizontal directions. Outside hurricane-prone regions, in special wind regions, or for special applications, ASCE 6.5.4.2 allows the use of basic wind speeds estimated from regional climatic data (see Sect. 2.4 of this book); 50-year speeds so estimated may be used in design even if they are lower than those of ASCE Fig. 6-1. In hurricane-prone regions, wind speeds may be estimated by using simulations acceptable to the authority having jurisdiction.

3.2.2 Importance Factor *I*

Buildings are divided into four categories (ASCE Table 1-1), depending on whether the hazard to human life in the event of failure is low (category I, including agricultural facilities, minor storage facilities, and certain temporary facilities, the meaning of the term *certain* being left by the Standard to the judgment of the user), substantial (category III, including buildings and other structures where more than 300 people congregate in one area[1]), or on whether the structures are designated as essential facilities (category IV). All structures not listed in categories I, III, and IV are defined as category II structures. The importance factors are (ASCE Table 6-1): *category I:* $I = 0.87$

[1] This criterion is usually interpreted as referring to places of assembly such as auditoria. A criterion for classification as category III buildings that is applicable to buildings whose overall occupancy exceeds 5000 people is specified in Table 1604.5 of the 2003 International Building Code, which is otherwise based primarily on ASCE 7-02. We are indebted to William F. Baker of Skidmore Owings and Merrill for drawing our attention to this criterion.

(I = 0.77 for hurricane-prone regions with $V > 100$ mph); *category II:* I = 1.00; *categories III and IV:* I = 1.15.

3.2.3 Exposure (ASCE 6.5.6; see also Sect. 2.3.4)

For each wind direction being considered, an exposure category extending in that direction upwind from the site must be determined as a function of the nature of the surface (e.g., water, open terrain, built-up terrain, wooded areas) and its changes, and of local topography.

ASCE 6.5.6.1 requires that for each direction being considered, the exposure category be obtained as follows. Determine the exposure category for two upwind sectors extending 45° on either side of the direction being considered. Use for both sectors the exposure category for the 45° sector that results in the largest wind loads (see also ASCE Commentary Fig. C6-3).

ASCE 6.5.6.4.1 requires that when designing mean wind force resisting systems (MWFRSs) based on ASCE Fig. 6-6 (Table 3.5.1), the design pressures used for each direction be based on the exposure obtained by this method. A similar requirement is specified for solid and freestanding walls and solid signs (ASCE 6.5.14). On the other hand, for the MWFRSs of low-rise buildings (see Sect. 3.3.6), for components and cladding, and for open buildings with monoslope, pitched, and troughed roofs, ASCE 6.5.6.4.2, 6.5.6.5, and 6.5.13.2 require that design pressures be based on the exposure that results in the highest wind loads *for any direction.* ASCE provides no requirements on exposure similar to those of ASCE 6.5.6.4.1 for MWFRS of buildings with domed roofs and arched roofs and other structures covered in ASCE 6.5.15. For these cases it is prudent to use design pressures based on exposure resulting in the highest wind loads for any direction.

3.2.3.1 Surface Roughness Categories (ASCE 6.5.6.2).

Surface roughness categories are based on measurements by meteorologists and assessments by wind engineers. They are commonly defined—and are also defined in the Standard—as follows:

- *Surface roughness B:* urban and suburban areas, wooded areas, or other terrain with numerous closely spaced obstructions having the size of single-family dwellings or larger
- *Surface roughness C:* open terrain with scattered obstructions generally less than 30 ft high, and flat open country, grasslands, and water surfaces in hurricane-prone regions[2]

[2] This definition is imprecise and contradicts measurement results in [2.6]. For this reason it has led to confusion and errors in ASCE Commentary material on Saffir–Simpson scale wind speeds over open terrain in hurricane-prone regions (see Table 2.2.1, footnote *c*, and Numerical Example 2.3.8).

- *Surface roughness D:* flat, unobstructed areas, including smooth mudflats, salt flats, and unbroken ice, and water surfaces outside hurricane-prone regions

An earlier classification included surface roughness A, now omitted due to the practical impossibility of defining reliably the surface roughness for centers of large cities.

3.2.3.2 Exposure Categories (ASCE 6.5.6.3). Exposure categories are defined as follows:

- *Exposure B* applies where surface roughness B prevails in the upwind direction for a distance of at least 2600 ft or 20 times the building height, whichever is greater, except that for buildings with a mean roof height of 30 ft or less, the upwind distance may be reduced to 1500 ft.
- *Exposure D* applies where surface roughness D prevails in the upwind direction for at least 5000 ft or 20 times the building height, whichever is greater. Exposure D extends inland from the shoreline for a distance of 660 ft or 20 times the building height, whichever is greater.
- *Exposure C* applies where exposures B and D do not apply and is commonly referred to as open terrain exposure.

As an aid to the user, the ASCE Commentary includes photographs of typical areas with various exposures, and typical schematic drawings for terrain with exposure B (ASCE Figs. C6-7 to C6-9). For a site located in the transition zone between exposure categories, the category resulting in the largest wind loads must be used except where an intermediate exposure is determined (see, e.g., ASCE Commentary Eqs. C6-7 to C6-10). For additional material on surface roughness, see Sect. 2.3.2.

3.2.4 Topographic Factor K_{zt} for Two-Dimensional Escarpments and Two- or Three-Dimensional Hills (ASCE 6.5.7)

Over the surface of a rising slope, speeds are higher, for any given height above ground, than their counterparts over the horizontal terrain upwind of the slope (see Sect. 2.5.2). ASCE provides a model for this increase (or *speed-up*) that is applicable to two-dimensional ridges and three-dimensional isolated hills, and two-dimensional escarpments, provided that all of the following conditions are satisfied: (1) no significant terrain features exist over sufficiently large distances (see ASCE 6.5.7.1 for details[3]; (2) the height of

[3] ASCE 6.5.7.1, item 2 refers to a 2-mile radius but does not define the point from which that radius is measured. According to K. C. Mehta, the intent of the standard is for that point to be the horizontal projection of the top of the topographic feature. We thank Dr. Mehta for providing this clarification.

the hill H exceeds 15 ft in exposures C and D and 60 ft in exposure B (see Fig. 3.2.1 for notations); and (3) $H/L_h \geq 0.2$.

The increased 3-s peak gust speed $V(x, z)$ depends on height z above ground at location x (see Fig. 3.2.1; ASCE Fig. 6-4). The ASCE model consists of a topographic factor $K_{zt} = [V(z, x)/V(z)]^2$, where $V(z)$ = 3-s peak gust speed at height z above ground in horizontal terrain with no topographic feature. The expression for K_{zt} is

$$K_{zt} = (1 + K_1K_2K_3)^2 \tag{3.2.1}$$

where the factor K_1 accounts for the shape of the topographic feature, K_2 accounts for the variation of the speed-up as a function of distance from the crest, and K_3 accounts for the variation of the speed-up as a function of height above the surface of the topographic feature. Expressions for K_1, K_2, and K_3 are given in ASCE Fig. 6.4. For example, for $H/L_h \leq 0.5$,

$$K_1 = \frac{aH}{L_h} \qquad K_2 = 1 - \frac{|x|}{\mu L_h} \qquad K_3 = \exp\left(\frac{-\gamma z}{L_h}\right) \tag{3.2.2a,b,c}$$

where for two-dimensional escarpments, $\gamma = 2.5$; $\mu = 1.5$ (upwind of crest), $\mu = 4.0$ (downwind of crest); and $a = 0.75$ (exposure B), $a = 0.85$ (exposure C), $a = 0.95$ (exposure D). These models are roughly consistent with those of [2.5].

Numerical Example 3.2.1. *Topographic factor.* A two-dimensional escarpment has exposure B and dimensions H = 100 ft and L_h = 400 ft. The

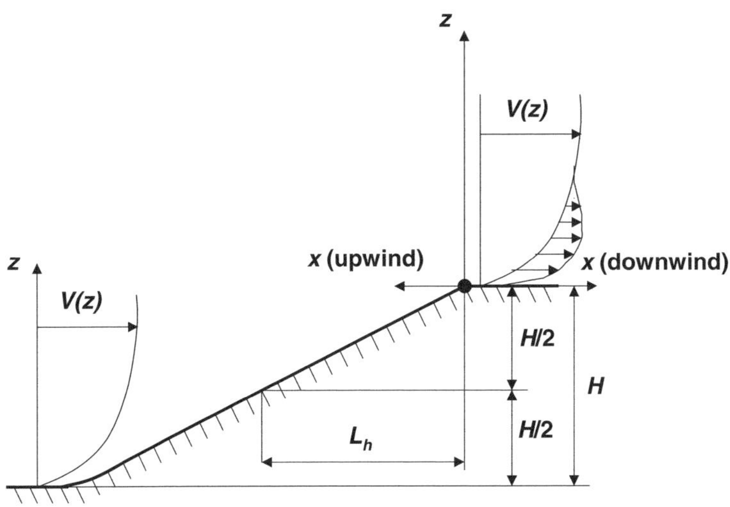

Figure 3.2.1. Notations.

topography upwind of the escarpment is assumed to satisfy condition 1. The building is located at the top of the escarpment, and the downwind distance (see Fig. 3.2.1) between the crest and the building's windward face is $x = 40$ ft. (In Fig. 3.2.1 the building would be located to the right of the crest.) We seek the quantity K_{zt} for elevation $z = 25$ ft above ground at $x = 40$ ft. Condition 2 is satisfied, since $H = 100$ ft > 60 ft, as is condition 3, since $H/L_h = 100/400 = 0.25 > 0.2$. Since $H/L_h < 0.5$, Eqs. 3.2.2 yield $K_1 = 0.75 \times 100/400 = 0.1875$, $K_2 = 1 - 40/(4.0 \times 400) = 0.975$, and $K_3 = \exp(-2.5 \times 25/400) = 0.855$. From Eq. 3.2.1, the topographic factor is $K_{zt} = (1 + 0.1875 \times 0.975 \times 0.855)^2 = 1.16^2 = 1.34$. This result implies that at $x = 40$ ft downwind of the crest and $z = 25$ ft above ground, the pressures are 1.34 times larger than those upwind of the escarpment, and the increased peak 3-s gust is 1.16 times larger than the peak 3-s gust at 25 ft above ground upwind of the escarpment.

3.2.5 Enclosure Classification (ASCE 6.2 and 6.5.9)

The enclosure classification of a structure controls the internal pressures and for open structures (defined subsequently) the external pressures as well. Internal pressures develop if air blown into a space cannot freely leave that space, in which case the internal pressures are positive; or if air sucked away from a space cannot be freely replaced, in which case the internal pressures are negative. The internal pressures thus depend on the way and the degree to which the building is enclosed, that is, on the amount and distribution of openings in the building envelope (see Sect. 2.5.4.2).

For the purposes of enclosure classification, glazing (i.e., glass or translucent plastic sheet in windows, doors, skylights, or curtain walls) is not defined as an opening except, under certain conditions (see item 5, this section), if the building is sited in a windborne debris region. The ASCE enclosure classification defines open, partially enclosed, and enclosed buildings and other structures as follows:

1. A building or other structure is *open* if in each wall that may be subjected to positive external pressure, the ratio of the area of the openings to the gross area of the wall is at least 80%.
2. A building is *partially enclosed* if (1) in a wall that experiences positive external pressure, the total area of openings exceeds 1.1 times the sum of the area of openings in the balance of the building envelope (walls and roof), *and* (2a) the total area of openings in that wall exceeds 4 ft^2 or 1% of the wall's gross area, whichever is smaller, *and* (2b) the percentage of openings in the balance of the building envelope does not exceed 20%.
3. A building that complies with both the definition of open buildings and the definition of partially enclosed buildings is classified as *open* (ASCE 6.5.9.4).

4. A building is *enclosed* if it is not open or partially enclosed.
5. *Windborne debris regions* are defined as hurricane-prone regions (a) within 1 mile of the coastal mean high water line where the basic wind speed is at least 110 mph, (b) in Hawaii, or (c) with a basic wind speed of at least 120 mph (ASCE 6.2). In such regions, for category 1 buildings glazing may be unprotected; for category II, III, and IV buildings, glazing may be unprotected if (1) the building being considered has no aggregate-covered roof surface (ASCE Commentary C6.5), and (2) the glazing elevation exceeds (a) 60 ft above ground and (b) 30 ft above roofs with aggregate-covered surfaces, located within 1500 ft of the building.[4] If conditions (1) and (2) are not satisfied, unprotected glazing is considered to be an opening for enclosure classification purposes. The protection must consist of an impact-resisting covering or the use of impact-resistant glazing, as required by approved test methods and performance criteria (see ASCE Commentary C6.5.9 for details and references).

Numerical Example 3.2.2. *Enclosure classification.* Consider a building with a rectangular shape in plan (60 ft $\times$ 30 ft), a flat roof, and eave height h = 20 ft. The exterior walls have gross areas $60 \times 20 = 1200$ ft^2 and $30 \times 20 = 600$ ft^2. Assume the area of the openings in each of the 1200-ft^2 walls is 240 ft^2. The 600-ft^2 walls have no openings, and the roof has two 1-ft^2 openings. If one of the 1200-ft^2 walls is considered, the sum of the areas of openings in the balance of the building envelope is 242 ft^2. Condition 2a for the classification of the building as partially enclosed is satisfied, that is, the total area of openings in the 1200-ft^2 wall (i.e., 240 ft^2) exceeds the smaller of the areas 4 ft^2 and $0.01 \times 1200 = 12$ ft^2. Condition 2b is also satisfied; that is, the percentage of openings in the balance of the building envelope is $242/(1200 + 2 \times 600 + 60 \times 30) < 20\%$. However, condition 1 is *not* satisfied; that is, 240 ft^2 $< 1.1(240 + 2 \times 1) = 266.2$ ft^2, so the building is not classified as partially enclosed. The building is not classified as open; that is, it does not satisfy the condition that for each wall, the ratio of the area of the wall's openings to the wall's gross area is at least 80%. Since the building is neither open nor partially enclosed, it is classified as enclosed.

3.3 BASIC MATERIAL COMMON TO SIMPLIFIED AND ANALYTICAL PROCEDURES, II

In this section we cover basic material pertaining to flexible and rigid structures (Sect. 3.3.1), pressure sign convention and net pressures (Sect. 3.3.2), pressure zone dimensions a (Sect. 3.3.3), effective areas (Sect. 3.3.4), mean

[4] ASCE Commentary C6.5.9 indicates that condition (b) may be relaxed if the roofs have "extremely high parapets," leaving the definition of this term to the judgment of the user.

roof and eave heights (Sect. 3.3.5), low-rise buildings (Sect. 3.3.6), and minimum loads (Sect. 3.3.7).

3.3.1 Flexible and Rigid Buildings and Other Structures (ASCE 6.2)

Buildings and other structures with a fundamental natural frequency of less than 1 Hz are defined as *flexible*. Buildings and other structures with a fundamental natural frequency of at least 1 Hz are defined as *rigid*. In practice, buildings with a ratio of height to least horizontal dimension not larger than 4 are assumed to be rigid, even in the absence of natural frequency calculations (ASCE Commentary C6.2). However, if there is concern that they may be flexible, natural frequency calculations should be performed. For more refined criteria, see Sect. 4A.1 and Numerical Example 4A.2.

3.3.2 Pressure Sign Convention, Net Pressures, and Single- and Double-Arrow Representation of Pressures

Positive and negative pressures are directed, respectively, toward and away from the surface on which they act. The net pressure is the sum of the external and internal pressures acting on a surface; for roof overhangs it is the sum of the external pressures acting on the upper and lower sides of the overhang; for parapets with no porosity it is the sum of the external pressures acting on the outer and inner parts of the parapet.

Numerical Example 3.3.1. *Sign convention and net pressures.* If the external pressure acting on the surface of a cladding panel is 40 psf (toward the exterior surface, since the external pressure is positive), and the internal pressure acting on that panel is 10 psf (toward the interior surface, since the internal pressure is positive), the net pressure on the panel is 40 − 10 = 30 psf (Fig. 3.3.1*a*). If the internal pressure were −10 psf (away from the internal

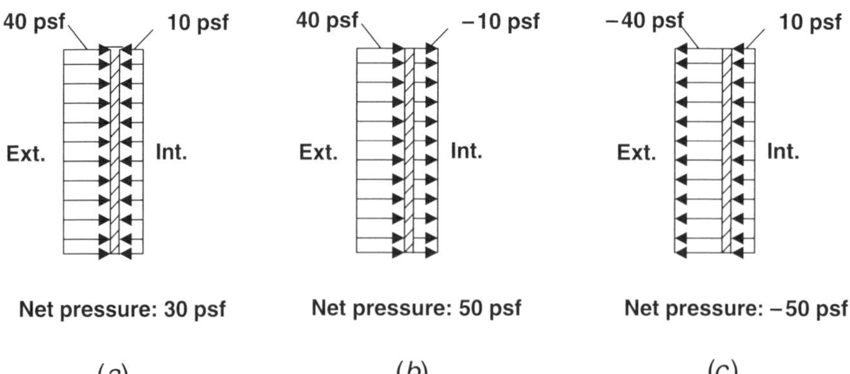

Figure 3.3.1. Pressure sign convention and net pressures, Numerical Example 3.3.1.

surface of the panel), the net pressure would be 40 − (−10) = 50 psf (Fig. 3.3.1*b*). If the external pressure were −40 psf and the internal pressure were 10 psf, the net pressure on the panel would be −50 psf (Fig. 3.3.1*c*).

3.3.3 Pressure Zone Dimensions *a*

Pressure zones specified in the ASCE Standard in terms of a dimension denoted by a (see, e.g., Figs. 3.4.1, 3.4.2, and 3.5.5 to 3.5.9) are governed by the following requirement: The dimension a is 10% of the least horizontal building dimension or $0.4h$ (h = mean roof height), whichever is smaller, but not less than 4% of the least horizontal building dimension or 3 ft.

Exceptions:

1. For multispan gable and sawtooth roofs, use the least horizontal dimension of a single-span module instead of the least horizontal building dimension (Figs. 3.5.7 and 3.5.8).
2. For Fig. 3.5.9 (ASCE Fig. 6-17), components and cladding, buildings with height $h > 60$ ft, a is 10% of the least horizontal dimension but not less than 3 ft.

3.3.4 Effective Wind Areas

Let the total wind forces on an area A_{eff1} and on a portion $A_{\text{eff2}} < A_{\text{eff1}}$ of that area be denoted by F_1 and F_2, respectively. It was indicated in Section 2.5.4.1 that, in general, $F_2/A_{\text{eff2}} > F_1/A_{\text{eff1}}$. This is the case because the extent to which pressures act in tune with each other (i.e., are mutually coherent) decreases as the area over which they act increases (see the pressures animation on the site http://www.nist.gov/wind). For this reason, ASCE specifies design pressures as functions of *effective wind area*. For example, for the design of cladding fasteners, the effective area is equal to the tributary area of an individual fastener. Should the effective area be chosen to be larger, the corresponding pressure coefficient would be too small and the resulting design would be unsafe. For the design of an exterior masonry wall supported by two successive floors, the effective wind area is the span length (the distance between the floors) times a width that must be neither too large, lest the resulting design be unsafe, nor too small, lest the resulting design be overconservative. A width equal to one-third the span length is viewed as a reasonable choice (ASCE 6.2). The choice of the effective area may in some cases require engineering judgment (see ASCE Commentary C6.2).

3.3.5 Mean Roof Height and Eave Height

Wind effects specified by the standard depend on a nominal building height denoted by h.

- For buildings with roof slopes $\theta > 10°$, h is defined as the *mean roof height,* that is, the vertical distance between ground level and the mean elevation of the roof (see, e.g., Fig. 3.4.2).
- For buildings with roof slopes $\theta \leq 10°$, h is defined as the *eave height,* that is, the distance between ground level and the eave (see, e.g., Fig. 3.4.1).

Note: For the simplified procedure the Standard inadvertently uses the criteria $\theta \geq 10°$ and $\theta < 10°$ instead of the criteria $\theta > 10°$ and $\theta \leq 10°$, respectively, used in the ASCE definition of mean roof height (ASCE 6.2) and in the analytical procedure.

For application to the Standard's Fig. 6-10, ASCE defines the eave height for a wall with nonhorizontal eave as the mean eave height for that wall. For a gable roof the mean eave height is the same as the mean roof height.

3.3.6 Low-Rise Buildings

Low-rise buildings are defined by ASCE as enclosed or partially enclosed buildings for which the height h defined as in Sect. 3.3.5 does not exceed (a) 60 ft and (b) the least horizontal dimension of the building.

3.3.7 Minimum Design Wind Loading

The minimum load cases are to be applied separately from and in addition to the other load cases specified in ASCE 6.

3.3.7.1 Main Wind-Force Resisting Systems (ASCE 6.1.4.1). For *enclosed or partially enclosed buildings or structures,* the minimum design force shall not be less than a net horizontal pressure of 10 psf (0.48 kN/m^2) times the area of the building or structure projected onto a vertical plane normal to the assumed wind direction (ASCE Commentary Fig. C6-1). For *open buildings and structures* the minimum design wind force shall not be less than 10 psf times the area used in the calculation of the wind load for the direction being considered (ASCE 6.1.4.1).

3.3.7.2 Components and Cladding (ASCE 6.1.4.2). Design net pressures shall be at least 10 psf acting in either direction normal to the surface.

3.4 SIMPLIFIED PROCEDURE (ASCE 6.4)

In this section we list the conditions that must be met for the simplified procedure to be applicable for components and cladding and for MWFRS

(Sects. 3.4.1.1 and 3.4.1.2), and describe the steps that must be followed to calculate the wind loading.

3.4.1 Conditions for Application of the Procedure

3.4.1.1 Components and Cladding (ASCE 6.4.1 and 6.4.1.2). For components and cladding, the simplified procedure may be applied provided that all the following conditions are satisfied:

1. The height $h \leq 60$ ft,[5] and the building has a *flat roof,* a *gable roof* with slopes of at most 45°, or a *hip roof* with slopes of at most 27°. The procedure may not be used for buildings with other roof types.
2. The building is *enclosed.* If it is located in a windborne debris region, it must satisfy the ASCE 6.5.9.3 requirements on protection against windborne debris (see Sect. 3.2.5, item 5).
3. The shape of the building has no unusual irregularity.
4. The building is not subject to dynamic vortex-induced effects, galloping or flutter effects, or significant channeling effects or buffeting in the wake of upwind obstructions.

If these four requirements are statisfied, the simplified procedure may be applied to components and cladding regardless of whether or not it is also applicable for MWFRS.

3.4.1.2 Main Wind-Force Resisting Systems (ASCE 6.4.1 and 6.4.1.1). For MWFRS the simplified procedure may be applied provided that except for the feature pertaining to hip roofs, the building has all the features of Sect. 3.4.1.1 and, in addition, all the following features:

1. A height $h \leq 60$ ft not exceeding the least horizontal building dimension (i.e., the building is a *low-rise building;* see Sect. 3.3.6).
2. Floor and roof diaphragms that transmit both the windward and leeward wind loads to the same MWFRS (i.e., there are no structural separations between the MWFRS that carries the windward loads and the MWFRS that carries the leeward wind loads).
3. An approximately symmetrical cross section in each direction.
4. If the building has a hip roof, it has roof slopes of at most 45°. The standard provides no specific indication on pressures for the design of

[5] This condition does not require that the building be a low-rise building (see Sect. 3.3.6). The statement in the ASCE Commentary C6.4 that the simplified procedure is for "the low-rise . . . building case" is not correct as far as the design of components and cladding is concerned.

hip roof MWFRS. However, ASCE Commentary Fig. C6-11 suggests, within the framework of the analytical procedure, a method for adapting to hip roofs the pressure coefficients provided in ASCE Fig. 6-10 for gable roofs.

5. Torsional loading need not be taken into account in design. This feature is satisfied by:
   a. One-story buildings with $h \leq 30$ ft, and one- or two-story buildings with light-framed construction (ASCE Fig. 6-10, note 5).
   b. Buildings with flexible roof and floor diaphragms.[6]
   c. Buildings whose design is not controlled by torsion as calculated using the analytical procedure (ASCE 6.4.1.1, item 8). Note the drawback of having to resort to the analytical procedure to ascertain whether the simplified procedure can be used.

Additional criteria for determining whether the design is controlled by wind-induced torsion are discussed in ASCE Commentary C6.4.

3.4.2 Design Procedure (ASCE 6.4.2)

The design procedure entails steps 1 to 6 below, as well as step 7 for MWFRS and step 8 for components and cladding.

1. Determine the ***basic wind speed V*** (Sect. 3.2.1; ASCE 6.5.4 and 6.5.4.2).
2. Determine the ***importance factor I*** (Sect. 3.2.2; ASCE Tables 1-1 and 6-1).
3. Determine the ***exposure category*** resulting in the highest wind loads for any direction at the site (Sect. 3.2.3; ASCE 6.5.6.4.2 and 6.5.6.5).
4. Determine the ***topographic factor*** $\boldsymbol{K_{zt}}$ (Sect. 3.2.4; ASCE 6.5.7). For horizontal terrain $K_{zt} = 1$.
5. Determine the ***height h*** (h = mean roof height, except that for roofs with slopes $\theta \leq 10°$, h = eave height; see note, Sect. 3.3.5).
6. Determine the ***adjustment factor*** $\boldsymbol{\lambda}$ for building height (h in feet) and exposure:

$$\lambda = \left(\frac{h}{60}\right)^{2/\alpha} \lambda_{60\ \mathrm{ft}} \quad \text{for } h \geq h_{\mathrm{E}} \qquad \lambda = \left(\frac{h_{\mathrm{E}}}{60}\right)^{2/\alpha} \lambda_{60\ \mathrm{ft}} \quad \text{for } h < h_{\mathrm{E}} \tag{3.4.1a,b}$$

[6]The definition of flexible diaphragms is not given in ASCE Sect. 6 (wind loads) but rather, in ASCE 9.5.2.3.1 (earthquake loads). If applied to the case of wind loading, it would require that the wind-induced in-plane deflection of the diaphragm itself exceed twice the average wind-induced horizontal deflection of the adjoining vertical MWFRS elements of the associated story.

where h_E is the height near the ground over which it is assumed that the wind profile is constant. The following parameters are specified:

a. *Exposure B:* $\alpha = 7$, $\lambda_{60\text{ ft}} = 1.22$, $h_E = 30$ ft

b. *Exposure C:* $\alpha = 9.5$, $\lambda_{60\text{ ft}} = 1.62$, $h_E = 15$ ft

c. *Exposure D:* $\alpha = 11.5$, $\lambda_{60\text{ ft}} = 1.87$, $h_E = 15$ ft

For example, for exposure C, $h_E = 15$ ft, $\lambda = (15/60)^{2/9.5} \times 1.62 = 1.21$. The factor λ can also be obtained from ASCE Figs. 6-2 and 6-3 instead of from Eqs. 3.4.1.

7. For ***main wind force-resisting systems*** (ASCE 6.4.2.1):

a. Determine the ***simplified design wind pressure p_{S30}*** corresponding to exposure B, $h = 30$ ft, $I = 1$, $K_{zt} = 1$:

$$p_{S30} = \left(\frac{V}{100}\right)^2 p_{S30}^{100} \tag{3.4.2}$$

where V is the basic speed in mph. It follows from Eq. 3.4.2 that p_{S30}^{100} is equal to the value of p_{S30} for $V = 100$ mph. Pressures p_{S30}^{100} for the design of MWFRS in the *direction normal to the ridge* (Fig. 3.4.1*a*) are listed in Table 3.4.1. (The entries in Table 3.4.1 are the same as the entries in ASCE Fig. 6-2 for $V = 100$ mph.) The pressure zones of Table 3.4.1 are shown in Fig. 3.4.1 (for sign conventions, see Sect. 3.3.2; for dimension a, see Sect. 3.3.3). The zones for pressures p_{S30}^{100} for the design of MWFRS in the *direction parallel to the ridge* (Fig. 3.4.1*b*) are designated by the letters A, C, E, F, G, and H; however, in this direction, *for all roof slopes θ*, the pressures p_{S30}^{100} have the values given in Table 3.4.1 for, respectively, zones A, C, E, F, G, and H of buildings with roof slopes $\theta = 0$ to 5°. Unnecessary duplicative pressure entries in ASCE Fig. 6-2 are not included in Table 3.4.1. The pressures p_{S30} may also be obtained from ASCE Fig. 6-2 for various values of V instead of from Eq. 3.4.2.

b. Determine the ***simplified design pressure*** (ASCE Eq. 6-1):

$$p_s = \lambda I K_{zt} p_{S30} \tag{3.4.3}$$

c. Determine the ***minimum simplified design wind pressures p_s*** (ASCE 6.4.2.1.1). The simplified design pressure p_s is not to be less than 10 psf for zones A, B, C, and D, while assuming no pressure is acting on zones E, F, G, and H.

d. Consider ***load pattern for each reference corner.*** The load patterns of Fig. 3.4.1 (ASCE Fig. 6-2) are associated with the reference corner indicated therein. Those patterns must be applied by considering in turn each corner of the building as a reference corner.

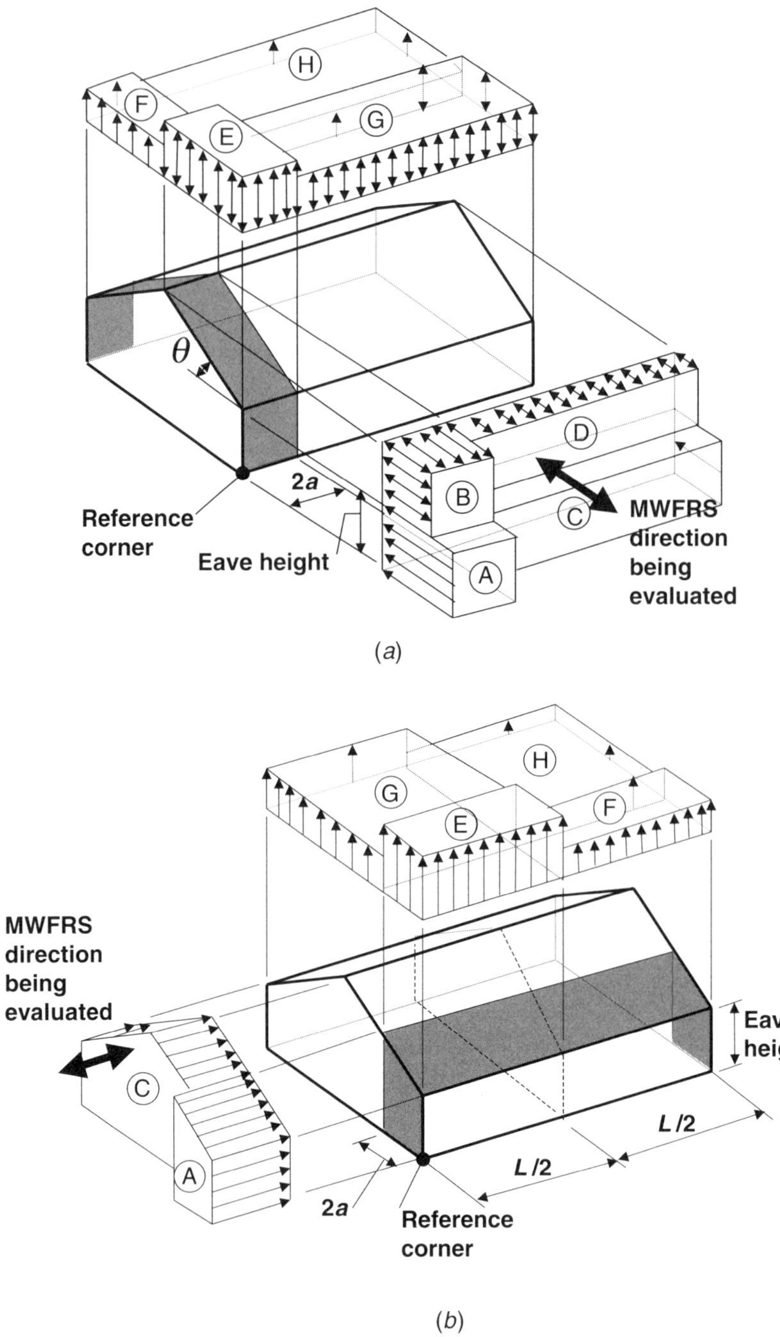

Figure 3.4.1. Simplified procedure; design wind pressures for design of *MWFRS, walls and roofs, enclosed structures,* $h \leq 60$ ft: (*a*) wind normal to a ridge or long building dimension; (*b*) wind parallel to a ridge or long building dimension.

TABLE 3.4.1. Pressures p_{S30}^{100} (psf)[a]

Roof Slope	Zone[b]									
	Horizontal Pressure				Vertical Pressure				Overhang	
θ (deg)	A	B	C	D	E	F	G	H	E_{OH}	G_{OH}
0–5	15.9	−8.2	10.5	−4.9	−19.1	−10.8	−13.3	−8.4	−26.7	−20.9
10	17.9	−7.4	11.9	−4.3	−19.1	−11.6	−13.3	−8.9	−26.7	−20.9
15	19.9	−6.6	13.3	−3.8	−19.1	−12.4	−13.3	−9.5	−26.7	−20.9
20	22.0	−5.8	14.6	−3.2	−19.1	−13.3	−13.3	−10.1	−26.7	−20.9
25	19.9	3.2	14.4	3.3	−8.8	−12.0	−6.4	−9.7	−16.5	−14.0
					(−3.4)	(−6.6)	(−0.9)	(−4.2)		
30–45	17.8	12.2	14.2	9.8	1.4	−10.8	0.5	−9.3	−6.3	−7.2
					or 6.9	or −5.3	or 5.9	or −3.8		

Source: Excerpted from ASCE Fig. 6-2, ASCE Standard 7-02, *Minimum Design Loads for Buildings and Other Structures,* 2003, American Society of Civil Engineers. Reproduced by permission of the publisher, ASCE. Unmodified in ASCE 7-05.

[a] Linear interpolation permitted for slopes not shown. Values in parentheses to be used only for interpolation between 25 and 30°. For example, for $\theta = 27.5°$, zone F, the interpolated pressures are $(-12.0 - 10.8)/2 = -11.4$ psf and $(-6.6 - 5.3)/2 = -5.95$ psf.

[b] For zone definitions, see Fig. 3.4.1 and Sect. 3.3.3.

e. Determine the ***horizontal pressures*** in Fig. 3.4.1 (ASCE Fig. 6-2), which represent sums of pressures acting on the building's windward and leeward sides. The total horizontal load shall not be less than the load obtained by assuming that the pressures in zones B and D are zero (i.e., if pressures in zones B and D are negative, they shall not be taken into account when calculating the overall horizontal wind load). Load paths of windward and leeward wall loads must be taken into account in design (e.g., in designing roof diaphragms with openings). For ***windward walls*** and ***leeward walls,*** respectively, the design pressures for zones A and C are 0.85 and -0.7 times the pressures p_s for those zones; for ***sidewalls*** they are -0.65 times the pressures p_s for zone C (ASCE Commentary C6.4).

f. Determine the ***vertical pressures*** on the roof in Fig. 3.4.1 (ASCE Fig. 6-2), which represent net pressures (sums of external and internal pressures), except for pressures on roof overhangs, which are sums of external pressures acting on the top and bottom sides of the overhang. The vertical presures are applied to the horizontal projection of the roof surface. For $\theta > 25°$ and zones E, F, G, and H, Table 3.4.1 has *two* entries for vertical pressures, which correspond to positive internal pressures (case 1) and negative internal pressures (case 2). For $\theta = 25°$ the second entry is shown in parentheses and is used only for interpolation from 25 to 30°. Members with spans from eave to ridge or longer may be designed for interior zone (rather than end zone) pressures (ASCE Commentary C6.4).

g. Determine the ***pressures on roof overhangs.*** Where zone E or G falls on a roof overhang on the windward side of the building, pressures listed in Table 3.4.1 for zones E_{OH} and G_{OH}, respectively, shall be used; for roof overhangs on the leeward side and on the building ends, the pressures for zones E and G shall be used. Negative pressures are directed upward.

Numerical Example 3.4.1. *Simplified procedure: MWFRS.* Consider a commercial building with a rectangular shape in plan (30 ft × 60 ft), 12.7-ft eave height, gable roof with slope 11°, and exterior masonry walls transmitting the horizontal loads to the ground-level floor and the roof diaphragm. The building is enclosed (Sect. 3.2.5) and is sited in flat, open terrain in Alabama. The terrain exhibits no surface roughness described as category B or D for more than 1 mile upwind of the building in all directions.

All the conditions of Sect. 3.4.1.2 are satisfied. (In accordance with Sect. 3.4.2, item 5, the mean roof height is $h = 12.7 + (30/4)\tan 11° = 14.2$ ft $<$ 60 ft; windborne debris provisions do not apply in Alabama, see Sect. 3.2.5, item 5.) The simplified method may therefore be used for the design of the MWFRS. We follow the steps indicated in Sect. 3.4.2.

The basic wind speed is 90 mph (ASCE Fig. 6-1). The building is category II, so the *importance factor* is $I = 1.00$ (Sect. 3.2.2). The building has *exposure C,* since the terrain fits neither the exposure B nor the exposure D description (Sect. 3.2.3). The *topographic factor* is $K_{zt} = 1$.

The adjustment factor λ for building height and exposure for $h = 14.2$ ft and exposure C is (Sect. 3.4.2, item 6, case $h \leq h_E$) $\lambda = (h_E/60)^{2/\alpha}\lambda_{60\,\text{ft}} = (15/60)^{2/9.5} \times 1.62 = 1.21$.

The simplified design pressures p_s for MWFRS are obtained as indicated in Sect. 3.4.2, item 7. (For zone designations, see Fig. 3.4.1.) Determine $p_{S30} = (V/100)^2 p_{S30}^{100} = 0.81 p_{S30}^{100}$ (since $V = 90$ mph), and $p_s = \lambda I K_{zt} p_{S30}$. From Table 3.4.1, for $\theta = 11°$, zone A, $p_{S30}^{100} = 18.3$ psf, so $p_{S30} = 0.81 \times 18.3 = 14.8$ psf and $p_s = 1.21 \times 1.0 \times 1.0 \times 14.8 = 17.9$ psf. Calculated pressures p_s in psf are shown below for zones A to H.

Zone	A	B	C	D	E	F	G	H
p_s	17.9	−7.1	11.9	−4.1	−18.7	−11.5	−13.0	−8.8

Minimum design pressures are +10 psf for zones A, B, C, and D, with no pressure on zones E, F, G, and H. Calculated pressures for *windward walls* are $0.85 \times 17.9 = 15.2$ psf (zone A) and $0.85 \times 11.9 = 10.1$ psf (zone C); for *leeward walls,* $-0.7 \times 17.9 = -12.5$ psf (zone A), $-0.7 \times 11.9 = -8.3$ psf (zone C); for *sidewalls,* $-0.65 \times 11.9 = -7.7$ psf.

The width of zones E and F is $2a$, where (Sect. 3.3.3) $a = 10\%$ of least horizontal building dimension (i.e., $a = 0.1 \times 30 = 3$ ft), or $a = 0.4h$ (i.e., $a = 0.4 \times 15 = 6$ ft), whichever is smaller, but $a \geq 4\%$ of least horizontal dimension (i.e., $a \geq 0.04 \times 30 = 1.2$ ft), so $a = 3$ ft, and $2a = 6$ ft. To establish load patterns, each building corner, in turn, is viewed as a reference corner.

Numerical Example 3.4.2. *Simplified procedure: MWFRS for roof overhangs.* Assume that the building of Numerical Example 3.4.1 has roof overhangs with a 3-ft horizontal projection outside the exterior walls. Since, as was shown in the previous example, the strip $2a = 6$ ft, zones E and G fall on a roof overhang. We now calculate the simplified wind pressures p_s on the roof overhang for MWFRS design purposes, see Sect. 3.4.2, item 7g.

For the leeward and side edges of the building, the pressures p_s on the roof overhangs are the same as the pressures p_s on the corresponding roof zones. However, for the windward side of the building, the simplified design pressures p_s on the roof overhang are determined for MWFRS design purposes by using for zone E_{OH} the pressures $p_{S30}^{100} = -26.7$ psf (Table 3.4.1, $\theta = 11°$); since $V = 90$ mph, $p_{S30} = (V/100)^2 p_{S30}^{100} = 0.81 p_{S30}^{100} = -21.6$ psf, and $p_s = \lambda I K_{zt} p_{S30} = 1.21 \times 1.0 \times 1.0 \times (-21.6) = -26.2$ psf. Similarly, for the overhang zone G_{OH}, $p_s = -20.5$ psf.

For the design of the overhangs, each building corner is to be viewed as a reference corner.

8. For ***components and cladding*** (ASCE 6.4.2.2):
   a. Determine the ***net pressure*** $\boldsymbol{p_{\text{net30}}}$ corresponding to exposure B, h = 30 ft, I = 1, and K_{zt} = 1:

$$p_{\text{net30}} = \left(\frac{V}{100}\right)^2 p_{\text{net30}}^{100} \tag{3.4.4}$$

   where V is the basic wind speed in mph. Net pressures p_{net30}^{100} are listed in Table 3.4.2 for roofs and walls and in Table 3.4.3 for roof overhangs. (The entries in Table 3.4.2 and 3.4.3 are the same as the entries in ASCE Fig. 6-3 for V = 100 mph.) For zone designations, see Fig. 3.4.2 (ASCE Fig. 6-3)[7] and Sect. 3.3.3. For effective areas (Sect. 3.3.4) not listed in the tables, linear interpolation is permitted; otherwise, the value associated with the lower effective area must be used. The pressure p_{net30} can be also be obtained directly for various speeds V from ASCE Fig. 6-3 instead of from Eq. 3.4.4.
   b. Determine the ***simplified net design pressure*** normal to the wall and roof surfaces, which has the expression (ASCE Eq. 6-2)

$$p_{\text{net}} = \lambda I K_{zt} p_{\text{net30}} \tag{3.4.5}$$

   Simplified net design pressures shall be at least 10 psf (positive pressures) and −10 psf (negative pressures) (ASCE 6.4.2.2.1).

Numerical Example 3.4.3. *Simplified procedure: components and cladding.* We consider again the building of Numerical Example 3.4.1 (λ = 1.21, I = 1.0, K_{zt} = 1.0). It can be verified that each of the conditions listed in Sect. 3.4.1.1 is satisfied, so the simplified method may be used for the design of components and cladding. For zone designations and dimensions, see Fig. 3.4.2 and Sect. 3.3.3.

Roof panel pressures. Assume a 15-ft^2 effective area of roof panels. From Table 3.4.2 (roof net pressures, $7° < \theta \leq 27°$), for an effective area of 15 ft^2, linear interpolation between entries for 10 and 20 ft^2 yields, for zone 1 (interior zone), p_{net30}^{100} = 9.9 psf and p_{net30}^{100} = −16.25 psf. The corresponding simplified design pressures are (Eqs. 3.4.4 and 3.4.5) p_{net} = 1.21 × 1.0 × 1.0 $(90/100)^2$ × 9.9 = 9.7 psf and p_{net} = −15.9 psf. For zone 2 (end zone), p_{net} = 9.7 psf and p_{net} = −27.0 psf. For zone 3 (corner zones), p_{net} = 9.7 psf and

[7]Pressure zones for cladding and components of flat roofs, gable roofs, and hip roofs for both the simplified and analytical procedures are defined in Fig. 3.4.2.

TABLE 3.4.2. Net Pressures p_{net30}^{100} (psf) on Components and Cladding for Roofs and Walls

Roof Net Pressure, $0° \le \theta \le 7°$												
Zone	1				2				3[a]			
Eff. area (ft²)	10	20	50	100	10	20	50	100	10	20	50	100
p_{net30}^{100}	7.3	6.9	6.3	5.8	7.3	6.9	6.3	5.8	7.3	6.9	6.3	5.8
	−18.0	−17.5	−16.9	−16.5	−30.2	−27.0	−22.3	−19.5	−45.4	−37.6	−27.3	−19.5
Roof Net Pressure, $7° < \theta \le 27°$												
Zone	1				2				3[b]			
Eff. area (ft²)	10	20	50	100	10	20	50	100	10	20	50	100
p_{net30}^{100}	10.4	9.4	8.2	7.3	10.4	9.4	8.2	7.3	10.4	9.4	8.2	7.3
	−16.5	−16.0	−15.4	−14.9	−28.7	−26.4	−23.3	−21.0	−42.4	−39.6	−36.0	−33.2
Roof Net Pressure, $27° < \theta \le 45°$												
Zone	1				2				3			
Eff. area (ft²)	10	20	50	100	10	20	50	100	10	20	50	100
p_{net30}^{100}	16.5	16.0	15.4	14.9	16.5	16.0	15.4	14.9	16.5	16.0	15.4	14.9
	−18.0	−17.1	−15.9	−14.9	−21.0	−20.1	−18.9	−18.0	−21.0	−20.1	−18.9	−18.0

Wall Net Pressure										
Zone	4					5				
Eff. area (ft²)	10	20	50	100	500	10	20	50	100	500
p_{net30}^{100}	18.0	17.2	16.1	15.3	13.4	18.0	17.2	16.1	15.3	13.4
	−19.5	−18.7	−17.6	−16.8	−14.9	−24.1	−22.5	−20.3	−18.7	−14.9

Source: Excerpted from Fig. 6-3, ASCE Standard 7-02, *Minimum Design Loads for Buildings and Other Structures,* 2003, American Society of Civil Engineers. Reproduced by permission of the publisher, ASCE. Unmodified in ASCE 7-05.

[a] For hip roofs, in zone 3 the values for zone 2 shall be used.

[b] For hip roofs with slope 25° or less, in zone 3 the values for zone 2 shall be used.

TABLE 3.4.3. Roof Overhang Net Pressures p_{net30}^{100} (psf) on Components and Cladding

	$0 < \theta \leq 7°$							
Zone	2				3[a]			
Eff. area (ft²)	10	20	50	100	10	20	50	100
p_{net30}^{100}	−25.9	−25.5	−24.9	−24.4	−42.7	−33.5	−21.4	−12.2
	$7 < \theta \leq 27°$							
Zone	2				3[b]			
Eff. area (ft²)	10	20	50	100	10	20	50	100
p_{net30}^{100}	−33.5	−33.5	−33.5	−33.5	−56.4	−50.9	−43.6	−38.1
	$27 < \theta \leq 45°$							
Zone	2				3			
Eff. area (ft²)	10	20	50	100	10	20	50	100
p_{net30}^{100}	−30.5	−29.6	−28.4	−27.4	−30.5	−29.6	−28.4	−27.4

Source: Excerpted from Fig. 6-3, ASCE Standard 7-02, *Minimum Design Loads for Buildings and Other Structures,* 2003, American Society of Civil Engineers. Reproduced by permission of the publisher, ASCE. Unmodified in ASCE 7-05.

[a]For hip roofs, in zone 3 the values for zone 2 shall be used.

[b]For hip roofs with slope 25° or less, in zone 3 the values for zone 2 shall be used.

$p_{net} = -40.2$ psf. The values $p_{net} = 9.7$ psf are superseded by the requirement that the minimum positive pressures be 10 psf.

Cladding pressures for the exterior walls. Assume that the span of the exterior wall is 15 ft. The effective wind area is $15 \times (1/3 \times 15) = 75$ ft² (Sect. 3.3.3). Linear interpolation between pressures for 50- and 100-ft² areas yields for zone 4 (interior zone) $p_{net30}^{100} = 15.7$ psf and $p_{net30}^{100} = -17.2$ psf (Table 3.4.2, wall net pressures). The simplified design pressures are $p_{net} = 1.21 \times 1.0 \times 1.0(90/100)^2 \times 15.7 = 15.4$ psf and $p_{net} = -16.9$ psf. For zone 5 (end zones), $p_{net} = 15.4$ psf and $p_{net} = -19.1$ psf.

3.5 ANALYTICAL PROCEDURE (ASCE 6.5)

In Sect. 3.5 we list the conditions that must be met for the analytical procedure to be applicable (Sect. 3.5.1), and describe the steps that must be followed to calculate the wind loading (Sect. 3.5.2). In Sect. 3.5.3 we provide details on the last of those steps: determining the external pressure and force coefficients, and calculating the design wind pressures and loads for most types of buildings and other structures covered by ASCE.

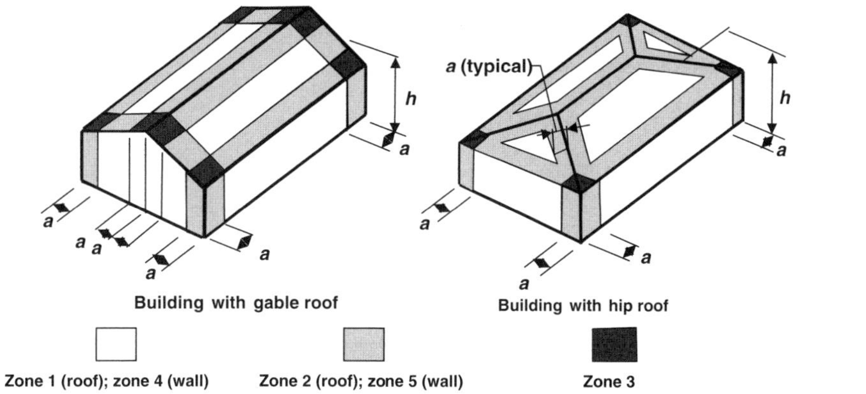

Notes
1. For *gable roofs with slopes 7° or less*, zone 2 and zone 3 adjacent to ridge shall be treated as zone 1 and zone 2, respectively.
2. For *flat roofs*, pressure zones shall be the identical to those for gable roofs with slopes 7° or less.
3. For *hip roofs with slopes 25° or less*, zone 3 shall be treated as zone 2.
4. h = mean roof height, except that for roofs with slopes $\theta \leq 10°$, h = eave height.
5. All dimensions a are horizontal dimensions.

Figure 3.4.2. Simplified procedure (*enclosed buildings*) and analytical procedure (*enclosed and partially enclosed buildings*); pressure zones for *components and cladding,* buildings with flat, gable, and hip roofs, $h \leq 60$ ft.

3.5.1 Conditions for Application of the Analytical Procedure (ASCE 6.5.1)

1. The building is located at a site with no significant channeling effects or buffeting in the wake of upwind obstructions.
2. The building has no unusual geometric irregularity.
3. The building is not subjected to vortex-induced, galloping, or flutter effects.

We note an inherent contradiction between the provision of a method for calculating along-wind response due to atmospheric turbulence fluctuations (ASCE 6.5.8.2) and the exclusion of buildings subjected to across-wind loading and vortex shedding (ASCE 6.5.1). The numerical example given in the ASCE Commentary C6.5.8 illustrates this contradiction. It involves a 600-ft-high building with a 100 ft × 100 ft square shape in plan. This building necessarily experiences vortex-induced effects and, according to ASCE 6.5.1, should therefore not be designed by the analytical procedure. The information on its along-wind response provided by ASCE as part of the analytical procedure is in general not sufficient for design purposes. Owing to the limitations of the standard with respect to the design of flexible buildings and structures for wind loads, this topic is treated separately in some detail in Sect. 4.3.

3.5.2 Design Procedure (ASCE 6.5.3)

The design procedure entails steps 1 to 11 below. Step 11 is described in detail in Sect. 3.5.3.

1. Determine the ***basic wind speed V*** (Sect. 3.2.1; ASCE 6.5.4).
2. Determine the ***wind directionality factor*** K_d (ASCE 6.5.4.4 and Table 6-4): $K_d = 0.85$ for buildings, arched roofs, signs, open lattice framework, and triangular, square, or rectangular trussed towers. For other types of structure, $K_d = 0.90$ or $K_d = 0.95$ (ASCE Table 6-4). *Note.* The wind directionality factor K_d is to be applied only when used in conjunction with the load combinations of ASCE 2.3 and 2.4 (ASCE 6.5.4.4).
3. Determine the ***importance factor I*** (Sect. 3.2.2; ASCE Tables 1-1 and 6-1).
4. Determine the ***exposure category*** for each direction being considered (Sects. 2.3.2.4 and 3.2.3; ASCE 6.5.6.4.1). However, the design of components and cladding, and MWFRSs for low-rise buildings, open buildings with monoslope, pitched, and troughed roofs, and freestanding walls and solid signs, shall be based on the exposure resulting in the highest wind loads for *any* exposure at the site (ASCE 6.5.6.4.2, 6.5.6.5, 6.5.13.2, and 6.5.14). ASCE provides no requirements on exposure similar to those of ASCE 6.5.6.4.1 for MWFRS of buildings with domed roofs and arched roofs, and other structures covered by ASCE 6.5.15. It is prudent to use for these cases design pressures based on exposure resulting in the highest wind loads for any direction.
5. Determine the ***topographic factor*** K_{zt} (Sects. 2.3.4 and 3.2.4; ASCE 6.5.7). For horizontal terrain $K_{zt} = 1$.
6. Determine the ***height h*** (h = mean roof height, except that for roofs with slope $\theta \leq 10°$, h = eave height).
7. Determine the ***gust effect factor*** (ASCE 6.5.8). For *rigid structures,* that is, structures with fundamental frequency of vibration equal to or greater than 1 Hz (ASCE 6.2 and 6.5.8.1) or with ratio $h/$(least horizontal dimension) < 4, the *gust effect factor* (or simply, the *gust factor*) $G = 0.85$ may be used. Alternatively, for structures that are, or may approximately be viewed as, prismatic or cylindrical,

$$G = 0.925\frac{1 + 5.78I_{\bar{z}}Q}{1 + 5.78I_{\bar{z}}} \qquad I_{\bar{z}} = c\left(\frac{33}{\bar{z}}\right)^{1/6}$$

$$Q = \frac{1}{\{1 + 0.63[(B + h)/\ell(\bar{z}/33)^{\bar{\varepsilon}}]^{0.63}\}^{1/2}} \qquad (3.5.1a,b,c)$$

where $\bar{z} = 0.6h$ or $z_{\min}$, whichever is larger; B = building width; h = mean roof height; and $z_{\min}$, c, ℓ, $\bar{\alpha}$, and $\bar{\varepsilon}$ are functions of exposure (ASCE Table 6-2).[8] For example, for exposure B, $z_{\min} = 30$ ft, $c = 0.30$, $\ell = 320$ ft, and $\bar{\varepsilon} = \frac{1}{3}$. *For flexible structures,* the gust effect factor, denoted by G_f, is specified in ASCE 6.8.2. An alternative and, in our opinion, more accurate expression is presented in Chapter 4, whose use is permitted by ASCE 6.5.8.3. ASCE 6.5.12.3 specifies an expression for estimating eccentricities for flexible structures. An alternative estimation method proposed by N. Isyumov of the University of Western Ontario is provided in Sect. 4A.3.

8. Determine the ***internal pressure coefficients*** (GC_{pi}) (Sect. 2.5.4; ASCE 6.5.11):
   a. ***Open buildings:*** $(GC_{pi}) = 0$.
   b. ***Partially enclosed buildings:*** $(GC_{pi}) = \pm 0.55$, except that if the building contains a large unpartitioned interior volume, (GC_{pi}) may be reduced as specified in ASCE 6.5.11.1.1.
   c. ***Enclosed buildings:*** $(GC_{pi}) = \pm 0.18$.

   In items 8b and c, G is given in combination with C_{pi} and is not to be obtained separately.
9. Determine the ***velocity pressure*** q_z (ASCE 6.5.10), calculated as follows:

$$q_z = 0.00256 I K_z K_d K_{zt} V^2 \quad \text{psf; } V \text{ in mph} \qquad (3.5.2a)$$

$$q_z = 0.613 I K_z K_d K_{zt} V^2 \quad \text{N/m}^2\text{; } V \text{ in m/s} \qquad (3.5.2b)$$

   where K_z is defined in item 10 below. The factors 0.00256 and 0.613 (corresponding to the units of Eq. 3.5.2*a* and *b*, respectively) are numerical values of the factor $\frac{1}{2}\rho$ (Sect. 2.5.1.1) for average conditions at zero altitude and may be modified if warranted by geographical considerations and climatic data.
10. Determine the ***velocity pressure exposure coefficient*** K_z (Eq. 3.5.2). At elevation z, K_z may be taken from ASCE Table 6-3. Alternatively, it can be calculated as follows:

[8] Except for slender towers, it is typically not necessary to use Eqs. 3.5.1 to calculate G; it is sufficient to use the value $G = 0.85$. Note that the expression for G (Eqs. 3.5.1) is derived by assuming wind loading on a wall normal to the wind direction and has in fact nothing whatsoever to do with pressures on the sidewalls, the leeward wall, and the roof. Its use in the ASCE Standard in conjunction with these pressures is conventional rather than being justified by physical considerations.

$$K_z = \begin{cases} 2.01\left(\frac{z}{z_g}\right)^{2/\alpha} & \text{for } z_1 \leq z \leq z_g \quad (3.5.3a) \\ 2.01\left(\frac{z_1}{z_g}\right)^{2/\alpha} & \text{for } z < z_1 \quad (3.5.3b) \end{cases}$$

For *exposure B*, z_g = 1200 ft, α = 7.0; for *exposure C*, z_g = 900 ft, α = 9.5; for *exposure D*, z_g = 700 ft, α = 11.5 (ASCE Table 6-2). In Eqs. 3.5.3, z_1 = 15 ft *except* that for exposure B, z_1 = 30 ft for components or cladding of all buildings and for MWFRS of low-rise buildings designed using Fig. 3.5.5 and Table 3.5.2 (ASCE Fig. 6-10). For details on the basis for calculating K_z, see Sect. 2.3.2.4.2.

The velocity pressures must be evaluated at elevations z, h, or at the elevation of the top of the parapet, as indicated for various types of buildings, structures, components, and cladding in Sect. 3.5.3; the respective values are denoted by q_z, q_h, or q_p.

11. Determine the ***external pressure or force coefficients,*** and the ***design wind pressures or loads,*** see Sect. 3.5.3. ASCE 6.5.2.1 allows *no pressure or load reductions* due to apparent shielding by buildings or other structures or terrain features.

3.5.3 Design Wind Pressures and Loads

Section 3.5.3 provides information on external pressure coefficients and design pressures or loads for MWFRS (Sect. 3.5.3.1) and components and cladding (Sect. 3.5.3.2) of enclosed and partially enclosed buildings, and for MWFRS and components and cladding of open buildings and structures (Sect. 3.5.3.3).

3.5.3.1 Enclosed and Partially Enclosed Buildings: Main Wind-Force Resisting Systems. Section 3.5.3.1 provides information on several types of building, on roof overhangs, and parapets, as follows:

- *Rigid buildings of all heights with flat, gable, hip, monoslope, and mansard roofs* (Fig. 3.5.1). Pressures are considered in Sect. 3.5.3.1.1. Design wind load cases, with and without torsion, are considered in Sect. 3.5.3.1.2 (Fig. 3.5.3).
- *Rigid buildings of all heights with domed roofs* (Sect. 3.5.3.1.3 and Fig. 3.5.4).
- *Arches* (Sect. 3.5.3.1.4).
- *Low-rise buildings* (Figs. 3.5.5, 3.5.6; Sect. 3.5.3.1.5). Low-rise buildings could be designed in accordance with the provisions for rigid buildings

of all heights. However, in an effort to achieve more economical designs, separate external pressure distributions were developed and specified in ASCE for low-rise buildings, on the basis of specially commissioned wind tunnel tests (see Sect. 4.2.1.1).

- *Roof overhangs* (Sect. 3.5.3.1.6).
- *Parapets* (Sect. 3.5.3.1.7).

3.5.3.1.1 Rigid Buidings of All Heights with Flat, Gable, Hip, Monoslope, and Mansard Roofs (Fig. 3.5.1; ASCE 6.5.12.2.1). Design wind pressures p are determined using the expression

$$p = qGC_p - q_i(GC_{pi}) \tag{3.5.4}$$

where C_p = external pressure coefficient, GC_{pi} = internal pressure coefficient (Sect. 3.5.2, item 8), the velocity pressures q and q_i are evaluated as indicated in items 9 and 10 of Sect. 3.5.2 at elevations specified subsequently in this section, and G = gust factor (Sect. 3.5.2, item 7).

Exception: For low-rise buildings the design pressures p may alternatively be determined as indicated in Sect. 3.5.3.1.5 instead of using Eq. 3.5.4.

The coefficient C_p is obtained from Table 3.5.1 (see ASCE Fig. 6-6).

The velocity pressure denoted by q in Eq. 3.5.4 is evaluated as follows: For *windward walls,* at elevation z [i.e., $q = q_z$ is an increasing function of

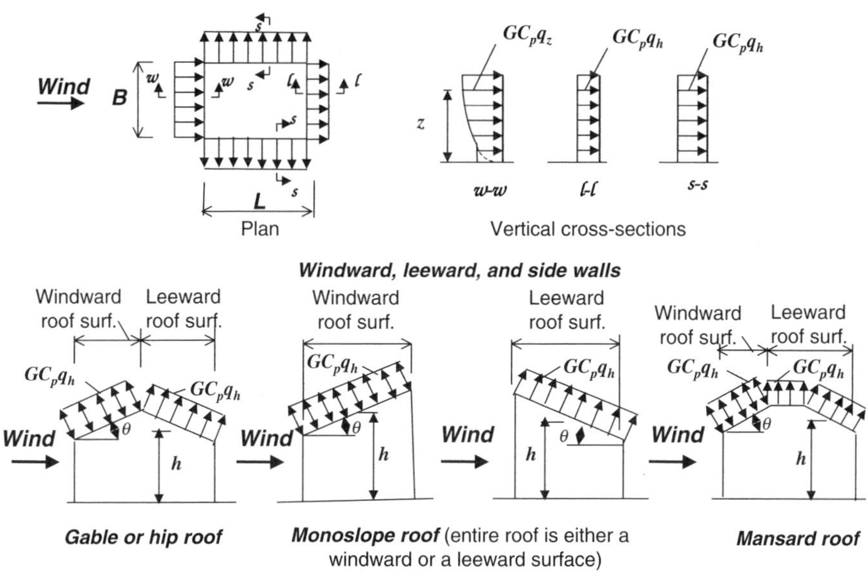

Figure 3.5.1. Analytical procedure, *enclosed and partially enclosed buildings of all heights;* external wind pressures for design of *MWFRS.*

TABLE 3.5.1. External Pressure Coefficients C_p, Enclosed and Partially Enclosed Buildings of All Heights with Roof Types of Fig. 3.5.1

Walls (Use with q_h Except Where Shown Otherwise)	
Windward wall	$C_p = 0.8$ (use with q_z)
Leeward wall	$0 < L/B \leq 1$: $C_p = -0.5$; $L/B = 2$: $C_p = -0.3$; $L/B \leq 4$: $C_p = -0.2$
Sidewalls	$C_p = -0.7$

Roof Slopes (Use with q_h)

		Windward Side					Leeward Side		
	$\theta°$	10	15	20	25	30	10	15	≥20
Normal to ridge θ ≥ 10°	$h/L \leq 0.25$	−0.7[a]	−0.5 0.0[b]	−0.3 0.2	−0.2 0.3	−0.2 0.3	−0.3	−0.5	−0.6
	$h/L = 0.5$	−0.9[a]	−0.7[a]	−0.4 0.0[b]	−0.3 0.2	−0.2 0.2	−0.5	−0.5	−0.6
	$h/L \geq 1.0$	−1.3[a,c]	−1.0[a]	−0.7[a]	−0.5 0.0[b]	−0.3 0.2	−0.7	−0.6	−0.6

		Horizontal distance u from Windward Edge	C_p
Normal to ridge for $\theta < 10°$, and parallel to ridge for all θ	$h/L \leq 0.5$	$0 < u \leq h$	−0.9[a]
		$h < u \leq 2h$	−0.5[a]
		$u > 2h$	−0.3[a]
	$h/L \geq 1.0$	$0 < u \leq h/2$	−1.3[a,c]
		$u > h/2$	−0.7[a]

Source: Excerpted from Fig. 6.3, ASCE Standard 7-95, *Minimum Design Loads for Buildings and Other Structures,* 1996, American Society of Civil Engineers. Reproduced by permission of the publisher, ASCE. Unmodified in ASCE 7-98, 7-02, and 7-05.

Notes:

1. L = horizontal dimension of building side parallel to wind direction. B = horizontal dimension of building side normal to wind direction.
2. For values of θ, L/B, and h/L not shown, linear interpolations are permitted between entries of the same sign. For ratios h/L, if the two entries used for interpolation have opposite signs, use 0.0 as one of the entries.
3. The horizontal component of the roof pressures is not to be used to reduce the total horizontal shear acting on the building, but is taken into account for members and connections near eaves.

[a] In addition to the C_p values listed in the table, ASCE 7-05 requires that the roof structure be designed for the value $C_p = -0.18$.

[b] Values listed for interpolation purposes only.

[c] May be reduced via multiplication by factor m obtained as follows: $Bh/2 \leq 100$ ft^2: $m = 1.0$; $Bh/2 = 200$ ft^2: $m = 0.9$; $Bh/2 \geq 1000$ ft^2: $m = 0.8$ (linear interpolation permitted).

height, except that at elevations $z < z_1$, $q = q_{z1}$ = const (see Eqs. 3.5.3)]; for *leeward walls, sidewalls, and roofs,* at elevation h (i.e., $q = q_h$).

The internal pressure q_i is evaluated as follows. For *enclosed buildings,* q_i is evaluated at elevation h (i.e., $q_i = q_h$). For *partially enclosed buildings,* q_i is evaluated at h, except that for positive internal pressures, q_i may be evaluated either at h or at the level of the highest opening that may affect positive internal pressures (e.g., if $h = 300$ ft, and that highest opening is at 50 ft above ground, the designer may evaluate q_i either at 300 ft or, less conservatively, at 50 ft).

The Standard provides no specific indication on pressures for the design of hip roof MWFRS. However, the ASCE Commentary C6.5.1.1 (Fig. C6-11) provides a method for adapting to hip roofs the pressure coefficients provided for gable roofs.

Numerical Example 3.5.1. *Analytical procedure. MWFRS for rigid building with flat roof.* This example illustrates in detail all the steps involved in calculating design pressures. To avoid unnecessary repetition, we focus in this example on determining pressures.

Consider an enclosed building with rectangular shape in plan (60 ft × 120 ft), eave height 95 ft, and flat roof. The building has *importance category II* ($I = 1.0$); is located in Florida's southern tip (*basic wind speed* $V = 150$ mph; see ASCE Fig. 6-1b); is *enclosed* (i.e., it satisfies the requirements of Sect. 3.2.5, items 1 to 5; item 5 applies because $V > 120$ mph); and has *exposure B in all directions*. The terrain is flat with *no topographic features* ($K_{zt} = 1.0$). The *wind directionality factor* is $K_d = 0.85$ (Sect. 3.5.2, item 2).

To obtain *external pressures* for the windward wall, the *velocity pressure*, and therefore the *coefficients* K_z, need to be evaluated at various elevations z. For the leeward walls, sidewalls, and roofs, they need to be evaluated at $z = h = 95$ ft. For example, from Sect. 3.5.2, item 10, Eq. 3.5.3, for $z \leq 15$ ft,

$$K_{z \leq 15\text{ft}} = 2.01 \times (15/1200)^{2/7.0} = 0.575$$

(K_z can also be obtained from ASCE Table 6-3). From Eqs. 3.5.2,

$$q_{z \leq 15\text{ ft}} = 0.00256 \times 1.0 \times 0.575 \times 0.85 \times 1.0 \times 150^2 = 28.2 \text{ psf}$$

From Eqs. 3.5.2 and 3.5.3, values of the velocity pressure q_z at elevations $z = 35$, 55, and 75 ft and $z = h = 95$ ft are obtained similarly. They are listed below along with the value at $z \leq 15$ ft:

z (ft)	≤15	35	55	75	95
q_z (psf)	28.2	35.8	40.8	44.6	47.7

(Note that the velocity pressure at $z = h = 95$ ft is denoted $q_z = q_h$.) The *external pressure coefficients* C_p are obtained from Table 3.5.1.

For wind direction parallel to the short building dimension (i.e., wind blowing in the transverse direction, Fig. 3.5.1, plan), $L = 60$ ft (L = building dimension parallel to wind direction, Fig. 3.5.1 and Table 3.5.1, note 1), $L/B = 60/120 < 1$, and $h/L = 95/60 > 1$. For the *windward wall*, $C_p = 0.8$. For the *leeward wall*, $C_p = -0.5$. For the *sidewalls*, $C_p = -0.7$. For the *roof*, since $\theta < 10°$ and $h/L > 1$, for a distance $h/2 = 47.5$ ft from the windward edge $C_p = -1.3$ (before reduction) and $C_p = -0.18$. The corresponding roof area is $Bh/2 = 120 \times 47.5 = 5700$ ft^2, so a 0.8 reduction factor (Table 3.5.1, footnote *c*) is applied to the coefficient $C_p = -1.3$. Therefore, $C_p = -1.3 \times 0.8 = -1.04$ and $C_p = -0.18$. For the roof area between 47.5 ft from windward edge and the leeward edge, $C_p = -0.7$ and $C_p = -0.18$.

The *gust effect factor* may be taken as $G = 0.85$ but can also be calculated using Eq. 3.5.1, as follows: $\bar{z} = 0.6h = 57$ ft; $c = 0.3$; $\ell = 320$ ft; $I_{\bar{z}} = 0.6(33/57)^{1/6} = 0.274$; $\bar{\varepsilon} = \frac{1}{3}$. Since $B = 120$ ft,

$$Q = \frac{1}{(1 + 0.63\{(120 + 95)/[320(57/33)^{1/3}]\}^{0.63})^{0.5}} = 0.83$$

$$G = 0.925 \frac{1 + 5.78 \times 0.274 \times 0.83}{1 + 5.78 \times 0.274} = 0.83$$

External pressures (i.e., term qGC_p on the right-hand side of Eq. 3.5.4) are as follows. For the *windward wall,* $z \leq 15$ ft,

$$p_{w\ \mathrm{ext}} = 28.2 \times 0.83 \times 0.8 = 18.7 \text{ psf}$$

(subscript w denoting "windward" and l denoting "leeward"). Values of $p_{w\ \mathrm{ext}}$ at elevations $z = 35$, 55, and 75 ft and $z = h = 95$ ft are obtained similarly. They are listed below along with the value at $z \leq 15$ ft:

z (ft)	≤15	35	55	75	95
$p_{w\ \mathrm{ext}}$ (psf)	18.7	23.8	27.1	29.6	31.7

For the *leeward wall,* since $C_p = -0.5$, $p_{l\ \mathrm{ext}} = -19.8$ psf at all elevations. For the *sidewalls,*

$$p_{s\ \mathrm{ext}} = 47.7 \times 0.83 \times (-0.7) = -27.7 \text{ psf at all elevations}$$

For the *roof,* between the windward edge and 47.5 ft from the windward edge,

$$p_{r\ \mathrm{ext}} = 47.7 \times 0.83 \times (-1.04) = -41.2 \text{ psf} \quad \text{and}$$

$$p_{r\ \mathrm{ext}} = 47.7 \times 0.83 \times (-0.18) = -7.1 \text{ psf}$$

For the balance of the roof (recall that $C_p = -0.7$ and $C_p = -0.18$), $p_{r\ \mathrm{ext}} = -27.7$ psf and -7.1 psf. Figure 3.5.2*b* shows external pressures on the windward and leeward walls and the roof for transverse wind loading.

Internal pressures $p_{\mathrm{int}} = q_i(GC_{pi})$ (i.e., the second term on the right-hand side of Eq. 3.5.4) are (Sect. 3.5.2, item 8)

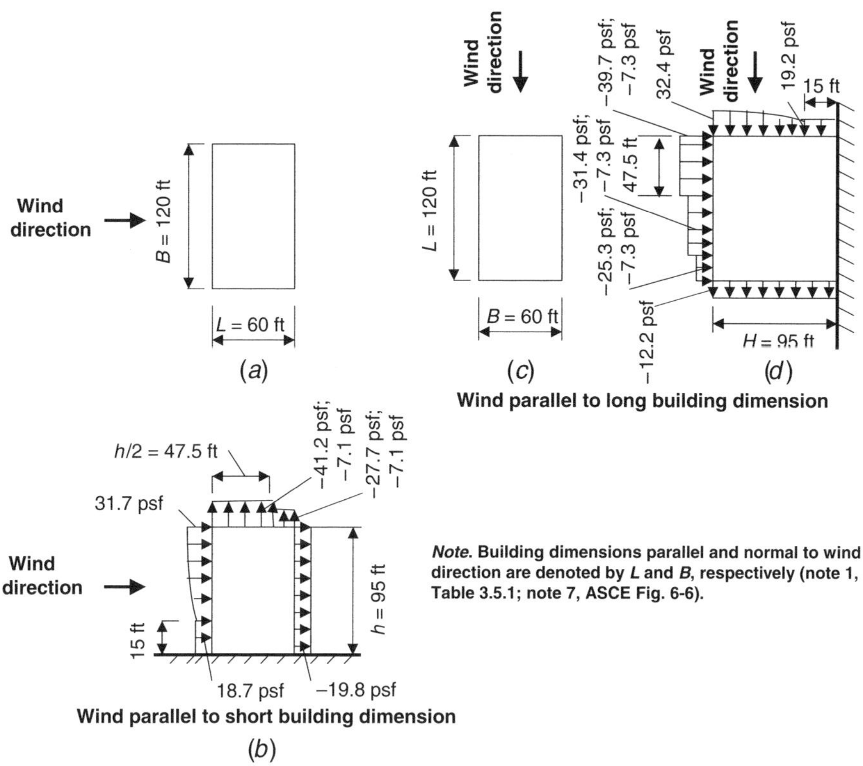

Note. **Building dimensions parallel and normal to wind direction are denoted by *L* and *B*, respectively (note 1, Table 3.5.1; note 7, ASCE Fig. 6-6).**

Figure 3.5.2. Wind pressures, Numerical Example 3.5.1.

$$p_{\text{int}} = 47.7 \times (\pm 0.18) = \pm 8.6 \text{ psf}$$

The *total* (*net*) *pressures* are $p = p_{\text{ext}} - p_{\text{int}}$ (Eq. 3.5.4). Positive internal pressures *increase* the upward loading on the roof from −41.2 psf to −(41.2 + 8.6) = −49.8 psf. Negative internal pressures create a net *downward* wind loading on the roof: −7.1 − (−8.6) = 1.5 psf.

For wind direction parallel to the long building dimension (i.e., wind blowing in the longitudinal direction, Fig. 3.5.2*c*), the calculations are similar, but L = 120 ft and B = 60 ft (L = building dimension parallel to wind direction, Fig. 3.5.1 and Table 3.5.1, note 1), $L/B = 120/60 = 2$, and $h/L = 95/120 = 0.79$. For the *windward wall,* $C_p = 0.8$. For the *leeward wall,* $C_p = -0.3$. For the *sidewalls,* $C_p = -0.7$. For the *roof,* $\theta < 10°$, and $h/L = 0.79$. The standard provides two options for the choice of C_p values. The first option consists of using, for each portion of the roof, the larger of the two values corresponding to the cases $h/L = 0.5$ and $h/L = 1.0$. The second option consists of using values obtained by interpolation between these two values. We list below the C_p values taken from Table 3.5.1, their maxima (in bold type), and the interpolated values (in italics):

	$0 < u \le h/2$	$h/2 < u \le h$	$h < u \le L$
$h/L \le 0.5$	−0.9	−0.9	−0.5
$h/L \ge 1.0$	$-0.9m = -1.04$	−0.7	−0.7
Interpolated	*−0.98*	*−0.78*	*−0.62*
Maximum	**−1.04**	**−0.9**	**−0.7**

(The roof area $Bh/2 = 60 \times 47.5 = 2850$ ft^2, so $m = 0.8$, and $-1.3m = -1.04$). For example, for $0 < u \le h/2$, the interpolated value is

$$C_p = -0.9 + [(0.79 - 0.5)/(1.0 - 0.5)][-1.04 - (-0.9)] = -0.98.$$

This example shows that, even if no interpolation is performed, Table 3.5.1 requires fairly complicated manipulations, which are unwarranted in view of the mediocre accuracy of the entries in the table (see Sect. 4.2), and should be avoided in a future version of the standard. In Eq. 3.5.3,

$$Q = \frac{1}{(1 + 0.63\{(60 + 95)/[320(57/33)^{1/3}]\}^{0.63})^{0.5}} = 0.86 \quad \text{and} \quad G = 0.85$$

External pressures (i.e., term qGC_p on the right-hand side of Eq. 3.5.4) are as follows. For the *windward wall,* $z \le 15$ ft,

$$p_{w\ \text{ext}} = 28.2 \times 0.85 \times 0.8 = 19.2 \text{ psf}$$

(the subscript w denoting "windward" and l denoting "leeward"). Values of $p_{w\ \text{ext}}$ at elevations of $z = 35$, 55, and 75 ft and $z = h = 95$ ft are obtained similarly. They are listed below along with the value at $z \le 15$ ft:

z (ft)	≤15	35	55	75	95
$p_{w\ \text{ext}}$ (psf)	19.2	24.3	27.7	30.3	32.4

For the *leeward wall,* since $C_p = -0.3$, $p_{l\ \text{ext}} = -12.2$ psf at all elevations. For the *sidewalls,*

$$p_{s\ \text{ext}} = 47.7 \times 0.85 \times (-0.7) = -28.4 \text{ psf at all elevations}$$

For the *roof,* between the windward edge and 47.5 ft from the windward edge,

$$p_{r\ \text{ext}} = 47.7 \times 0.85 \times (-0.98) = -39.7 \text{ psf} \quad \text{and}$$

$$p_{r\ \text{ext}} = 47.7 \times 0.85 \times (-0.18) = -7.3 \text{ psf}$$

For the area of the roof between 47.5 and 95 ft from the windward edge,

$$p_{r\ \mathrm{ext}} = 47.7 \times 0.85 \times (-0.78) = -31.4 \text{ psf} \quad \text{and}$$

$$p_{r\ \mathrm{ext}} = 47.7 \times 0.85 \times (-0.18) = -7.3 \text{ psf}$$

For the area of the roof between 95 ft from the windward edge and the leeward edge,

$$p_{r\ \mathrm{ext}} = 47.7 \times 0.85 \times (-0.62) = -25.3 \text{ psf} \quad \text{and}$$

$$p_{r\ \mathrm{ext}} = 47.7 \times 0.85 \times (-0.18) = -7.3 \text{ psf}$$

Figure 3.5.2*d* shows external pressures on the windward and leeward walls and roof for transverse wind loading.

Internal pressures $p_{\mathrm{int}} = q_i(GC_{pi})$ (i.e., the second term on the right-hand side of Eq. 3.5.4) are (Sect. 3.5.2, item 8)

$$p_{\mathrm{int}} = 47.7 \times (\pm 0.18) = \pm 8.6 \text{ psf}$$

The *total* (*net*) *pressures* are $p = p_{\mathrm{ext}} - p_{\mathrm{int}}$ (Eq. 3.5.4).

3.5.3.1.2 Design Wind Load Cases: Wind Acting on Two or Four Building Faces, with and Without Torsion (Fig. 3.5.3; ASCE 6.5.12.3). All enclosed and partially enclosed buildings for which wind loads are determined in accordance with Sects. 3.5.3.1.1 are designed for the loading cases 1a and 1b, 2a and 2b, 3, and 4 of Fig. 3.5.3 (a more compact version of ASCE Fig. 6-9 that can easily be computerized for office use). In Fig. 3.5.3 the pressures are windward and leeward wall pressures, *except* that one-story buildings with $h \leq 30$ ft and one- or two-story buildings with light-framed construction or flexible diaphragms need to be designed for cases 1 and 3 only (ASCE 6.5.12.3). For flexible buildings, see Sect. 4A.3.

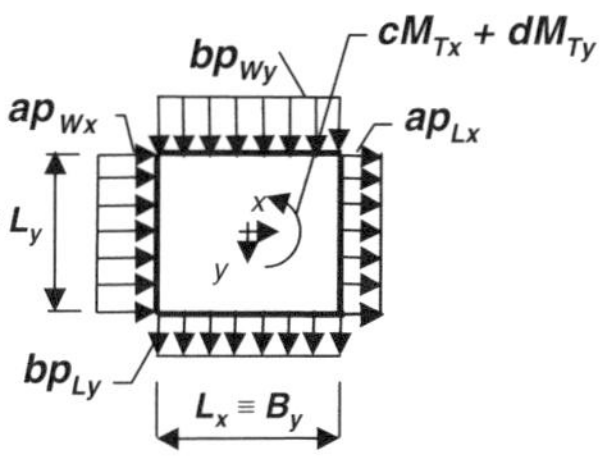

p_{Wx} = windward face external pressure in *x*-direction
p_{Lx} = leeward face external pressure in *x*-direction
p_{Wy} = windward face external pressure in *y*-direction
p_{Ly} = leeward face external pressure in *y*-direction

$M_{Tx} = \pm 0.1125(p_{Wx} + p_{Lx})B_x^2$

$M_{Ty} = \pm 0.1125(p_{Wy} + p_{Ly})B_y^2$

Case 1a: $a = 1$; $b = 0$; $c = 0$; $d = 0$
Case 1b: $a = 0$; $b = 1$; $c = 0$; $d = 0$
Case 2a: $a = 0.75$; $b = 0$; $c = 1$; $d = 0$
Case 2b: $a = 0$; $b = 0.75$; $c = 0$; $d = 1$

Case 3: $a = 0.75$; $b = 0.75$; $c = 0$; $d = 0$ *Case 4:* $a = 0.563$; $b = 0.563$; $c = 0.75$; $d = 0.75$

Figure 3.5.3. Analytical procedure, *enclosed and partially enclosed buildings of all heights;* design wind load cases for *MWFRS*.

Numerical Example 3.5.2. *Analytical procedure: design wind load cases with and without torsion, MWFRS.* Consider the building of Numerical Example 3.5.1. The external pressures calculated at $h = 95$ ft on the windward and leeward walls were shown to be $p_w = 31.7$ psf, $p_l = -19.8$ psf in the transverse direction (i.e., parallel to the short sides of the building, denoted the x-direction), and $p_w = 32.4$ psf, $p_l = -12.2$ psf for the longitudinal direction, denoted the y- direction (we omit the subscript "ext"). For the elevation $h = 95$ ft, the loading cases of Fig. 3.5.3 are as follows:

- *Case 1a:* $31.7 + 19.8 = 51.5$ psf (x-direction).
- *Case 1b:* $32.4 + 12.2 = 44.6$ psf (y-direction).
- *Case 2a:* $0.75 \times 51.5 = 38.7$ psf (x-direction) and, simultaneously, torsional moment $M_T = M_{Tx} = \pm 0.1125 \times 51.5$ psf $\times\ 120^2$ ft$^2 = \pm 83.4$ kip-ft/ft (moment per lineal foot of height).
- *Case 2b:* 0.75×44.6 psf (y-direction) $= 33.4$ psf and, simultaneously, torsional moment $M_T = M_{Ty} = \pm 0.1125 \times 44.6$ psf $\times\ 60^2$ ft$^2 = \pm 18.1$ kip-ft/ft.

Note: For wind in the x-direction, the dimension normal to the wind direction, denoted by B in Table 3.5.1 (ASCE Fig. 6-6), is denoted for clarity in Fig. 3.5.3 by B_x ($B_x = 120$ ft); similarly, for wind in the y-direction, the dimension normal to the wind direction is denoted by B_y ($B_y = 60$ ft).

- *Case 3:* $0.75 \times 51.5 = 38.7$ psf (x-direction) and, simultaneously, $0.75 \times 44.6 = 33.5$ psf (y-direction).
- *Case 4:* $0.563 \times 51.5 = 29.0$ psf (x-direction) and, simultaneously, 0.563×44.6 psf $= 25.1$ psf (y-direction), torsional moment $M_T = \pm 0.75(M_{Tx} + M_{Ty}) = \pm 76.1$ kip-ft/ft.

These results correspond to elevation $h = 95$ ft. Similar results must be obtained for other elevations of the windward and leeward walls. They depend on elevation, owing to the variation of the windward wall pressures with height. The resulting overall forces and moments are used for the design of the MWFRS. Internal pressures do not contribute to the overall transverse and longitudinal forces and torsional moments and may be disregarded in the calculations, as was done in this example.

3.5.3.1.3 Domed Roofs, MWFRS (Fig. 3.5.4; ASCE Fig. 6-7). Design wind pressures are given by Eq. 3.5.4, where q and q_i are evaluated at the top of the dome (i.e., at elevation $h_D + f$; see Fig. 3.5.4a). To determine C_p, two cases, referred to as cases 1 and 2, are considered.[9] *Case 1* is used to check

[9]Cases 1 and 2 are denoted in ASCE as cases A and B, respectively. Our notation avoids the confusion between the notations A and B used in ASCE for the two loading cases just described, points A and B, and the lines pertaining to those points in Fig. 3.5.4.

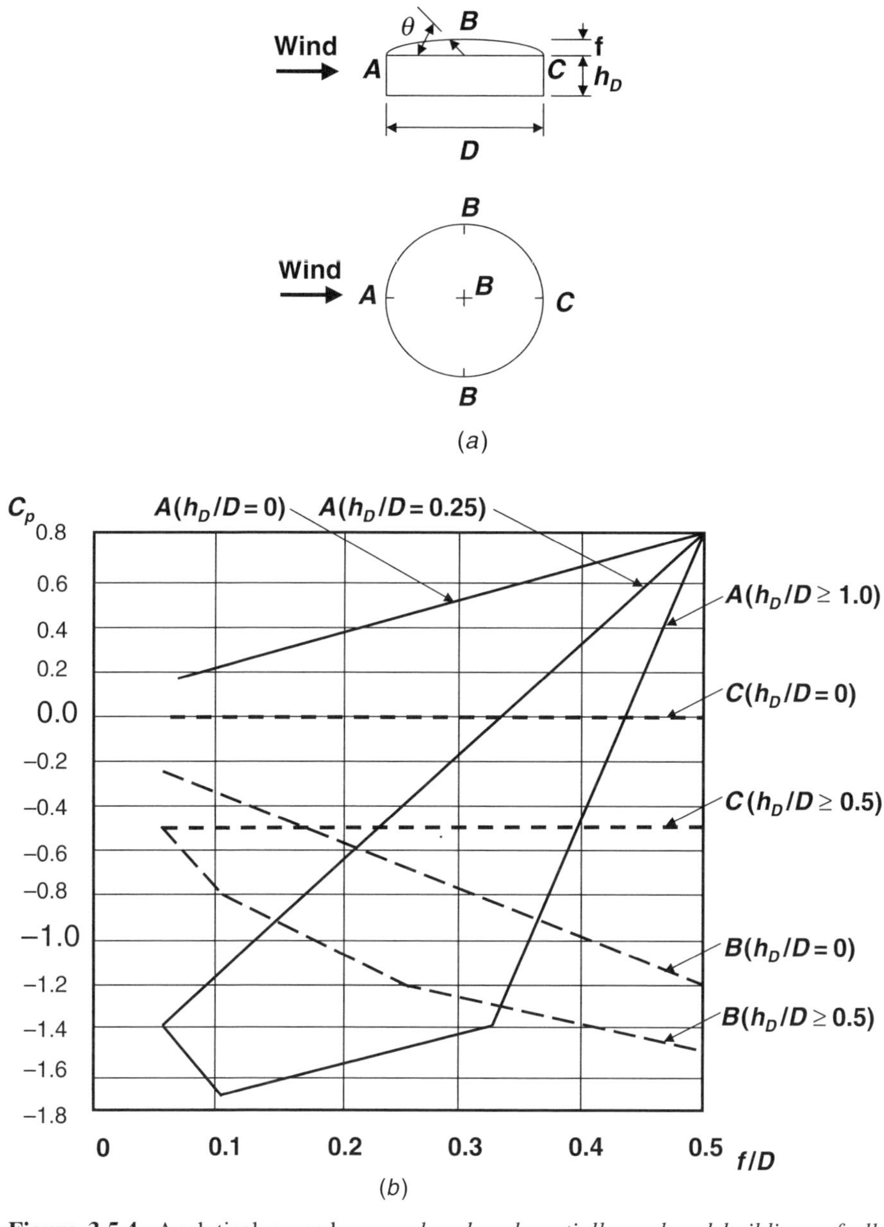

Figure 3.5.4. Analytical procedure, *enclosed and partially enclosed buildings of all heights with domed roofs:* (*a*) elevation and plan of a dome with a circular base; (*b*) plot for the estimation of the pressure coefficient C_p (courtesy of Eurocode).

maximum uplift. C_p values are obtained from Fig. 3.5.4*b* by linear interpolation between lines *A* and *B* for the windward side of the dome and between lines *B* and *C* for the leeward side. *Case 2* is used to check the maximum base shear. For $0 \leq \theta \leq 25°$, C_p is constant and equal to the value at point *A*. For $25° < \theta \leq 90°$, C_p is determined by linear interpolation between lines *A* and *B*, and for $90° < \theta \leq 180°$, by linear interpolation between lines *B* and *C*. Interpolations may be based on horizontal distances. For ratios h_D/D not shown in Fig. 3.5.4*b*, linear interpolation between lines *A*, between lines *B*, and between lines *C* is permitted. C_p is constant on the dome surface on arcs normal to the wind direction (on the arc *B–B–B* and all arcs parallel to *B–B–B* in Fig. 3.5.4*a*).

The total horizontal force may not be reduced by accounting for wind forces on the roof. For $f/D < 0.05$, use Table 3.5.1 (ASCE Fig. 6-6).

Numerical Example 3.5.3. *Analytical procedure. External pressure coefficients C_p for domed roofs, MWFRS.* Let $D = 150$ ft, $h_D = 37.5$ ft, $f = 30$ ft; $h_D/D = 0.25$, $f/D = 0.2$. For $A(h_D/D = 0.25)$, $C_p = -0.67$. For $B(h_D/D = 0)$, $C_p = -0.58$, for $B(h_D/D = 0.5)$, $C_p = -1.07$, so for $B(h_D/D = 0.25)$, $C_p = -0.83$. For $C(h_D/D = 0.25)$, $C_p = -0.25$.

Case 1: Obtain pressure coefficients by linear interpolation, (1) between points *A* and *B*: point *A*, $C_p = -0.67$; point *B*, $C_p = -0.83$; point at, e.g., 30 ft horizontal distance downwind from point *A*, $C_p = -0.67 - [30/(150/2)](0.83 - 0.67) = -0.73$; (2) between points *B* and *C*: point *B*, $C_p = -0.83$; point *C*, $C_p = -0.25$; point at, e.g., 30 ft distance downwind from line *B–B–B*, $C_p = -0.25 - [(150/2 - 30)/(150/2)](0.83 - 0.25) = -0.60$.

Case 2: Horizontal distance downwind of *A* over which pressures are the same as at *A*: $x = 44.2$ ft (x depends on the curvature of the dome surface and was determined graphically). Pressure coefficients: For *A* and the entire dome surface downwind of *A* up to horizontal coordinate x: $C_p = -0.67$. For *B*: $C_p = -0.83$. For point at 20 ft downwind of coordinate x: $C_p = -0.67 - [20/(150/2 - 44.2)](0.83 - 0.67) = -0.77$. Between *B* and *C*, pressures are the same as for case 1.

3.5.3.1.4 Arched Roofs. C_p varies with rise-to-span ratio r and position (ASCE Fig. 6-8).

3.5.3.1.5 Low-Rise Buildings (Sect. 3.3.6, Figs. 3.5.5, 3.5.6, ASCE 6.5.12.2.2). Design wind pressures p may be determined as in Sect. 3.5.3.1.1 (rigid buildings), or as follows:

$$p = q_h(GC_{pf}) - (GC_{pi}) \tag{3.5.5}$$

where q_h = velocity pressure (Eqs. 3.5.2, 3.5.3), GC_{pi} = internal pressure coefficient (Sect. 3.5.2, item 8), and (GC_{pf}) = external pressure coefficient from Table 3.5.2 (see ASCE Fig. 6-10) for the zones indicated in Fig. 3.5.5,

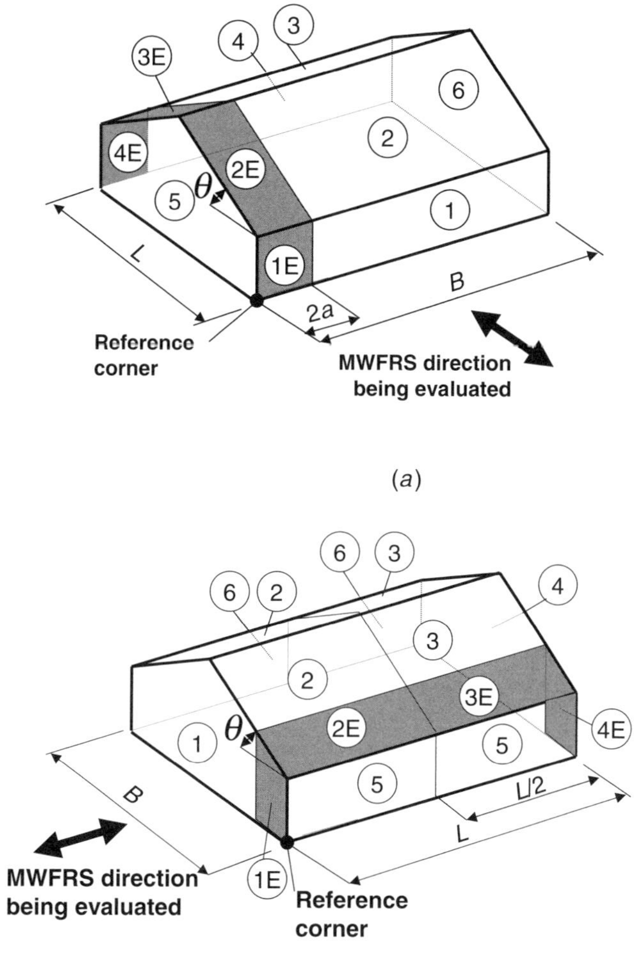

Figure 3.5.5. Analytical procedure, *enclosed and partially enclosed buildings, $h \leq 60$* ft; pressure zones for basic loading on *walls and roofs:* (*a*) wind normal to a ridge or long building dimension; (*b*) wind parallel to a ridge or long building dimension.

in which a is defined in Sect. 3.3.3 (linear interpolation is permitted for roof angles not shown in Table 3.5.2).

Note 8 of ASCE Fig. 6-10 requires that if for zone 2 or 2E the pressure coefficient GC_{pf} is negative, it is applied from the edge of the roof for a distance equal to half the dimension of the building parallel to the direction of the MWFRS being designed, as shown in Fig. 3.5.5*b*, or from the edge of the roof for a distance equal to 2.5 times the windward wall eave height, whichever is smaller. For wind normal to the ridge, if it is the latter distance

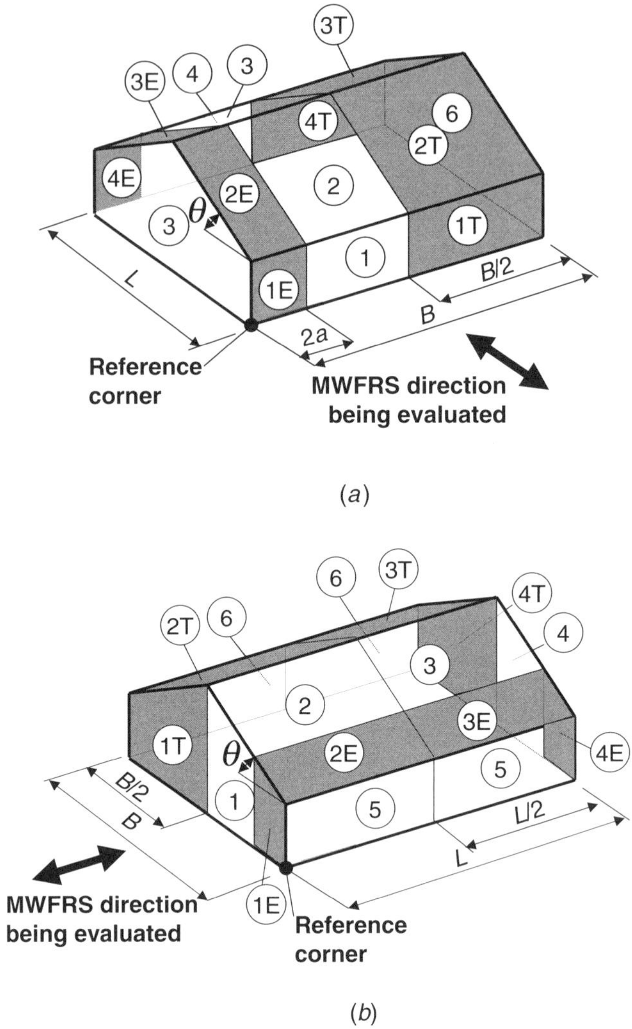

Figure 3.5.6. Analytical procedure, *enclosed and partially enclosed low-rise buildings;* pressure zones for torsional loading: (*a*) wind normal to a ridge or long building dimension; (*b*) wind parallel to a ridge or long building dimension.

that is smaller, the remaining portion of zone 2 or 2E extending to the ridge line is to be treated as zone 3 or 3E, respectively. For gable end walls, the average value of the eave height is used (Sect. 3.3.5, ASCE 6.2).

Note 7 of ASCE Fig. 6-10 requires that for the MWFRS in the direction parallel to the ridge, or for flat roofs, GC_{pf} be based on an assumed angle $\theta = 0°$. It requires further that the boundary between zones 2 and 3 be located at midlength of the building. This is inconsistent with note 8 of ASCE Fig.

TABLE 3.5.2. External Pressure Coefficients GC_{pf}, Low-Rise Buildings

	Zone (Fig. 3.5.5)									
Roof Slope	1	2	3	4	5	6	1E	2E	3E	4E
$0 \le \theta \le 5°$	0.40	−0.69	−0.37	−0.29	−0.45	−0.45	0.61	−1.07	−0.53	−0.43
$\theta = 20°$	0.53	−0.69	−0.48	−0.43	−0.45	−0.45	0.80	−1.07	−0.69	−0.64
$30 \le \theta \le 45°$	0.56	0.21	−0.43	−0.37	−0.45	−0.45	0.69	0.27	−0.53	−0.48
$\theta = 90°$	0.56	0.56	−0.37	−0.37	−0.45	−0.45	0.69	0.69	−0.48	−0.48

Source: From ASCE Standard 7-02, *Minimum Design Loads for Buildings and Other Structures,* 2003, American Society of Civil Engineers. Reproduced by permission of the publisher, ASCE. Unmodified in ASCE 7-05, except for a typographical error in ASCE 7-02, where the external pressure coefficients are incorrectly referred to as C_{pf} instead of GC_{pf}.

6-10, which, as was just shown, may allow zone 2 to extend for a distance from the roof edge of *less* than half the dimension of the building parallel to the direction of the ridge. Until this inconsistency is resolved it is prudent to assume in design that note 7 is applicable to the MWFRS in the direction of the ridge (or of the longer dimension of a flat roof) and that note 8 is applicable to the MWFRS only in the direction normal to the ridge (or the direction normal to the longer dimension of a flat roof).

Except for one-story buildings with $h \leq 30$ ft and one- or two-story buildings with light-framed construction or flexible diaphragms, *torsional loading* is considered in design by applying to zones 1T, 2T, 3T, and 4T shown in Fig. 3.5.6 (ASCE Fig. 6-10) 25% of the design pressures calculated by Eq. 3.5.5 for zones 1, 2, 3, and 4, respectively, while the design pressures for the balance of the building are given by Eq. 3.5.5.

The load patterns of Figs. 3.5.5 and 3.5.6 for winds acting transversely and longitudinally are associated with the reference corner shown therein. They are applied by considering each building corner in turn as the reference corner.

Reduction of the total horizontal shear on the building by roof pressures is neglected, except near the eaves, as indicated in note 3, Table 3.5.1, Sect. 3.5.3.1.1 for buildings of all heights.

Numerical Example 3.5.4. *Analytical procedure: low-rise building, MWFRS.* A rectangular building has dimensions in plan 45 ft $\times$ 40 ft, eave height 15 ft, gable roof with slope $\theta = 30°$, and mean roof height $h = 15 \text{ ft} + \frac{1}{2}(\frac{1}{2} \times 40 \text{ ft}) \sin 30° = 20 \text{ ft} < 40$ ft. We follow the procedure of Sect. 3.5.2. The building is located in Arkansas (basic wind speed $V = 90$ mph, see ASCE Fig. 6-1); its wind directionality factor is $K_d = 0.85$; it is classified as category II ($I = 1.0$); it is enclosed, so $GC_{pi} = \pm 0.18$; and it has exposure B in all directions. The building is located at the top of a two-dimensional escarpment, and the downwind distance between the escarpment's crest and the building's 45-ft-long face is $x = 40$ ft. From Numerical Example 3.2.1, the topographic factor is $K_{zt} = 1.34$. The velocity pressure exposure coefficient is $K_h = 2.01(30/1200)^{2/7} = 0.70$ (Eq. 3.5.3*a*; alternatively, see ASCE Table 6-3). The velocity pressure at height h is (Eq. 3.5.2*a*)

$$q_h = 0.00256 \times 1.0 \times 0.70 \times 0.85 \times 1.34 \times 90^2 = 16.5 \text{ psf}$$

The design pressures are given by Eq. 3.5.5 for the zones of Fig. 3.5.5. In the *transverse direction:*

- *Zone 1:* $p = 16.5[0.56 - (\pm 0.18)] = 6.3$ psf or 12.2 psf.
- *Zone 2:* $p = 16.5[0.21 - (\pm 0.18)] = 0.5$ psf or 6.4 psf.
- *Zone 3:* $p = 16.5[-0.43 - (\pm 0.18)] = -10.1$ psf or -4.1 psf.
- *Zone 4:* $p = 16.5[-0.37 - (\pm 0.18)] = -9.1$ psf or -3.1 psf.
- *Zones 5 and 6:* $p = 16.5[-0.45 - (\pm 0.18)] = -10.4$ psf or -4.5 psf.

Similarly:

- *Zone 1E:* p = 4.5 psf or 14.3 psf; *Zone 2E:* p = 1.5 psf or 7.4 psf; *Zone 3E:* p = −11.7 psf or −5.8 psf; *Zone 4E:* p = −10.9 psf or −5.0 psf.

Design pressures for zones 1T, 2T, 3T, and 4T (Fig. 3.5.6) are 25% of those in zones 1, 2, 3, and 4; that is, zone 1T: $p_T = 0.25 \times 6.3 = 1.6$ psf or $0.25 \times 12.2 = 3.1$ psf; zone 2T: $p_T = 0$ or 1.6 psf; zone 3T: $p_T = -2.5$ psf or −1.0 psf; zone 4T: $p_T = -2.3$ psf or −1.1 psf. Note that unlike in Sect. 3.5.3.1.2 and Fig. 3.5.5 (ASCE Fig. 6-9), these calculations must involve the design pressures rather than just the external pressures.

Zones 2 and 2E extend for a distance from the edge of the roof equal to 0.5 times 40 ft (i.e., 20 ft). Zones 3 and 3E extend from the edge of zones 2 and 2E to the leeward edge of the roof.

In the *longitudinal direction,* pressures are based on slope $\theta = 0°$; for example, for zones 2, 3, and 2E, 3E:

- *Zone 2:* $p = 16.5[-0.69-(\pm 0.18)] = -14.3$ psf or −6.8 psf.
- *Zone 3:* $p = 16.5[-0.37-(\pm 0.18)] = -9.1$ psf or −3.1 psf.
- *Zone 2E:* $p = 16.5[-1.07-(\pm 0.18)] = -20.6$ psf or −14.7 psf.
- *Zone 3E:* $p = 16.5[-0.53-(\pm 0.18)] = -11.7$ psf or −5.8 psf.

The boundary between zones 2 and 3, and 2E and 3E, is at 45/2 = 22.5 ft from the outer faces of the end walls.

All the pressures calculated above are based on the use of the factor $K_{zt} = 1.34$. This is correct for wind blowing uphill in a direction normal to the crest line. However, these pressures would be conservative for winds blowing downhill in a direction normal to the crest line and for winds blowing in a direction parallel to the crest line. Pressures due to these winds could be reduced by using in the calculations the factor $K_{zt} = 1.0$ instead of $K_{zt} = 1.34$.

3.5.3.1.6 Roof Overhangs (ASCE 6.5.11.4.1). A positive pressure coefficient $C_p = 0.8$ applicable to the bottom overhang surface is to be used in addition to the pressure coefficient for the top surface from Table 3.5.1 (ASCE Fig. 6-6) or 3.5.2 (ASCE Fig. 6-10).

Numerical Example 3.5.5. *Analytical procedure: roof overhang, MWFRS.* The low-rise building of Numerical Example 3.5.4 has a roof with overhang projecting 3 ft outside the longitudinal exterior walls. The building has exposure B from all directions. In this example we consider for the top area of the overhang the most strongly loaded zone only, that is, zone 2E, for an angle $\theta = 0$. Therefore (see Numerical Example 3.5.4), the net pressure on the overhang is $p_{\text{net}} = 16.5(-1.07 - 0.8) = -30.9$ psf. Similar calculations are performed for other types of building.

3.5.3.1.7 Parapets (ASCE 6.5.12.2.4). For all enclosed or partially enclosed buildings, the effect on the MWFRS of wind acting on the parapets is based on the total pressures

$$p_p = q_p(GC_{pn}) \tag{3.5.6}$$

where q_p is the velocity pressure (Eqs. 3.5.2 and 3.5.3) evaluated at the top of the parapet, and $(GC_{pn}) = 1.5$ for windward parapets and -1.0 for leeward parapets.

Numerical Example 3.5.6. *Analytical procedure: parapets, MWFRS.* Assume that the building of Numerical Example 3.5.1 has a 3-ft-high parapet on all roof sides. At the top of the parapet (98 ft elevation) the velocity pressure exposure coefficient is $K_z = 2.01(98/1200)^{2/7} = 0.98$, and the velocity pressure is $q_p = 0.00256 \times 1.0 \times 0.98 \times 0.85 \times 1.0 \times 150^2 = 48.0$ psf (Eqs. 3.5.3 and 3.5.2; 0.85 is the wind directionality factor K_d). The total pressures on the windward and leeward parapets are $p_p = 48.0 \times 1.5 = 72.0$ psf and $p_p = 48.0 \times (-1.0) = -48.0$ psf (directed toward and away from parapet), respectively.

3.5.3.2 Enclosed and Partially Enclosed Buildings: Components and Cladding. Components and cladding are designed for the calculated maximum positive and negative net pressures. If the tributary area exceeds 700 ft^2, the design provisions for MWFRSs may be used. For effective areas of components and cladding, see Sect. 3.3.4. This section provides information on pressures for:

- *Buildings with height $h \leq 60$ ft* and gable, hip, multispan gable, monoslope, and sawtooth roofs (Sect. 3.5.3.2.1)
- *Buildings with $h > 60$ ft* (Sect. 3.5.3.2.2)
- *Buildings with 60 ft $< h <$ 90 ft* for which alternative design pressures are permitted (Sect. 3.5.3.2.3)
- *Buildings with stepped flat roofs (all heights)* (Sect. 3.5.3.2.4)
- *Domed roofs (all heights)* (Sect. 3.5.3.2.5)
- *Arched roofs* (Sect. 3.5.3.2.6)
- *Parapets* (Sect. 3.5.3.2.7)

3.5.3.2.1 Buildings with Height $h \leq 60$ ft: Components and Cladding (ASCE 6.5.12.4.1). The pressures are

$$p = q_h(GC_p) - (GC_{pi}) \tag{3.5.7}$$

q_h = velocity pressure at height h (Sect. 3.5.2, item 9; Eq. 3.5.2), (GC_{pi}) = internal pressure coefficient (Sect. 3.5.2, item 8).

For *walls* (Fig. 3.4.2), *gable roofs* (Figs. 3.4.2), *hip roofs* (Fig. 3.4.2), *multispan gable roofs* (Fig. 3.5.7), *monoslope roofs* (Figs. 3.5.8*a*, *b*), and *sawtooth roofs* (Fig. 3.5.8*c*), the external pressure coefficients (GC_p) depend on effective area A_{eff} as follows:

$$A_{\text{eff}} < A_{\text{eff 1}}\text{: } (GC_p) = (GC_p)_1 \tag{3.5.8a}$$

$$A_{\text{eff}} > A_{\text{eff 2}}\text{: } (GC_p) = (GC_p)_2 \tag{3.5.8b}$$

$$A_{\text{eff 1}} \leq A_{\text{eff}} < A_{\text{eff 2}}\text{: } (GC_p) = (GC_p)_1 + g \log_{10} \frac{A_{\text{eff}}}{A_{\text{eff 1}}} \tag{3.5.8c}$$

The areas $A_{\text{eff 1}}$, $A_{\text{eff 2}}$, and the values $(GC_p)_1$, $(GC_p)_2$, and g are listed in Table 3.5.3. Because in Eq. 3.5.8*c* (and in similar equations presented subsequently) the logarithmic function depends on nondimensional ratios, these values are independent of units (i.e., they are the same for English and metric units of area). This advantage could be used in a future, more compact version of the standard. Equations 3.5.8 and similar equations presented subsequently can easily be computerized for office use.

Addenda to Table 3.5.3

1. For *walls of buildings with roof slopes* $\theta \leq 10°$, values of (GC_p) are reduced by 10% (ASCE Fig. 6-11A, note 5).
2. For *gable roof overhangs,* $\theta \leq 7°$ (ASCE Fig. 6-11B):
   a. For zones 1 and 2:

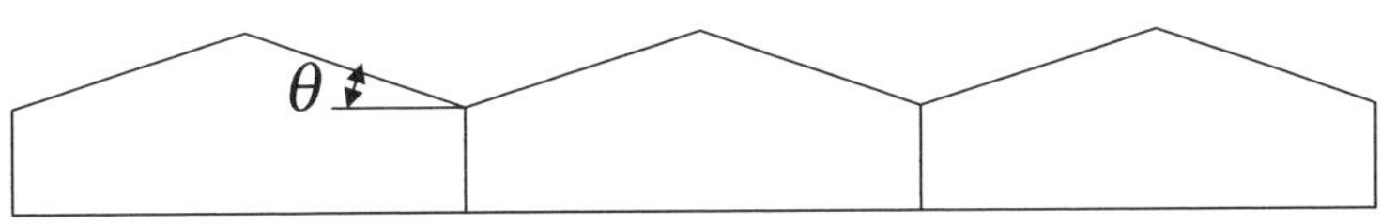

Figure 3.5.7. Analytical procedure, *enclosed and partially enclosed buildings with multispan gable roofs, h* ≤ 60 ft, elevation of building (two or more spans).

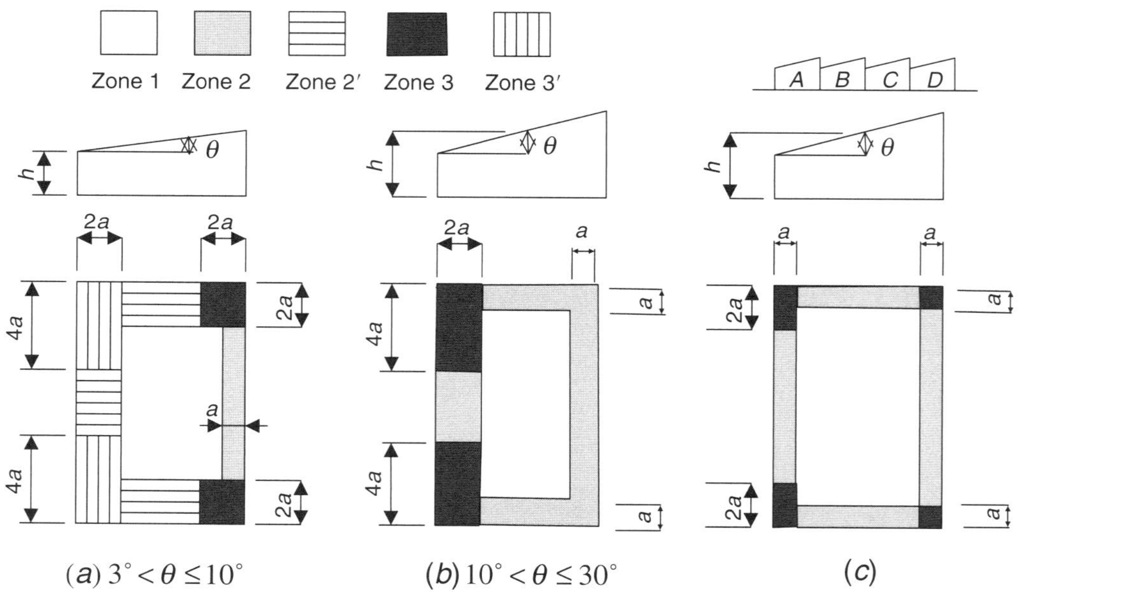

Monoslope roofs
Elevation (top) and plan (bottom)

Sawtooth roofs (two or more spans)
Elevation of building with four spans (top)
Elevation (middle) and plan (bottom) of single-span module

Note. For monoslope roofs with $\theta \leq 3°$ and sawtooth roofs with $\theta \leq 10°$ use pressure coefficients for gable roofs with $\theta \leq 7°$ (Table 3.5.4 and Fig. 3.4.2).

Figure 3.5.8. Analytical procedure, *enclosed and partially enclosed buildings, monoslope roofs, and sawtooth roofs,* $h \leq 60$ ft; pressure zones for *components and cladding.*

TABLE 3.5.3. Effective Areas $A_{eff\,1}$, $A_{eff\,2}$, (sq ft), Coefficients $(GC_p)_1$, $(GC_p)_2$, and Coefficients g: Buildings with $h \leq 60$ ft

	Zone(s)	Pressure	$A_{eff\,1}$	$A_{eff\,2}$	$(GC_p)_1$	$(GC_p)_2$	g
Walls[a]	4, 5	Positive			1.0	0.7	−0.177
Fig. 3.4.2	4	Negative	10	500	−1.1	−0.8	0.177
(ASCE Fig. 6-11A)	5	Negative			−1.4	−0.8	0.353
Gable roofs	1, 2, 3	Positive			0.3	0.2	−0.1
$\theta \leq 7°$[a]	1	Negative			−1.0	−0.9	0.1
Fig. 3.4.2	2	Negative	10	100	−1.8	−1.1	0.7
(ASCE Fig. 6-11B)	3	Negative			−2.8	−1.1	1.7
	Overhang, 1, 2	Negative			−1.7	−1.6	0.1
	Overhang, 3	Negative			−2.8	−0.8	2.0
Gable and hip roofs	1, 2, 3	Positive			0.5	0.3	−0.2
$7° \leq \theta \leq 27°$	1	Negative			−0.9	−0.8	0.1
Fig. 3.4.2	2	Negative	10	100	−1.7	−1.2	0.5
(ASCE Fig. 6-11C)	3	Negative			−2.6	−2.0	0.6
	Overhang, 2	Negative			−2.2	−2.2	0.0
	Overhang, 3	Negative			−3.7	−2.5	1.2
Gable roofs	1, 2, 3	Positive			0.9	0.8	−0.1
$27° < \theta \leq 45°$	1	Negative	10	100	−1.0	−0.8	0.2
Fig. 3.4.2	2, 3	Negative			−1.2	−1.0	0.2
(ASCE Fig. 6-11D)	Overhang, 2, 3	Negative			−2.0	−1.8	0.2
Multisp. gable roofs[a]	1, 2, 3	Positive			0.6	0.4	−0.2
$0° < \theta \leq 30°$	1	Negative	10	100	−1.6	−1.4	0.2
Figs. 3.4.2, 3.5.7	2	Negative			−2.2	−1.7	0.5
(ASCE Fig. 6-13)	3	Negative			−2.7	−1.7	1.0
Multisp. gable roofs[a]	1, 2, 3	Positive			1.0	0.8	−0.2
$30° < \theta \leq 45°$	1	Negative	10	100	−2.0	−1.1	0.9
Figs. 3.4.2, 3.5.7	2	Negative			−2.5	−1.7	0.8
(ASCE Fig. 6-13)	3	Negative			−2.6	−1.7	0.9

TABLE 3.5.3. (*Continued*)

	Zone(s)	Pressure	$A_{\text{eff }1}$	$A_{\text{eff }2}$	$(GC_p)_1$	$(GC_p)_2$	G
Monoslope roofs	All zones	Positive	10	100	0.3	0.2	−0.1
$3° < \theta \leq 10°$	1	Negative			−1.1	−1.1	0.0
Fig. 3.5.8*a*	2	Negative			−1.3	−1.2	0.1
(ASCE Fig. 6-14A)	2′	Negative			−1.6	−1.5	0.1
	3	Negative			−1.8	−1.2	0.6
	3′	Negative			−2.6	−1.6	1.0
Monoslope roofs	All zones	Positive	10	100	0.4	0.3	−0.1
$10° < \theta \leq 30°$	1	Negative			−1.3	−1.1	0.2
Fig. 3.5.8*b*	2	Negative			−1.6	−1.2	0.4
(ASCE Fig. 6-14B)	3	Negative			−2.9	−2.0	0.9
Sawtooth roofs[a]	1	Positive	10	500	0.7	0.4	−0.177
Fig. 3.5.8*c*	2	Positive	10	100	1.1	0.8	−0.3
(ASCE Fig. 6-15)	3	Positive	10	100	0.8	0.7	−0.1
	1	Negative	10	500	−2.2	−1.1	0.647
	2	Negative	10	500	−3.2	−1.6	0.942
	3, span A	Negative	10	100	−4.1	−3.7	0.4
	3, spans B, C, D	Negative	100	500	−2.6	−1.9	1.0

Source: Based on ASCE Standard 7-02, *Minimum Design Loads for Buildings and Other Structures*, 1996, American Society of Civil Engineers. Reproduced by permission of the publisher, ASCE. Unmodified in ASCE 7-05.

[a] See Addenda to Table 3.5.3: item 1 for walls of buildings with roof slopes $\theta \leq 10°$; item 2a for overhangs of gable roofs with slopes $\theta \leq 7°$; item 2b for parapet effects on pressures for roofs with slopes $\theta \leq 7°$; item 3 for sawtooth roofs; item 4 for multispan gable roofs, and for multispan gable and sawtooth roofs with slopes $\theta \leq 10°$.

$A_{\text{eff }2} = 100 \text{ ft}^2 < A_{\text{eff}} \leq 500 \text{ ft}^2$:

$$(GC_p) = -1.6 + 0.715 \log_{10} \frac{A_{\text{eff}}}{A_{\text{eff }2}} \qquad (3.5.8e)$$

$$A_{\text{eff}} > 500 \text{ ft}^2\text{: } (GC_p) = -1.1 \qquad (3.5.8f)$$

b. If a *parapet* of height 3 ft or higher is provided around the perimeter of a gable roof with $\theta \leq 3°$, the negative values of (GC_p) in zone 3 shall be equal to those of zone 2, and the positive values of (GC_p) in zones 2 and 3 shall be equal to those in zones 4 and 5 for walls (Fig. 3.4.2), respectively.

3. For *sawtooth roofs,* zone 3, span A (ASCE Fig. 6-15):

$A_{\text{eff }2} = 100 \text{ ft}^2 < A_{\text{eff}} \leq 500 \text{ ft}^2$:

$$(GC_p) = -3.7 + 2.29 \log_{10} \frac{A_{\text{eff}}}{A_{\text{eff }2}} \qquad (3.5.8g)$$

$$A_{\text{eff}} > 500 \text{ ft}^2, (GC_p) = -2.1 \qquad (3.5.8h)$$

4. For *multispan gable roofs,* pressure zones shown in Fig. 3.4.2 for gable roofs apply to all spans. For roof slopes $\theta \leq 10°$, the values listed in Table 3.5.4 for gable roofs shall be used instead of those listed for multispan gable and sawtooth roofs (ASCE Figs. 6-13 and 6-15, note 5).

Numerical Example 3.5.7. *Analytical procedure: external pressure coefficients (GC_p) for components and cladding, walls of building with height $h \leq$ 60 ft.* See Fig. 3.4.2, Table 3.5.3, and Eqs. 3.5.8. For walls, $h \leq 60$ ft:

TABLE 3.5.4. Effective Areas $A_{\text{eff }1}$, $A_{\text{eff }2}$ (ft^2), Corresponding External Pressure Coefficients $(GC_p)_1$, $(GC_p)_2$, and Coefficients g: Buildings with $h > 60$ ft[a]

	Zone(s)		$A_{\text{eff }1}$	$A_{\text{eff }2}$	$(GC_p)_1$	$(GC_p)_2$	g
Walls; flat and	4, 5	Positive	20	500	0.9	0.6	−0.215
gable roofs	4	Negative	20	500	−0.9	−0.7	0.143
$\theta \leq 10°$	5	Negative	20	500	−1.8	−1.0	0.572
Fig. 3.5.9	1	Negative	10	500	−1.4	−0.9	0.294
(ASCE Fig. 6-17)	2	Negative	10	500	−2.3	−1.6	0.412
	3	Negative	10	500	−3.2	−2.3	0.53

Source: Based on Fig. 6-17, ASCE Standard 7-02, *Minimum Design Loads for Buildings and Other Structures,* 1996, American Society of Civil Engineers. Reproduced by permission of the publisher, ASCE. Unmodified in ASCE 7-05.

[a] If a 3-ft-high or higher parapet is provided around the perimeter of roofs with slope $\theta \leq 10°$, zone 3 becomes zone 2.

1. Zones 4 and 5 (positive pressures):

$$A_{\text{eff}} \leq 10 \text{ ft}^2\text{: } (GC_p) = 1.0 \qquad \text{(Eq. 3.5.8}a\text{);}$$

$$A_{\text{eff}} = 100 \text{ ft}^2\text{: } (GC_p) = 1.0 - 0.177 \log_{10} \frac{100}{10} = 0.82 \qquad \text{(Eq. 3.5.8}b\text{);}$$

$$A_{\text{eff}} \geq 500 \text{ ft: } (GC_p) = 0.7 \qquad \text{(Eq. 3.5.8}c\text{).}$$

2. Zone 4 (negative pressures):

$$A_{\text{eff}} \leq 10 \text{ ft}^2\text{: } (GC_p) = -1.1 \qquad \text{(Eq. 3.5.8}a\text{);}$$

$$A_{\text{eff}} = 100 \text{ ft}^2\text{: } (GC_p) = -1.1 + 0.177 \log_{10} \frac{100}{10} = -0.92 \qquad \text{(Eq. 3.5.8}b\text{);}$$

$$A_{\text{eff}} \geq 500 \text{ ft: } (GC_p) = -0.8 \qquad \text{(Eq. 3.5.8}c\text{).}$$

3.5.3.2.2 Buildings with $h > 60$ ft: Components and Cladding (ASCE 6.5.12.4.2). The pressures are

$$p = q(GC_p) - q_i(GC_{pi}) \tag{3.5.9}$$

For windward walls, $q = q_z$ (Eqs. 3.5.2); for leeward walls, sidewalls, and roofs, $q = q_h$; for internal pressures, $q_i = q_h$; for positive internal pressures of *partially enclosed* buildings, the value $q_i = q_{ho}$ may be used in lieu of $q_i = q_h$, where h_o is the elevation of the highest opening that may affect positive internal pressures. For dimensions a see Sect. 3.3.3.

For (GC_{pi}), see Sect. 3.5.2, item 8. For walls and for roofs with slopes $\theta \leq 10°$, (GC_p) in Eq. 3.5.9 is based on Table 3.5.4, Fig. 3.5.9, and Eqs. 3.5.8*a*–*c*. For gable and hip roofs with slopes $\theta > 10°$, (GC_p) in Eq. 3.5.9 is based on Table 3.5.3, Fig. 3.4.2, and Eqs. 3.5.8*a*–*c* (ASCE Fig. 6-17, note 6).

3.5.3.2.3 Buildings with 60 ft $< h <$ 90 ft: Components and Cladding (ASCE 6.5.12.4.3). As an alternative to Sect. 3.5.3.2.2, if the height/width ratio is 1 or less, Eq. 3.5.7 (normally applicable to buildings with height $h \leq$ 60 ft) may be used, with pressure coefficients based on Table 3.5.3 and Eqs. 3.5.8. *Exception:* If the roof slope is $\theta > 10°$, Eq. 3.5.9 is used, with coefficients from Table 3.5.3 (gable and hip roofs, see Figs. 3.4.2) and Eqs. 3.5.8*a*–*c*, even if the height/width ratio is larger than 1; the exposure is determined as in paragraph 2 of Sect. 3.2.3.

3.5.3.2.4 Buildings with Stepped Flat Roofs: Components and Cladding (All Heights; ASCE Fig. 6-12). Provisions are restricted to buildings with dimensions $h_1 \geq 10$ ft, $0.3 < h_i/h < 0.7$, and $0.25 < W_i/W < 0.75$ $(i = 1,2)$

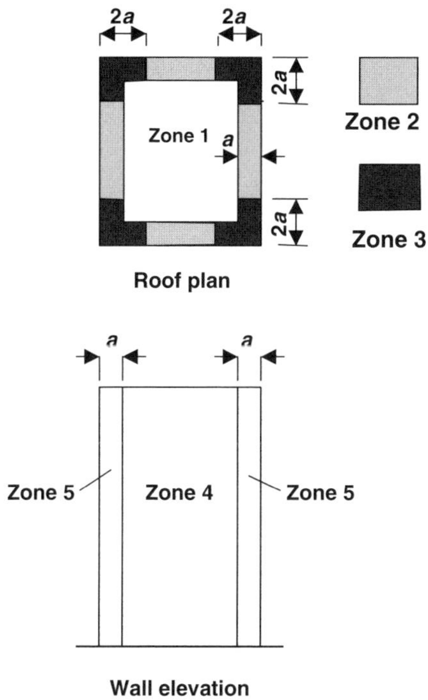

Figure 3.5.9. Analytical procedure, *enclosed and partially enclosed buildings, $h > 60$ ft*; pressure zones for *components and cladding, walls, and roofs.*

(see notations in ASCE Fig. 6-12). For the lower level of roof, zone designations and coefficients (GC_p) for gable roofs with slopes $\theta \leq 7°$ apply (see Table 3.5.3), except that at roof–upper wall intersections, zone 3 is treated as zone 2, and zone 2 is treated as zone 1. The width of the crosshatched areas of ASCE Fig. 6-12 is $b = 1.5h_1$ or 100 ft, whichever is less. The positive pressures (GC_p) on those areas shall be

$$A_{\text{eff}} < 10 \text{ ft}^2, \ GC_p = 1.0$$

$$10 \leq A_{\text{eff}} < 500 \text{ ft}^2, \ GC_p = 1.0 - 0.177 \log_{10} \frac{A_{\text{eff}}}{10}$$

$$A_{\text{eff}} > 500 \text{ ft}^2, \ GC_p = 0.7$$

No pressures are specified for walls and the upper level of roof, but coefficients may be used for walls and gable roofs with $\theta \leq 7°$ from Table 3.5.3 and Eqs. 3.5.8*a*–*c*.

3.5.3.2.5 Domed Roofs: Components and Cladding (All Heights; ASCE Fig. 6-16). Using the notations of Fig. 3.5.4*a*,

$$p = q_{h_D+f}(GC_p) - (GC_{pi}) \tag{3.5.10}$$

where q_{h_D+f} = velocity pressure (Sect. 3.5.2, item 9) at elevation $h_D + f$ (i.e., at the top of the dome); $(GC_p) = -0.9$ for $0° < \theta \leq 90°$, $(GC_p) = 0.9$ for $0° < \theta \leq 60°$, and $(GC_p) = 0.5$ for $60° < \theta \leq 180°$. For internal pressure coefficient (GC_{pi}), see Sect. 3.5.2, item 8. The use of these values is restricted to domes for which $0 < h_D/D \leq 0.5$ and $0.2 \leq f/D \leq 0.5$.

3.5.3.2.6 Arched Roofs: Components and Cladding (ASCE Fig. 6-8, note 4). At the roof perimeter use the external pressure coefficients of Table 3.5.3 for gable roofs with slope equal to the spring-line slope of the arched roof. For remaining roof areas, use external pressure coefficients of ASCE Fig. 6-8 divided by 0.87. (ASCE mistakenly uses "multiplied" instead of "divided"; in fact, wind loads on components and cladding have smaller effective areas than loads on MWFRSs, which, as noted in ASCE Fig. 6-8, note 1, are average loads; so the external pressures for components and cladding should be larger, not smaller, than the average loads used for MWFRSs.)

3.5.3.2.7 Parapets: Components and Cladding (ASCE 6.5.12.4.4). Pressures on components of parapets are

$$p = q_p(GC_p) - (GC_{pi}) \tag{3.5.11}$$

where q_p is the velocity pressure (Eqs. 3.5.2) evaluated at the top of the parapet, and (GC_{pi}) is taken from Sect. 3.5.2, item 8. Positive and negative internal pressures, if any, are to be based on the porosity of the parapet envelope and depend on information supplied by the manufacturer. Two load cases shall be considered. Load case A pertains to parapets on the building's windward side: the positive wall pressure (from Table 3.5.3 or 3.5.4, as appropriate) at the top of the parapet is applied to the parapet's windward (outer) face, and the negative (edge or corner zone) roof pressure is applied to the parapet's leeward (inner) face. Load case B pertains to parapets on the building's leeward side: the positive wall pressure (from Table 3.5.3 or 3.5.4) at the top of the parapet is applied to the parapet's windward (inner) face, and the negative wall pressure at the top of the parapet is applied to the parapet's leeward (outer) face.

Numerical Example 3.5.8. *Analytical procedure: parapet components and cladding.* The building of Numerical Example 3.5.1 (h = 95 ft, flat roof) is assumed to have 3-ft-high nonporous parapets (i.e., there are no internal pressures within the parapets). The components are assumed to be 3 ft high and

2 ft wide, so $A_{\text{eff}} = 3 \text{ ft} \times 2 \text{ ft} = 6 \text{ ft}^2$. The external pressure coefficients are taken from Table 3.5.4 and Fig. 3.5.9.

For *case A* (windward parapet), the positive external coefficient for walls $(GC_p) = 0.9$. The negative zone 2 (edge) roof pressure coefficient is $(GC_p) = -2.3$, and the negative zone 3 (corner) roof pressure is $(GC_p) = -3.2$. According to the footnote of Table 3.5.4, if the building has a 3-ft parapet around the roof perimeter, zone 3 becomes zone 2, so zone 2 pressures are also applicable for the roof corners, meaning that the parapet protects the roof from experiencing at the corners very high (zone 3) pressures, corresponding to $(GC_p) = -3.2$.

For *case B,* the positive wall pressure coefficient $(GC_p) = 0.9$. The negative wall pressure coefficient for zone 4 $(GC_p) = -0.9$, and for zone 5 (corners) $(GC_p) = -1.8$.

The velocity pressures are at the top of the parapet, $q_p = 48.0$ psf (Numerical Example 3.5.6), and for the roof, $q_h = 47.7$ psf (Numerical Example 3.5.1). (Given the small difference between q_p and q_h, some designers may prefer to use q_p for the roof pressure as well.)

For *case A,* the design pressure corresponding to the zone 2 portion of the roof is $p = 48[0.9 - (-2.3)] = 153.6$ psf.

For *case B,* the design pressures are $48[0.9 - (-0.9)] = 86.4$ psf for zone 4 and $48[0.9 - (-1.8)] = 129.6$ psf for zone 5.

3.5.3.3 Open Buildings and Structures (ASCE 6.5.13). In this section we discuss design wind pressures or forces on MWFRSs and components and cladding of:

- *Monoslope, pitched, and troughed roofs of open buildings* (Sect. 3.5.3.3.1)
- *Solid walls and solid signs* (Sect. 3.5.3.3.2)
- *Open signs, lattice frameworks, and trussed towers* (Sect. 3.5.3.3.3)
- *Chimneys, tanks, rooftop equipment, and similar structures* (Sect. 3.5.3.3.4)

Velocity pressures q_z and q_h are evaluated by Eqs. 3.5.2 and 3.5.3 at height z and height h, respectively. The exposure resulting in the highest load for any direction is used.

It was noted earlier that the expression for gust factors given by Eqs. 3.5.1 (ASCE Eq. 6-4) was derived for rigid rectangular buildings. That expression is typically not applicable to open buildings and structures other than free-standing solid walls and trussed towers.

3.5.3.3.1 Design Wind Pressures on Open Buildings with Monoslope, Pitched, or Troughed Roofs: MWFRS (ASCE 6.5.14). The net design pressure is

$$p = q_h G C_N \tag{3.5.12}$$

where q_h is the velocity pressure (Sect. 3.5.2, item 9) at elevation h (Fig. 3.5.10 and Sect. 3.3.5), and the gust factor is $G = 0.85$ (Sect. 3.5.2, item 7). For $\gamma = 0°$ and $\gamma = 180°$, $\theta \le 45°$, and $0.25 \le h/L \le 1.0$, the net pressure coefficient (sum of coefficients for the top and bottom surfaces) C_N is taken from Tables 3.5.5 and 3.5.6 (ASCE Fig. 6-18A and B), in which $C_N = C_{NW}$ for the windward half of the roof and $C_N = C_{NL}$ for the leeward half. Linear interpolation is permitted, except that for $\theta < 7.5°$ the values listed for monoslope roofs for $\theta = 0°$ are used.

Note 5 of ASCE Table 6-18D specifies that for monoslope roofs with slopes $\theta < 5°$ and $0.05 \le h/L \le 0.25$, for $\gamma = 0°$ the values of C_N are taken from that table (i.e., from Table 3.5.7), but no indication is given on the values of C_D for $\gamma = 180°$. However, since for monoslope roofs with $0.25 \le h/L \le 1.0$ and $\theta < 7.5°$ there is no differentiation between the directions $\gamma = 0°$ and $\gamma = 180°$, it is reasonable to assume that this is also true for monoslope roofs with $\theta < 5°$ and $0.05 \le h/L \le 0.25$ (see the footnote to Table 3.5.7). Cases A and B of Tables 3.5.5, 3.5.6, and 3.5.7 are both considered in design.

For all roofs with slopes $\theta \le 45°$, the net pressure coefficients C_N for wind parallel to the direction of the roof's horizontal edges ($\gamma = 90°$ or $\gamma = 270°$)

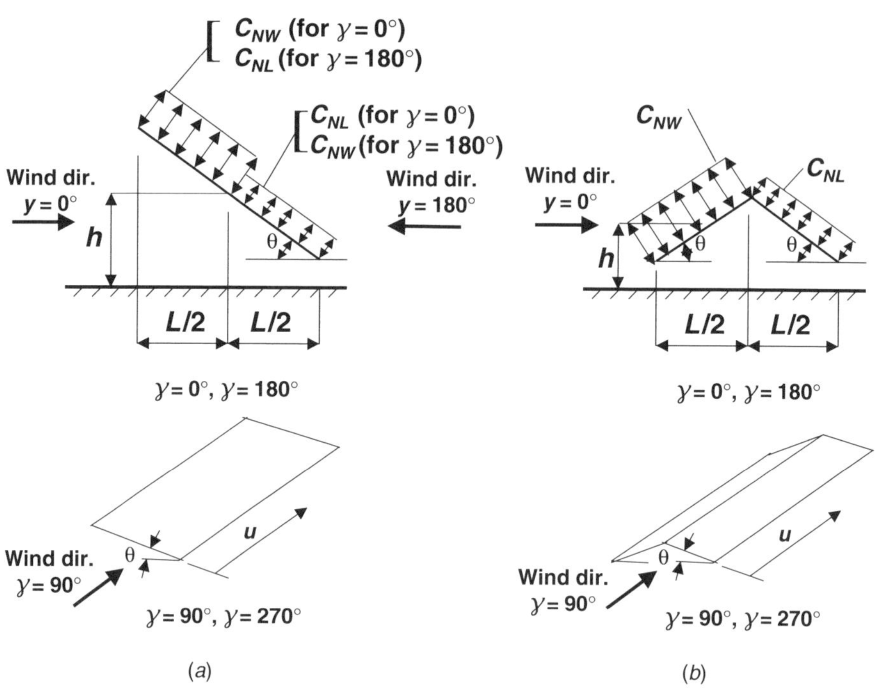

Figure 3.5.10. Analytical procedure, *open buildings*, $0.25 \le h/L \le 1.0$, $\theta \le 45°$; *MWFRS*, net pressure coefficients C_{NW} and C_{NL}: (*a*) monoslope roofs; (*b*) pitched roofs.

TABLE 3.5.5. Net Pressure Coefficients, Monoslope Roofs, $0.25 \leq h/L \leq 1.0$ (MWFRS), for Selected Roof Slopes[a]

		Wind Direction $\gamma = 0°$				Wind Direction $\gamma = 180°$			
		Clear Flow		Obstructed Flow		Clear Flow		Obstructed Flow	
Roof Angle, θ (deg)	Load Case	C_{NW}	C_{NL}	C_{NW}	C_{NL}	C_{NW}	C_{NL}	C_{NW}	C_{NL}
0	A	1.2	0.3	−0.5	−1.2	1.2	0.3	−0.5	−1.2
	B	−1.1	−0.1	−1.1	−0.6	−1.1	−0.1	−1.1	−0.6
7.5	A	−0.6	−1.0	−1.0	−1.5	0.9	1.5	−0.2	−1.2
	B	−1.4	0.0	−1.7	−0.8	1.6	0.3	0.8	−0.3
15	A	−0.9	−1.3	−1.1	−1.5	1.3	1.6	0.4	−1.1
	B	−1.9	0.0	−2.1	−0.6	1.8	0.6	1.2	0.3
22.5	A	−1.5	−1.6	−1.5	−1.7	1.7	1.8	0.5	−1.0
	B	−2.4	−0.3	−2.3	−0.9	2.2	0.7	1.3	0.0

Source: Based on Fig. 6-18, ASCE Standard 7-05, *Minimum Design Loads for Buildings and Other Structures,* 1996, American Society of Civil Engineers. Reproduced by permission of the publisher, ASCE.

[a]Cases A and B and wind directions $\gamma = 0°$ and $\gamma = 180°$ are all considered in design.

depend on the horizontal distance u from the windward roof edge as shown in Fig. 3.5.10 and Table 3.5.7.

Fascia panels of roofs with $\theta \leq 5°$ are considered as inverted parapets with contributions to the MWFRS loads calculated as in Sect. 3.5.3.1.7, in which q_h is used instead of q_p.

For *components and cladding,* the pressure coefficients C_N are

TABLE 3.5.6. Net Pressure Coefficients, Pitched Roofs, $0.25 \leq h/L \leq 1.0$ (MWFRS), for Selected Roof Slopes[a]

		Pitched Roofs, Wind Direction $\gamma = 0°$, $\gamma = 180°$			
		Clear Flow		Obstructed Flow	
Roof Angle, θ (deg)	Load Case	C_{NW}	C_{NL}	C_{NW}	C_{NL}
7.5	A	1.1	−0.3	−1.6	−1.0
	B	0.2	−1.2	−0.9	−1.7
15	A	1.1	−0.4	−1.2	−1.0
	B	0.1	−1.1	−0.6	−1.6
22.5	A	1.1	0.1	−1.2	−1.2
	B	−0.1	−0.8	−0.8	−1.7

Source: Based on Fig. 6-18, ASCE Standard 7-05, *Minimum Design Loads for Buildings and Other Structures,* 1996, American Society of Civil Engineers. Reproduced by permission of the publisher, ASCE.

[a]For troughed roofs, see ASCE Fig. 6-18C.

TABLE 3.5.7. Net Pressure Coefficients C_N, Monoslope, Pitched, and Troughed Roofs, $0.25 \le h/L \le 1.0$, $0 < \theta \le 45°$, $\gamma = 90°$ or $270°$ (MWFRS)[a]

Horizontal Distance, u	Load Case	C_N, Clear Flow	C_N, Obstructed Flow
$u \le h$	A	−0.8	−1.2
	B	0.8	0.5
$h < u \le 2h$	A	−0.6	−0.9
	B	0.5	0.5
$u > 2h$	A	−0.3	−0.6
	B	0.3	0.3

Source: Based on Fig. 6-18, ASCE Standard 7-05, *Minimum Design Loads for Buildings and Other Structures,* 1996, American Society of Civil Engineers. Reproduced by permission of the publisher, ASCE.

[a]Values from this table are also to be used for directions $\gamma = 0°$ and $\gamma = 180°$ for monoslope roofs with slopes $\theta < 5°$ and ratios $0.05 \le h/L \le 0.25$.

$$p = q_h G C_N \tag{3.5.13}$$

where $G = 0.85$ and the C_N are given as functions of effective area A_{eff} (see Sect. 3.3.4) in Table 3.5.8 and Fig. 3.5.11 (ASCE Figs. 6-19). For the width a in Fig. 3.5.11, see Sect. 3.3.3.

Note. For both MWFRSs and components and cladding, pressure coefficients are given for *clear* wind flow (50% or less obstructed vertical area under the roof) and for *obstructed* wind flow (more than 50% obstructed vertical area). Both positive and negative pressures are used in design.

Numerical Example 3.5.9. *Analytical procedure: monoslope roof of an open building.* Assume that for a monoslope roof (Fig. 3.5.10): $h = 15$ ft, $L = 24$ ft ($h/L = 0.625$), and $\theta = 15°$; and the wind flow is clear. For the MWFRS, Tables 3.5.5 and 3.5.7 yield:

Wind Direction, γ	Load Case	C_{NW}	C_{NL}
0°	A	−0.9	−1.3
	B	−1.9	0.0
180°	A	1.3	1.6
	B	1.8	0.6

Wind Direction, γ	Distance u from Windward Edge	Load Case	C_N
90°	$u \le 15$ ft	A	−0.8
		B	0.8
	15 ft $< u \le 24$ ft	A	−0.6
		B	0.5

TABLE 3.5.8. Net Pressure Coefficients C_N, Monoslope, and Pitched Roofs, $0.25 \le h/L \le 1.0$ (Components and Cladding)[a,b]

Roof	Flow	Zone	A_{eff}	$\theta = 0°$	$\theta = 7.5°$	$\theta = 15°$	$\theta = 30°$	$\theta = 45°$
Monoslope	Clear	1	All A_{eff}	1.2;−1.1	1.6;−1.4	1.8;−1.9	2.6;−2.5	2.6;−2.3
		2	$A_{eff} \le 4a^2$	1.8;−1.7	2.4;−2.1	2.7;−2.9	3.9;−3.8	3.9;−3.5
			$A_{eff} > 4a^2$	1.2;−1.1	1.6;−1.4	1.8;−1.9	2.6;−2.5	2.6;−2.3
		3	$A_{eff} \le a^2$	2.4;−3.3	3.2;−4.2	3.6;−3.8	5.2;−5.0	5.2;−4.6
			$a^2 < A_{eff} \le 4a^2$	1.8;−1.7	2.4;−2.1	2.7;−2.9	3.9;−3.8	3.9;−3.5
			$A_{eff} > 4a^2$	1.2;−1.1	1.6;−1.4	1.8;−1.9	2.6;−2.5	2.6;−2.3
	Obstr.	1	All A_{eff}	0.5;−1.2	0.8;−1.7	1.2;−2.1	1.6;−2.3	2.1;−1.9
		2	$A_{eff} \le 4a^2$	0.8;−1.8	1.2;−2.6	1.8;−3.2	2.4;−3.5	3.2;−2.9
			$A_{eff} > 4a^2$	0.5;−1.2	0.8;−1.7	1.2;−2.1	1.6;−2.3	2.1;−1.9
		3	$A_{eff} \le a^2$	1.0;−3.6	1.6;−5.1	2.4;−4.2	3.2;−4.6	4.2;−3.8
			$a^2 < A_{eff} \le 4a^2$	0.8;−1.8	1.2;−2.6	1.8;−3.2	2.4;−3.5	3.2;−2.9
			$A_{eff} > 4a^2$	0.5;−1.2	0.8;−1.7	1.2;−2.1	1.6;−2.3	2.1;−1.9
Pitch	Clear	1	All A_{eff}	1.2;−1.1	1.1;−1.2	1.1;−1.1	1.3;−0.9	1.1;−0.8
		2	$A_{eff} \le 4a^2$	1.8;−1.7	1.7;−1.8	1.7;−1.7	2.0;−1.4	1.7;−1.2
			$A_{eff} > 4a^2$	1.2;−1.1	1.1;−1.2	1.1;−1.1	1.3;−0.9	1.1;−0.8
		3	$A_{eff} \le a^2$	2.4;−3.3	2.2;−3.6	2.2;−2.2	2.6;−1.8	2.2;−1.6
			$a^2 < A_{eff} \le 4a^2$	1.8;−1.7	1.7;−1.8	1.7;−1.7	2.0;−1.4	1.7;−1.2
			$A_{eff} > 4a^2$	1.2;−1.1	1.1;−1.2	1.1;−1.1	1.3;−0.9	1.1;−0.8
	Obstr.	1	All A_{eff}	0.1;−1.2	0.5;−1.7	0.5;−1.6	0.5;−1.2	0.5;−1.2
		2	$A_{eff} \le 4a^2$	0.8;−1.8	0.8;−2.6	0.8;−2.4	0.8;−1.8	0.8;−1.8
			$A_{eff} > 4a^2$	0.5;−1.2	0.5;−1.7	0.5;−1.6	0.5;−1.2	0.5;−1.2
		3	$A_{eff} \le a^2$	1.0;−3.6	1.0;−5.1	1.0;−3.2	1.0;−2.4	1.0;−2.4
			$a^2 < A_{eff} \le 4a^2$	0.8;−1.8	0.8;−2.6	0.8;−2.4	0.8;−1.8	0.8;−1.8
			$A_{eff} > 4a^2$	0.5;−1.2	0.5;−1.7	0.5;−1.6	0.5;−1.2	0.5;−1.2

Source: Based on Fig. 6-19, ASCE Standard 7-05, *Minimum Design Loads for Buildings and Other Structures,* 1996, American Society of Civil Engineers. Reproduced by permission of the publisher, ASCE.

[a] For troughed roofs, see ASCE Fig. 6-19C.

[b] Note the elimination in this table of tens of redundant entries in ASCE Figs. 6-19.

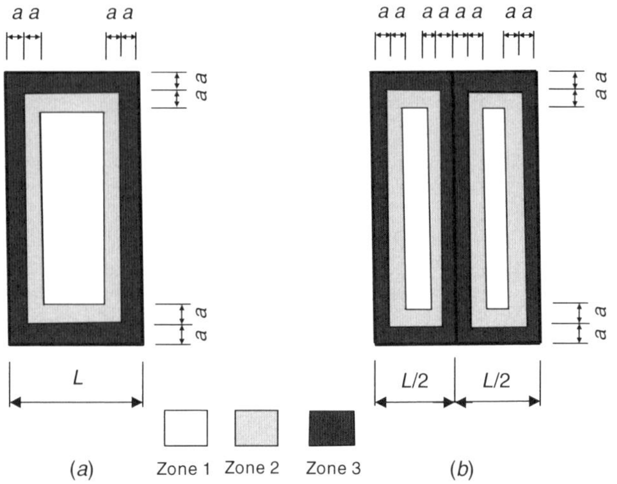

Figure 3.5.11. Analytical procedure, *open buildings,* $0.25 \leq h/L \leq 1.0$, $\theta \leq 45°$; pressure zones for *components and cladding:* (*a*) monoslope roofs and pitched roofs with $\theta \leq 10°$ (ridge of pitched roof not shown; (*b*) pitched roofs with $\theta > 10°$.

The net pressure coefficients (components and cladding) for the wind direction 270° are obtained from the list of coefficients for 90° by using symmetry. We assume that the effective area for the components is 9 ft^2. The width a is $0.1 \times$ least horizontal dimension $= 0.1 \times 24 = 2.4$ ft or $0.4 \times h = 0.4 \times 15 = 6$ ft, whichever is smaller, but not less than 3 ft, so $a = 3$ ft. The effective area $A_{\text{eff}} = 9 \text{ ft}^2 = a^2$, so the pressure coefficients are:

- *Central zone* (*zone 1*): $C_N = 1.8$ or -1.9
- *Intermediate zone* (*zone* 2): $C_N = 2.7$ or -2.9
- *Periphery zone* (*zone 3*): $C_N = 3.6$ or -3.8

3.5.3.3.2 Signs, Lattice Frameworks, and Trussed Towers. The design force is

$$F = q_z G C_f A_f \tag{3.5.15}$$

where q_z is the velocity pressure at the height z of the area A_f and G is specified for rigid structures in Sect. 3.5.2, item 7 (ASCE 6.5.8.1) and for flexible structures in ASCE 6.5.8.2.

For *open signs* (defined as signs with a ratio of solid area to gross area $\varepsilon \leq 0.3$) and *lattice frameworks,* A_f is the area of all exposed members and elements projected on a plane normal to the wind direction, C_f is the net force coefficient from ASCE Fig. 6-22, and D is the diameter of a typical round member. (Two entries are listed for rounded members for each ratio ε

in the tower segment under consideration, owing to the dependence of the drag force on speed for cylindrical members; see Sects. 2.5.5.1 and 2.5.1.2.) The force F is assumed to act in the wind direction.

For *solid signs* and *solid freestanding walls,* see ASCE 6.5.14.

For *trussed towers,* A_f is the solid area of a tower projected on the plane of the face under consideration and C_f is the net force coefficient for towers with structural angles or similar flat-sided members (ASCE Fig. 6-23). For trusses with *rounded members,* the forces on those members may be multiplied by the factor $0.51\varepsilon^2 + 0.57$, where ε is the ratio of solid area to gross area of one tower face for the tower segment being considered, provided that the factor does not exceed unity. For square towers the force along a tower diagonal is obtained through multiplication of the force based on ASCE Fig. 6-23 by the factor $1 + 0.75\varepsilon$ or 1.2, whichever is smaller. Wind forces on tower appurtenances such as ladders, conduits, lights, elevators, and so on, are calculated using appropriate force coefficients (see, e.g., [2.5, p. 421]). Loads due to ice accretion are accounted for in accordance with ASCE Sect. 11.

3.5.3.3.3 Chimneys, Tanks, Rooftop Equipment, and Similar Structures. The design force is

$$F = q_z G C_f A_f \qquad (3.5.16)$$

with the notations of Sect. 3.5.3.3.2; C_f is taken from ASCE Fig. 6-21, in which h is the height of structure; D is the diameter of a circular cross section, the side of a square cross section, or the depth of a hexagonal or octagonal cross section (i.e., cross-sectional dimension normal to two parallel sides); and D' is the depth of protruding elements such as ribs and spoilers.

If the rooftop equipment is placed on top of a building with $h \leq 60$ ft, the coefficients C_f of ASCE Fig. 6-21 are to be multiplied by a factor that is dependent on the ratio A_f/Bh, where B is the horizontal dimension of the building normal to the wind flow. If A_f/Bh is less than 0.1, the factor used is 1.9. The factor may be reduced linearly from 1.9 to 1.0 as A_f/Bh increases from 0.1 to 1. In addition, uplift forces for the equipment, left by ASCE to the judgment of the user, must be provided for in design.

APPENDIX 3A: MAIN CHANGES AND ADDITIONS IN THE ASCE 7-05 STANDARD AND COMMENTARY WITH RESPECT TO ASCE 7-02

Section 6.1

- Added material on minimum wind loading on MWFRS, including Fig. C6-1, to Commentary Sect. C6.1.4.1.

Section 6.2

- Definition of "Building, simple diaphragm," made more precise.
- Added definition of eave height.
- Revised Commentary Sect. C6.2 to clarify the definition of effective area for design of windows and doors.

Section 6.3

- Added definition of symbol A_s, deleted symbols M, N, and n, and modified definition of ε.

Section 6.4

- Removed ASCE 7-02 restriction of method 1 to buildings not subjected to effects of topography.
- Revised footnotes 1 and 4, Fig. 6-2.
- Revised footnote 1, Fig. 6-3.
- Excluded from method 1 buildings whose design is controlled by torsional loads as determined for method 2 (note 5, Fig. 6-10).
- Revised Commentary Sect. C6.4.

Section 6.5.4

- Sect. 6.5.4.2: Added material on estimation of basic wind speeds from regional data in regions not prone to hurricane winds.
- Commentary Sect. C6.5.4: Added comment on winds in parts of Alaska.
- Commentary Sect. C6.5.4: Added material on relation between peak 3-s gust at 10 m above open terrain, corresponding to Saffir–Simpson scale wind speeds over open water. (*Note:* This material is internally inconsistent and therefore in error; see Table 2.1.1, note c and Numerical Example 2.3.8.)

Section 6.5.6

- Sect. 6.5.6.3: Modified fetch requirements for definition of exposure categories.
- Modified comments on exposure categories in Commentary Sect. C6.5.6.
- Provided approximate relation between logarithmic law roughness length parameter z_0 and power law exponent α, and approximate model for modification of wind profile in intermediate exposure zones as a function of fetch, in Commentary C6.5.6.4.

Section C6.5.8

- Added material on fundamental frequency of vibration and damping.

Section 6.5.9

- Sect. 6.5.9.3: Modified to require glazing in windborne debris regions to be protected, with specified exceptions.

Section 6.5.11

- Modified Commentary Sect. C6.5.11.1 material on glazing in regions with debris hazards.
- Revised note 8 of Fig. 6-10 to account for eave height definition.
- Revised note 5 of Fig. 6-11B.
- Revised Commentary C6.5.11.5 on parapets.

Section 6.5.12

- Sect. 6.5.12.4.4: Revised parapet provisions for components and cladding.

Section 6.5.13

- Replaced ASCE 7-02 Fig. 6-18 with ASCE 7-05 Figs. 6-18A through 6-18D (MWFRSs); added Figs. 6-19A through 6-19C (components and cladding); renumbered Figs. 6-19 through 6-22 to Figs. 6-20 through 6-23, respectively. Changed references to current Figs. 6-19 and 6-22.
- Edited provisions of ASCE 7-02 Sect. 6.5.13, and renumbered the section as Sect. 6.5.14.
- Modified Commentary C6.5.9 on Sect. 6.5.9.
- Edited ASCE 7-02 Sect. 6.5.13 and changed its numbering to ASCE 7-05 Sect. 6.5.14.
- Added new ASCE 7-05 Sect. 6.5.13 on monoslope, pitched, and troughed roofs.
- Added new commentary in ASCE 7-05 Commentary C6.5.13.

APPENDIX 3B: EXAMPLES OF PROVISIONS IN THE ASCE 7-05 STANDARD WHOSE IMPROVEMENT SHOULD BE CONSIDERED

- To ascertain whether the simplified procedure may be used, in at least one instance the designer is required to perform calculations based on the analytical procedure (ASCE 6.4.1.1, item 8), which is precisely what the simplified method is purported to avoid. Also, the Standard is so structured that to use the simplified procedure (ASCE 6.4), the designer must obtain information (e.g., on basic wind speeds, importance factors, terrain exposure, and topographic effects) included in the Standard as part of the analytical procedure (ASCE 6.5). More clearly and logically, the information common to both procedures can be included in separate, preliminary sections.

- The terms *flexible diaphragms* and *drag struts* used in ASCE Sect. 6.5.12.3, Table 6-10 (note 5) and ASCE Commentary C6.4 are not defined in ASCE Sect. 6 (wind loads). Rather, they are defined in ASCE Sect. 9 (earthquake loads), with no cross-reference being provided for the benefit of the ASCE Sect. 6 user.
- For MWFRS design, ASCE 6.5.6.4.1, 6.5.6.4.2, and 6.5.13.2 require that: (a) if the loads are determined from ASCE Fig. 6-6, the exposure category appropriate for each wind direction must be used, whereas (b) if the loads are determined from ASCE Fig. 6-10, or for open buildings with monoslope, pitched, or troughed roofs, the exposure category must be determined for the most unfavorable wind direction at the site. What exposure category should be used for MWFRSs of structures other than those covered by ASCE Figs. 6-6, Fig. 6-10, and 6.5.13.2 is not specified by the Standard.
- The title of ASCE Sect. 6.5.12.4.1 is redundant. Instead of referring to "Low-Rise Buildings and Buildings with $h \leq 60$ ft" it is enough to refer to "Buildings with $h \leq 60$ ft" since, by definition (ASCE Sect. 6.2), for low-rise buildings $h \leq 60$ ft.
- The notation (GC_{pi}) is unnecessarily awkward. A single-letter subscripted symbol would suffice.
- The notation GC_p is in most cases awkward. The gust factor G is based on first principles only in the case of winds normal to a wall surface. In almost all situations (e.g., pressures on side walls, or on roofs) it is conventionally assumed in the Standard to be 0.85. Instead of the symbol GC_p a single-letter subscripted symbol would suffice.
- The definition of strip width for buildings with height $h \leq 60$ ft is given in the Standard about 15 times, and the sign convention for pressures acting toward and away from the surfaces is repeated about 20 times. This kind of repetition is unnecessary.
- Pressure zones for components and cladding of buildings with flat, gable, and hip roofs are defined by ASCE in about 10 drawings (ASCE Figs. 6-3, 6-11A, B, C, D), of which four are provided twice. In fact, those zones may be defined by just two drawings (Fig. 3.4.2).
- ASCE Figs. 6-2 and 6-3 contain large tables where pressures are listed separately for 12 different wind speeds. With a view to reducing the bulk

of the standard, it may be desirable to list this information for one wind speed (V_0) only and obtain it for any other speed V through multiplication by $(V/V_0)^2$. This would have the added advantage of eliminating the need for interpolations for speeds not listed in the current tables.

- Notes 7 and 8 of ASCE Fig. 6-10 contain an inconsistency, as indicated in Sect. 3.5.3.1.5.
- ASCE Figs. 6-11, 6-13, 6-14, 6-15, and 6-17 contain about 15 plots, from which the user needs to obtain pressures as functions of effective areas, in general by unwieldy "eyeball" interpolations from logarithmic plots. Simple, easily programmable formulas can be used instead that eliminate the need for interpolations and save a large amount of space (Eqs. 3.5.8 and Tables 3.5.3 and 3.5.4).
- ASCE figures contain a maze of footnotes perceived by some users as veritable minefields. The reduction in the number of footnotes and the presentation of their content in a clearer and more logical fashion would render the standard much easier to use.
- Conversions of 1-min speeds over open water to peak 3-s gust speeds over open terrain are based in the ASCE Commentary on erroneous values of surface roughness, thereby resulting in the large overestimation of the wind speeds over open terrain equivalent to Saffir–Simpson scale speeds over open water specified in the ASCE Commentary itself (see Table 2.2.1, note *c*, and Numerical Example 2.3.8).

CHAPTER 4

RIGID AND FLEXIBLE BUILDINGS

4.1 INTRODUCTION

The total response of buildings to wind loads consists of three parts: (1) the mean response induced by the mean wind load (i.e., the *static* part of the response), (2) the fluctuating response due to wind force components whose frequencies differ significantly from the building's natural frequencies of vibration (i.e., the *quasi-static* or *background* part of the response), and (3) the fluctuating response due to wind force components with frequencies equal or close to the building's natural frequencies of vibration (i.e., the response associated with dynamic amplification effects, called the *resonant* part of the response).

Rigid buildings (Sect. 4.2) are defined by the property that their wind-induced resonant response is negligible. For practical purposes, buildings may be considered rigid if their fundamental natural frequency of vibration is at least 1 Hz (ASCE 7 Standard, Sect. 6.2), or their height-to-least-width ratio is less than 4 (ASCE 7 Commentary, Sect. C6.2).[1]

Comprehensive information on design wind loads and torsional moments for rigid buildings is provided in the ASCE 7 Standard [4.1]. (See Chapter 3 of this book for details and numerical examples) and other standards (e.g., the Eurocode [4.2], the British Standard [4.3], the Australian/New Zealand Standard [4.4].)[2] However, inherent in standard provisions for wind loads on

[1] Chapter and section numbers pertain to chapters and sections in this book unless it is specifically indicated that they pertain to the ASCE 7 Standard or its Commentary.

[2] Standards other than the ASCE 7 Standard provide design pressures for some cases not covered by ASCE 7 (e.g., the British Standard provides design wind pressures for buildings with reentrant corners, on which ASCE 7 provides no information).

rigid buildings are errors that can be significant. These are due in part to the reduction of vast amounts of aerodynamic data to a few numbers contained in tables and plots—a reduction imposed in the past by the use of the slide rule for structural engineering calculations. Like poor construction or design practices, inadequate specification of wind pressures can lead to poor building performance under wind (Fig. 4.1.1).

For buildings whose design warrants the use of more accurate procedures than those based on standard tables and plots, such errors can be reduced significantly by using database-assisted design (DAD) procedures. Essentially, DAD consists of determining wind effects by using large sets of electronically recorded aerodynamic pressure data measured in the wind tunnel at large numbers of points on the building surfaces, either for typical building configurations or for the specific building being designed. DAD is particularly suited for estimates of mean recurrence intervals of wind effects that take into account directional properties of the extreme wind speeds and the building aerodynamics. DAD is also useful for conducting nonlinear analyses providing information on ultimate capacities under wind and gravity loads. The use of DAD procedures is permitted by the ASCE 7 Standard (Sect. 6.6, and Sect. 6.6.2, item 3) and is evolving into an effective, user-friendly tool (see DAD software for rigid buildings on the site http://www.nist.gov/wind).

Figure 4.1.1. Damage to community center building, Gulfport, Mississippi, following passage of hurricane Katrina (2005). Photograph by L. T. Phan, National Institute of Standards and Technology.

Flexible buildings are discussed in Sect. 4.3. To estimate wind effects on flexible buildings it is necessary to conduct dynamic analyses, that is, analyses that account for resonant response. The estimation of wind effects is based on data obtained in the wind tunnel by using one of three approaches: (1) The database-assisted design (DAD) approach, which uses rigid wind tunnel models; (2) the high-frequency force balance (HFFB) approach, which also uses rigid models; and (3) aeroelastic modeling. DAD is superior to HFFB from the point of view of accuracy, transparency, and ease of use by the structural designer. It also has the advantage of supplying not only information for the design of structural members, but also the pressure data needed for cladding design, since in the absence of DAD testing, separate tests are normally required for cladding pressure measurements. On the other hand, HFFB has the advantage of allowing faster testing of alternative building configurations, used for comparisons among the respective aerodynamic performances. Aeroelastic modeling, the most elaborate of the three approaches, is conducted in the relatively rare cases in which the building may be expected to exhibit significant aeroelastic effects.

To estimate the response of flexible buildings it is necessary to account in an integrated manner for the directional dependence of the extreme wind speeds, the aerodynamic response, *and* the dynamic response. How this is done directly influences the quality of the response estimates. The fact that estimates of the New York City World Trade Center towers' response to wind, performed in 2002 by two independent North American wind tunnel laboratories, differed from each other by 40% was due to in part to the treatment of wind directionality effects [2.1].

4.2 RIGID BUILDINGS

The ASCE 7 Standard defines the specialized term *low-rise buildings* (Sect. 3.3.6), which refers to a class of rigid buildings for which (a) the height $h \leq$ 60 ft and (b) $h \leq$ least horizontal building dimension. The term is tied to specific ASCE 7 Standard provisions and should therefore be used primarily within their context. In this section we use the more generic term *rigid buildings* as defined earlier in this chapter.

When subjected to wind blowing from any one direction, buildings experience external pressures on all faces. The external pressures' variation in time and space is illustrated strikingly in the animation—constructed from actual wind tunnel measurements—included on the site http://www.nist.gov/wind.[3] The pressures produce wind loads on components and cladding and wind forces and torsional moments acting on the main wind-force resisting system (MWFRS).

[3] We thank Andrea Grazini of the University of Perugia, who created the animation from data provided by the University of Western Ontario.

The codification of pressures in standard provisions is aimed at achieving the best possible balance between the conflicting aims of developing simple, easy-to-use design tools and representing faithfully the actual aerodynamic pressures. In the United States, the ASCE 7 Standard's simplified or analytical procedure should be used for all routine building designs (Chapter 3).

However, as mentioned earlier, the ASCE 7 Standard provisions do not adequately reflect the actual complexity of the loading, despite their complicated structure. For this reason they can differ significantly—in some instances by more than 50%—from the loads calculated by fully accounting for the time and space dependence of the pressures as reflected in wind tunnel measurements [4.5]. Such errors are far greater than errors inherent in structural analyses based on specified loads, and for buildings that warrant more accurate designs, they can be reduced substantially. This can be done by using records of simultaneous wind tunnel pressure measurements at large numbers of taps on the building surface. The technique for doing so, database-assisted design (DAD), is permitted, within the framework of the wind tunnel method, by Sect. 6.6.2, item 3 of the ASCE 7-05 Standard. DAD can also be used for improving on current standard provisions on wind-induced torsional moments. Finally, DAD is suited for use with pressures obtained by computational fluid dynamics methods, which in the not too distant future are likely to be powerful enough to compete with wind tunnel measurements for routine structural engineering applications.

In Section 4.2.1 we describe the DAD approach and in Section 4.2.2 discuss the estimation of wind effects with specified mean recurrence intervals by accounting for the dependence on direction of both aerodynamic pressures and wind speeds. In Section 4.2.3 we present material on the estimation of wind load factors, which provide safety margins that account for errors inherent in the estimation of wind load effects. In Section 4.2.4 we discuss briefly the estimation of ultimate structural capacities under extreme fluctuating wind loads.

4.2.1 Database-Assisted Design

4.2.1.1 Methodology Used to Develop ASCE 7 Standard Provisions for Wind Loads on Low-Rise Buildings. Database-assisted design has evolved from the methodology used in the late 1970s by the University of Western Ontario to develop ASCE 7 Standard wind loads for low-rise structures. The methodology is illustrated in Fig. 4.2.1, in which the spatial variation of the pressures at a fixed instant in time is represented for portions of lines 3 and r_2. At any fixed point on the building envelope, the pressures vary in time. Pressures are measured in the wind tunnel at a number of pressure taps, of which those affecting frame 2 are represented in Fig. 4.2.1 as black dots.

Consider, for example, the bending moment at time t induced at a cross section j of frame 2 by wind blowing from a specified direction. The moment is caused by wind forces acting on the frame at time t. To show how these forces are approximated for each of the taps, we refer, for specificity, to the

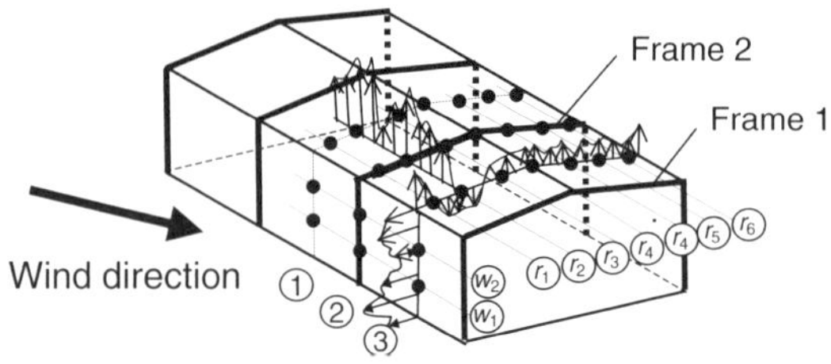

Figure 4.2.1. Building model with pressure taps tributary to frame 2.

taps located at the intersections of lines 2 and w_1 and lines 2 and w_2. The following steps are required:

1. Multiply the pressure measured at each tap at time t by the area tributary of the tap. This yields the approximate total wind force acting at time t at the center of that tributary area. For the pressures at taps $2 - w_1$ and $2 - w_2$, $p_{2-w1}(t)$ and $p_{2-w2}(t)$, the tributary areas are $A_{2-w1} = (d_{13}/2)(h_2/2)$ and $A_{2-w2} = (d_{13}/2)(h_e - h_1)/2$, where d_{13} is the distance between rows 1 and 3, h_1 and h_2 are the elevations of taps $1 - w_2$ and $2 - w_2$, respectively, and h_e is the elevation of the eave. The corresponding approximate total wind forces are $F_{2-w1}(t) = p_{2-w1}(t)\, A_{2-w1}$ and $F_{2-w2}(t) = p_{2-w2}(t)\, A_{2-w2}$, respectively.
2. The roof and wall systems distribute to the frame the forces calculated as in item 1. For example, assuming that girts are simply supported on the frames, the calculated wind forces associated with taps $1 - w_1$, $2 - w_1$, and $3 - w_1$ acting on frame 2 are $F_{1-w1}(t) \times 0.5$, $F_{2-w1}(t) \times 1.0$, and $F_{3-w1}(t) \times 0.5$, respectively.
3. The forces $0.5F_{1-w1}(t)$, $F_{2-w1}(t)$, and $0.5F_{3-w1}(t)$ (or similar forces corresponding to the support conditions for the purlins and girts) induce at the cross section j of frame 2 the bending moments $0.5F_{1-w1}(t)m_{2-w1,j}$, $F_{2-w1}(t)m_{2-w1,j}$, and $0.5F_{3-w1}(t)m_{2-w1,j}$, respectively, where the influence coefficient $m_{2-w1,j}$ is the moment induced at cross section j by a unit load acting at the intersection of frame 2 with line w_1.
4. The total bending moment at time t at cross section j of frame 2 is equal to the sum of contributions similar to those discussed in item 3 from all the taps affecting the frame.
5. The preceding steps allow the calculation of bending moments and other internal forces induced by pressures measured in the wind tunnel at any time t within an interval t_f corresponding to the duration of a typical storm ($t_f = 20$ min to 1 h, say). For any internal force record obtained as indicated in this section, the value of interest is the *peak value* during that record.

The application of this methodology in the 1970s was affected by the limitations of the computational and measurement technology capabilities available at the time of its development. The steps described in items 1 through 5 above were carried out on line following a spatial averaging of pressures by the now largely obsolete pneumatic averaging technique. The density of the taps was typically on the order of 1 tap per 10 m^2, that is, about one order of magnitude less than current technology allows. Typically, tests were conducted for a total of eight wind directions (45° increments), that is, at a directional resolution almost one order of magnitude lower than in current wind tunnel practice. As noted earlier, the internal forces being obtained were based on generic sets of influence coefficients, that is, influence coefficients obtained for a few building types purported to be typical, but that in practice could differ significantly from the buildings being designed. Similarly, a distance between frames purported to be typical was considered in the online calculations, even though the distance between frames, which depends on the particular design being considered, affects the extent to which loads tributary to the frames are mutually correlated. Finally, so-called *pseudo-loads* (this is the term used in the ASCE 7 Standard's Commentary) were developed, largely by eye, for use in the ASCE 7 Standard. These were intended to induce in frames peak moments, horizontal thrusts, and uplift forces comparable to those obtained by using the five steps listed earlier.

The adoption of the methodology just described was an innovative and useful step forward in the codification of wind loads. However, even assuming that the wind tunnel measurements and the performance of the pneumatic averaging technique were satisfactory, errors inherent in that methodology arise owing to (1) the relatively small number of taps and small number of wind directions for which measurements were made; (2) the choice of frame properties and locations, deemed by the code writer to be typical but which can in fact vary fairly widely for any given building configuration; (3) the use for codification purposes of observed peaks, which owing to the random nature of the peak response can differ widely from test to test; and (4) the use of pseudo-loads, that is, of "eyeball" envelopes of measured wind effects for the various wind directions considered in the tests. As mentioned earlier, such errors have been shown to be as high as 50% and can be either on the conservative or unconservative side.

4.2.1.2 Description of the Database-Assisted Design Approach. The principle of DAD is similar to the principle of the methodology described in Sect. 4.2.1.1. However, in its practical application DAD differs from that methodology in that it takes advantage of the following advances:

1. A far larger number of pressure taps—by a factor of 10 or more—can be accommodated on a building envelope than was possible in the 1970s.
2. Pressure time histories can be measured simultaneously at all the pressure taps affecting the loading of components and main wind force–resisting systems.

3. The simultaneous pressure time histories can be stored electronically for use in calculations.
4. The influence coefficients required to calculate the internal forces can be obtained by standard structural analysis programs for the structure of interest rather than from generic structures purported to be typical.
5. Time histories of internal forces can be obtained in a routine, user-friendly manner by using software whose input consists of pressure tap locations, stored pressure data, frame locations, and influence coefficients obtained for the structure being designed.
6. DAD software includes a subroutine for calculating *statistics of peak values* by using the information contained in the entire time history of the internal force of interest. Such statistics (e.g., the peaks' 50th or 90th percentile) are needed because by virtue of the random nature of the wind loading, the observed peak of the internal force varies from record to record. A single observed value of the peak is therefore unrepresentative. The subroutine is based on a generalized *Rice method* [4.6], which allows the estimation of any desired percentile of the peak and is included in the DAD software provided on the site http://www.nist.gov/wind.
7. DAD software also includes software that effects interpolations based on data for buildings with geometric characteristics close to those of the building being designed (Sect. 4.2.1.3).

Database-assisted design software and examples of its application are included in the site http://www.nist.gov/wind. The software is aimed at implementing the following sequence of operations. A preliminary structural design is developed and used to obtain influence coefficients. Spatial coordinates of the pressure taps and measured pressure time series, provided by the wind tunnel operator or available on file, are used in conjunction with information on the positions of the structural frames and the requisite structural influence coefficients to obtain peak values of the response for each wind direction. It is convenient to evaluate peak responses corresponding to a unit wind speed. The resulting direction-dependent response quantities are referred to as *directional influence factors* (DIFs). Since linear structural responses being sought are unaffected by dynamic effects, the peak response for any arbitrary wind speed can be obtained by multiplying the corresponding DIF by the square of that wind speed. In this manner, DIFs can be combined with databases of directional extreme wind speeds from a sufficiently large number of storm events to compute values of the internal forces having a specified mean recurrence interval (see Sect. 4.2.2.2). These internal forces can then be used to redesign the structure. Influence coefficients are computed for the revised structural design, and the analysis procedure can be repeated as necessary, until satisfactory convergence is achieved.

The ASCE Standard provisions for low-rise buildings were developed on the basis of tests on a modest number of basic configurations—about 10—

that do not represent all the building configurations covered by the provisions. Also, interpolations had to be performed by eye. To date, a considerably larger number of building configurations have been covered for DAD development purposes by tests, conducted mainly at the University of Western Ontario. For database-assisted design to be a routine tool resulting in safer and more economical design of buildings of a wide variety of types it is necessary that the acquisition, certification, and archiving of aerodynamic data be expanded in the future. Whenever warranted by economic considerations, ad hoc wind tunnel tests can be performed for the building being designed.

Numerical Example 4.2.1. *Internal forces in a portal frame of an industrial metal building.* Consider a rectangular building with dimensions 200 ft × 100 ft in plan, eave height 20 ft, and gable roof with slopes 1/24, located in terrain with exposure B in the Miami, Florida area. Pressures taps were placed on the model in accordance with the schematic of Fig. 4.2.2. The main wind force–resisting system consists of portal frames (Fig. 4.2.3*a*). The frames are shown in plan in Fig. 4.2.3*b*. Steps similar to those described in Sect. 4.2.1.1, items 1 through 5, resulted in the time history shown in Fig. 4.2.4, representing the bending moment at a knee (cross section 2) of frame 2 (basic wind speed: $V = 140$ mph peak 3-s gust at 33 ft above ground over terrain with exposure C, blowing into the frame in a direction parallel to the ridge; for use in conjunction with wind tunnel data, that speed must be converted to a mean hourly speed at the wind tunnel reference height; see Sect. 2.3.1.2). Calculations described in Sect. 4.2.1.2, item 6, yielded the results of Table 4.2.1.

Table 4.2.1 suggests that statistical inferences based on records shorter than 1 h (e.g., 30 min) differ from inferences based on 1-h records by relatively small amounts that contribute negligibly to the overall uncertainty in the estimation of peak wind effects. According to these results, it would be possible to use for peak estimation purposes records of, say, 30 min length (about 30 s at model scale), instead of 1 h, as is common practice at present.

We note that DAD can also be used for components (e.g., purlins).

4.2.1.3 Interpolation. In practice, it is not possible to provide aerodynamic databases for all possible sets of building dimensions corresponding to a given building configuration. Therefore, if pressure records are available for two buildings having the same configuration but somewhat different dimensions, internal forces on a similar building with intermediate dimensions can be obtained by interpolation. We now describe a simple and effective interpolation procedure [4.9].

Assume that pressure data are available for two buildings, denoted by A and C, whose dimensions (and/or roof slopes) are relatively close to and, respectively, larger and smaller than those of the building of interest, denoted by B. The coordinates of the pressure taps of buildings A and C are scaled to match the dimensions of building B. Using these scaled coordinates, DIFs are computed exactly as described in Sect. 4.2.1.2, treating the measured

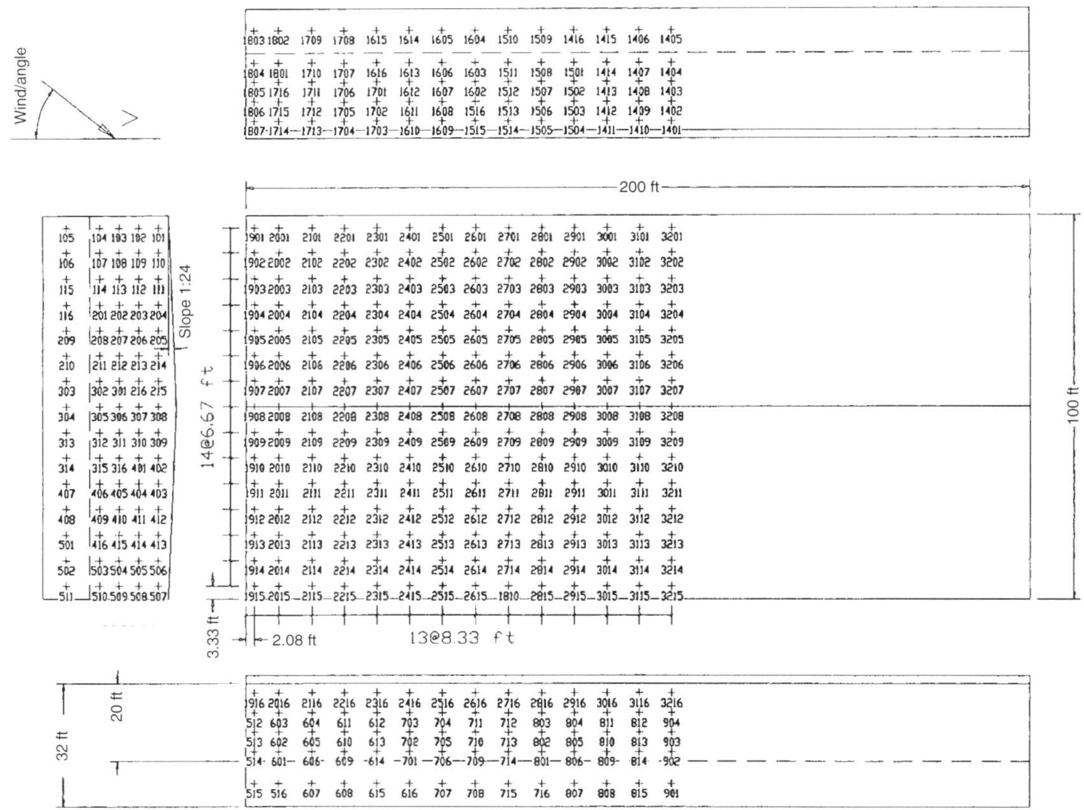

Figure 4.2.2. Building model with locations of pressure taps. Tap pressure measurements are made simultaneously. (From [4.7]; courtesy of the University of Western Ontario.)

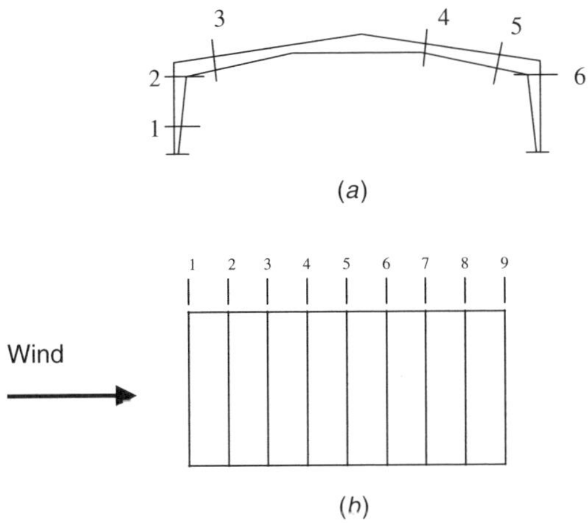

Figure 4.2.3. (*a*) Frames and (*b*) their locations in the plan for the building of Fig. 4.2.2. (Courtesy of Gill Harris, CECO Building Systems.)

pressures on buildings A and C as if they had been recorded on a model having the same dimensions as building B. This procedure can be applied using measurements from only one wind tunnel model (i.e., A or C), but improved accuracy can generally be achieved by making use of measurements from several models, having dimensions that bound those of the structure of interest. For each of the models the tap coordinates are scaled and DIFs are computed. The DIFs for the structure of interest are then estimated by taking a weighted average of the DIFs corresponding to each of the wind tunnel models, with greater weight being given to results from models that more

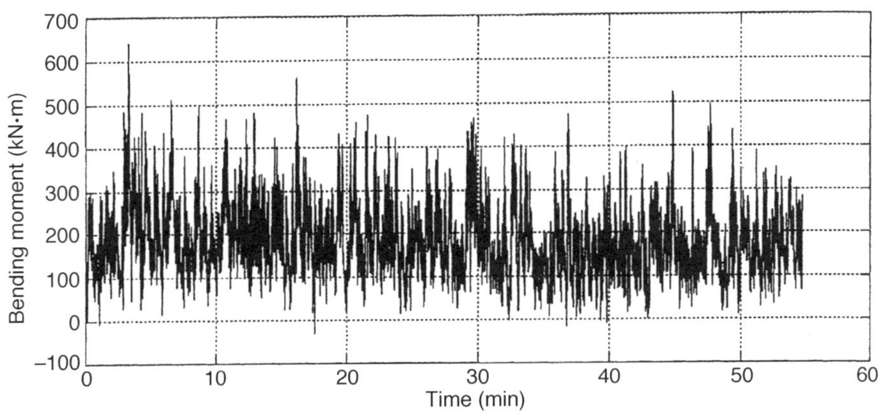

Figure 4.2.4. Time history of the bending moment at the windward knee joint of frame 2, wind normal to plane of frame.

TABLE 4.2.1. Observed Peaks and Estimated Mean Values, Standard Deviations, and Percentiles Peaks (Bending Moments in ft-kips)

Record Length	Observed Peak	Mean	Standard Deviation	84th Percentile	97.5th Percentile
1 h	642	643	55	695	905
30 min	642	639	54	691	897
20 min	642	639	53	690	896

Source: After [4.8].

closely match the dimensions of the structure of interest. The structure is designed on the basis of those DIFs.

The requisite software is provided on the site http:/www.nist.gov/wind, along with an example of its application to a case where the results of the interpolation are found to match closely results based on data obtained directly in the wind tunnel for the building of interest.

Alternative interpolation procedures were proposed in [4.38, 4.39]. In applying any interpolation procedure it should be noted that changes in roof slope can cause relatively sharp changes in aerodynamic behavior [4.40]. For this reason it appears that interpolations should only be performed between buildings with slopes differing by at most 5 to 10°.

4.2.1.4 Errors and Uncertainties Associated with Wind Tunnel Simulations. Errors and uncertainties associated with wind tunnel simulations are due to the violation of the Reynolds number in the laboratory (Sects. 2.5.1.2 and 2.5.6), flow simulation problems, possible blockage problems (i.e., distortions of the wind tunnel flow velocity due to the presence of structures being modeled), and possible measurement problems, especially of mean wind speeds at reference heights (e.g., eave heights) of low-height buildings.

A number of attempts to correct wind tunnel pressures to match approximately full-scale measurements have been reported. A systematic compendium of requisite corrections based on the state of the art remains to be developed, however. A recent study of wind effects based on measurements performed at six wind tunnels around the world showed that peak pressures at a roof corner and peak bending moments near portal frame knees for the building of Numerical Example 4.2.1 varied widely among wind tunnels, the coefficients of variation being as high as about 30% [2.25]; similar discrepancies were reported in [2.26] (Sect. 2.5.6.3).

4.2.2 Wind Directionality Effects

For rigid structures with linearly elastic behavior, the wind effect $Q(\theta)$ (e.g., a bending moment or an axial load) induced by a reference wind speed $v(\theta)$ blowing from direction θ may be written as

$$Q(\theta) = \kappa(\theta)v^2(\theta) \tag{4.2.1}$$

The directional effect coefficient $\kappa(\theta)$ is the wind effect induced by a unit wind speed at the building location blowing from direction θ, that is, the directional influence factor (DIF; see Sect. 4.2.1.2). For the speed $v(\theta)$, both the averaging time and the height above ground must be defined (see Sects. 2.3.1.1 and 2.3.2). In this section we discuss several approaches used in standards or by wind engineering consultants to account for the directionality of both the aerodynamic effects and the extreme wind speeds.

The approach specified in the analytical procedure of the ASCE 7 Standard consists of using a *wind directionality factor* applied to the wind effect calculated by disregarding wind directionality (Sect. 4.2.2.1). This approach is the simplest, but as shown subsequently, can either overestimate or underestimate the response.

The approach based on the *one-dimensional sample of wind effects* (Sect. 4.2.2.2) incorporates the directional aerodynamic and wind climate effects in a simple, transparent, and rigorous manner. In particular, this approach is applicable within a DAD framework [4.10].

The *outcrossing* approach (Sect. 4.2.2.3) is, to our knowledge, used currently by one or two laboratories and has been implemented in various ways using in-house models. The models include in the data samples nonextreme wind speed data such as speeds recorded at 1-h intervals [4.11, p. 167], or low speeds occurring in peripheral hurricane zones. This tends to result in the underestimation of wind effects. In fact, the peaks-over-threshold approach to the estimation of extremes is motivated by the need to avoid the use of nonextreme data—to "let the peaks speak for themselves" (see Sect. 2.4.3.1). The concern that the outcrossing approach may result in the underestimation of extreme wind effects has been explicitly expressed in [4.12].

The *sector-by-sector* approach (Sect. 4.2.2.4) is based on the separate calculation of extreme wind effects for each of equally spaced (e.g., 8 or 16) azimuthal sectors and has two variants. One variant, currently in use by some wind tunnel operators, is unconservative (on the unsafe side). A second variant yields conservative results.

The *simulated tropical storm passages* approach [4.12] is similar in principle, if not in its details, to the approach described in Sect. 4.2.2.2.

4.2.2.1 Approach Based on Use of the ASCE Standard Wind Directionality Factor. The ASCE Standard provisions are based on the assumption that in Eq. 4.2.1, the wind speeds and the directional effect coefficients are independent of direction and are equal, respectively, to $V = \max_\theta[v(\theta)]$ and $K = \max_\theta[\kappa(\theta)]$. (Conceptually, this is tantamount to rotating the building about its vertical axis so that the aerodynamically most unfavorable direction coincides with the direction of the largest wind speed during a year or a storm, and then multiplying the wind effect so obtained by K.) For buildings, the corresponding wind effects are assumed in the Standard to be

$$Q = K_d \max_\theta[\kappa(\theta)] \{\max_\theta[v(\theta)]\}^2$$
$$= 0.85KV^2 \tag{4.2.2}$$

where $K_d = 0.85$ is a wind directionality factor (see Sect. 3.5.2, item 2). If the wind effect in question consists of peak external pressures, $\kappa(\theta)$ is proportional to the peak pressure coefficient $C_{\text{pk}}(\theta)$ (the proportionality factor being $\rho/2$, where ρ is the specific mass of air), so K is proportional to $\max_\theta[C_{\text{pk}}(\theta)]$. For specificity we consider henceforth the case of peak external pressures, but any wind effect may be considered similarly.

To illustrate the approach implicit in Eq. 4.2.2 we show in Table 4.2.2 the largest yearly directional wind speeds for two years of record. In the table the largest yearly speed regardless of direction, V, is indicated in bold type for each year; that is, $V = 100$ mph for year 1 and $V = 107$ mph for year 2. (As shown in Sect. 2.4, some extreme wind speed samples do not consist of extreme yearly wind speeds but rather, of speeds in hurricane events or speeds exceeding a specified threshold. For such samples the index of the storm event should be used instead of the index of the year of occurrence: e.g., hurricane 1, hurricane 2, . . . instead of year 1, year 2, ...)

For a square building with specified orientation, let $\kappa(\theta)/(\rho/2)$ represent the peak pressure coefficients $C_{\text{pk}}(\theta)$ on a cladding element (Table 4.2.3). The values $K/(\rho/2)] = \max_\theta[\kappa(\theta)]/(\rho/2) = \max_\theta[C_{\text{pk}}(\theta)]$ for peak negative pressures and for peak positive pressures are indicated in bold italic and bold roman type, and are -0.6 and 0.9, respectively (for negative quantities we apply the term *maximum* to the absolute values).

Table 4.2.4 lists the terms of the samples $0.85 \max_\theta[C_{\text{pk}}(\theta)] \max_\theta[v^2(\theta)]$ obtained from Tables 4.2.2 and 4.2.3. Such terms belong to samples used for inferences on extreme wind effects with various nominal mean recurrence intervals (MRIs) considered in design. The quantity $\max_\theta[C_{\text{pk}}(\theta)]$ is independent of direction, so instead of the terms listed in Table 4.2.4 it is possible to use the terms of the time series $V = \max_\theta[v(\theta)]$ and estimate from that series the values of the speeds V corresponding to the requisite MRI values. The squares of those values are then multiplied by the quantity $0.85(\rho/2)$ $\max_\theta[C_{\text{pk}}(\theta)]$ to yield the design wind pressures with those MRIs. The approach based on the use of the ASCE Standard wind directionality factor is assessed in Sect. 4.2.2.2.

Note: The building orientation affects the dependence upon azimuth of the directional effect coefficients but it does *not* affect their maxima used in the

TABLE 4.2.2. Largest Yearly Wind Speeds $v(\theta)$ (mph)

Azimuth	NE	E	SE	S	SW	W	NW	N
Direction Index	1	2	3	4	5	6	7	8
Year 1	90	**100**	81	96	87	89	93	78
Year 2	78	94	83	**107**	80	83	72	76

TABLE 4.2.3. Pressure Coefficients $C_{pk}(\theta)$

Azimuth	NE	E	SE	S	SW	W	NW	N
Direction Index	1	2	3	4	5	6	7	8
$C_{pk}(\theta)$	0.7	−0.4	***−0.6***	−0.3	***−0.6***	−0.4	0.7	**0.9**

terms $0.85 \max_\theta[C_{pk}(\theta)] \max_\theta[v^2(\theta)]$. ASCE 7 Standard estimates of wind effects are therefore independent of building orientation.

4.2.2.2 Approach Based on One-Dimensional Sample Incorporating Directional Aerodynamic and Extreme Wind Effects: Comparison with ASCE 7-Based Results. This approach is practicable in conjunction with the wind tunnel procedure (see Sect. 3.1.1) or within the framework of database-assisted design. In Sect. 4.2.2.2.1 we present by means of a numerical illustration the definition and construction of one-dimensional time series and compare the results obtained by this approach with those obtained by the ASCE 7 Standard approach described in Sect. 4.2.2.1. In Sect. 4.2.2.2.2 we define samples of equivalent wind speeds that incorporate the effects of the dependence on directionality of both the aerodynamic response and the extreme wind speeds.

The conceptual framework of the approach discussed in this section is simple. However, so much confusion has been generated and so many divergent results have been obtained by the application of various approaches to the wind directionality problem in the past that in our opinion, the material that follows warrants special attention.

4.2.2.2.1 Definition and Construction of One-Dimensional Samples of Directional Wind Effects. To introduce the simple concepts that underlie the approach described in this section, we consider the largest yearly directional wind speeds $v(\theta)$ of Table 4.2.2 and the peak pressure coefficients $C_{pk}(\theta)$ of Table 4.2.3. We then calculate for each direction the terms $C_{pk}(\theta)v^2(\theta)$. The results are shown in Table 4.2.5.

Not all the quantities $C_{pk}(\theta)v^2(\theta)$ listed in Table 4.2.5 for each year of record are of interest for design purposes. Rather, we are interested only in the quantities

$$\max_\theta[C_{pk}(\theta)v^2(\theta)] \tag{4.2.3}$$

TABLE 4.2.4. Terms $0.85 \max_\theta[C_{pk}(\theta)] \max_\theta[v^2(\theta)] \equiv 0.85KV^2$ (mph)^2

Year	Negative Terms	Positive Terms
1	$0.85 \times 100^2 \times (-0.6) = -5100$	$0.85 \times 100^2 \times 0.9 = 7650$
2	$0.85 \times 107^2 \times (-0.6) = -5840$	$0.85 \times 107^2 \times 0.9 = 8760$

TABLE 4.2.5. Terms $C_{pk}(\theta)v^2(\theta)$ (mph)^2

Azimuth	NE	E	SE	S	SW	W	NW	N
Direction Index	1	2	3	4	5	6	7	8
Year 1	5670	−4000	−3935	−2765	***−4540***	−3170	**6055**	5475
Year 2	4260	−3535	***−4135***	−3435	−3840	−2755	3630	**5200**

which are proportional to the largest yearly wind *effects* being sought (rather than to the largest yearly wind *speeds*). These maximum negative and positive quantities are indicated in Table 4.2.5 in bold italic and bold roman type, respectively. The quantities $\max_\theta[C_{pk}(\theta)v^2(\theta)]$ for year 1, 2, . . . can be used to form samples from which inferences can be made on pressures corresponding to various mean recurrence intervals. The terms of the samples for the two years covered in Table 4.2.5 are shown in Table 4.2.6. The two samples $\max_\theta[C_{pk}(\theta)v^2(\theta)]$ of Table 4.2.6 constitute *one-dimensional samples of directional wind effects*. These samples fully incorporate the directional information concerning the wind climate and the aerodynamic response for a building with specified orientation and aerodynamic properties. The time series are called one-dimensional because they reduce a two-dimensional matrix (with lines associated with year 1, 2, . . . and columns associated with wind directions) to a vector whose components are associated with year 1, 2, ...

Table 4.2.6 suggests that for *some* building orientations and in some wind climates, the ASCE 7 approach overestimates the wind effects. However, this is not true for all building orientations. Also, in many instances the overestimation of wind effects corresponding to relatively short mean recurrence intervals (e.g., 50 years) is considerably reduced for wind effects with longer mean recurrence intervals (e.g., 1000 years) [4.10]. This issue is discussed briefly in Sect. 4.2.2.2.2.

4.2.2.2.2 Estimation of $\bar{N}$-Year Directional Wind Effects: Equivalent Wind Speeds. In the directional approach just described, it is convenient to make statistical inferences from time series whose terms are the quantities

$$v_{eq} = \{\max_\theta[C_{pk}(\theta)v^2(\theta)]\}^{1/2} \tag{4.2.4}$$

TABLE 4.2.6. Terms $\max_\theta[C_{pk}(\theta)v^2(\theta)]$ vs. 0.85 $\max_\theta[C_{pk}(\theta)]\max_\theta[v^2(\theta)]$ (mph)^2

Year	Negative	Positive
1	−4540 vs. −5100	6055 vs. 7650
2	−4135 vs. −5840	5200 vs. 8760

called *equivalent wind speeds,* rather than from the quantities $\max_\theta[C_{pk}(\theta)v^2(\theta)]$. The wind effects corresponding to the MRIs of interest can be obtained from the equivalent wind speeds with the same MRIs simply by squaring the latter and then applying to the result the proportionality constant that converts quantities of the type $C_{pk}(\theta)v^2(\theta)$ to wind pressures or other wind effects. The estimation methods applied to equivalent wind speeds are identical to those described in Sect. 2.4.3 for application to wind speeds.

By construction, the sample mean and sample standard deviation of the time series of the equivalent wind speeds are typically lower and higher, respectively, than their counterparts for the nondirectional time series. This statement is illustrated by the following example.

Numerical Example 4.2.2. *Sample means for the sample* v_{eq} *and the sample* $\{max_\theta[C_{pk}(\theta)]\ max_\theta[v^2(\theta)]\}^{1/2} \equiv \{max_\theta[C_{pk}(\theta)]\}^{1/2}\ max_\theta[v(\theta)]$. For the purpose of this example it is sufficient to consider just two angles, θ_1 and θ_2, the respective coefficients $C_{pk}(\theta)$, and wind speeds (mph) for three years of record. Data and calculations follow.

	θ_1	θ_2		θ_1	θ_2
$C_{pk}(\theta)$	0.8	1.0			
$v_1(\theta)$	**70**	60	$[C_{pk}(\theta)v_1^2(\theta)]^{1/2}$	**63**	60
$v_2(\theta)$	50	**65**	$[C_{pk}(\theta)v_2^2(\theta)]^{1/2}$	45	**65**
$v_3(\theta)$	**60**	55	$[C_{pk}(\theta)v_3^2(\theta)]^{1/2}$	54	**55**

As indicated by the bold numbers, the values of $\{\max_\theta[C_{pk}(\theta)]\}^{1/2}\ \max_\theta[v(\theta)]\} = 1.0 \times \max_\theta[v(\theta)]$ are $\max_\theta[v_1(\theta)] = 70$ mph, $\max_\theta[v_2(\theta)] = 65$ mph and $\max_\theta[v_3(\theta)] = 60$ mph. The values of the equivalent speeds $[C_{pk}(\theta)v^2(\theta)]^{1/2}$ are $v_{eq\ 1} = 63$ mph, $v_{eq\ 2} = 65$ mph, and $v_{eq\ 3} = 55$ mph. The means and standard deviation are 65 and 5.0 mph for the $\max_\theta[v(\theta)]$ sample, and 61 and 5.3 mph for the v_{eq} sample.

Given a sample of equivalent wind speeds, the estimated equivalent wind speed with a specified mean recurrence interval (MRI) $\bar{N}$ is a sum of two terms. The first term is the sample mean, and the second term is proportional to the standard deviation and increases as $\bar{N}$ increases (see Eqs. 2.4.3 and 2.4.4). The estimated nondirectional wind speed with MRI $\bar{N}$ obtained from the sample $\max_\theta[v(\theta)]$ of maximum annual speeds is obtained similarly. If the two samples have distributions with about the same tail length, as was found to be the case for a number of sites, for short MRIs $\bar{N}$ (e.g., 50 years) the first term tends to dominate, and nondirectional speeds may yield a larger estimate of the $\bar{N}$-year wind effect than equivalent speeds do. Owing to the relative magnitudes of the means and of the standard deviations of the nondirectional and the directional wind speeds, the differences tend to decrease, however, for relatively large $\bar{N}$.

4.2.2.3 Approach Based on Outcrossing of the Limit-State Boundary. This approach is applied by some wind engineers within the framework of the wind tunnel procedure. As mentioned earlier, the approach typically includes in the statistical samples nonextreme wind speed data, such as largest speeds recorded during successive 1-h intervals [4.11] or low speeds occurring in peripheral hurricane zones. *We do not recommend its use* unless detailed documentation is provided in a transparent manner so that it can be understandable to and clearly assessed by the structural engineer in charge of the project. As noted earlier, the outcrossing approach tends to underestimate wind effects corresponding to specified MRIs, a concern that was also noted in [4.12].

4.2.2.3.1 Statement of the Outcrossing Problem for Structures Experiencing Directional Wind Effects. Let us represent the wind velocity as a vector **v** with origin at a point O, and let the wind speed (i.e., the magnitude of the vector **v** corresponding to the wind direction θ) be denoted by $v(\theta)$. To every angle θ there corresponds a wind speed $v_R(\theta)$ that induces in the structure or member being considered the limit state R of interest. The limit-state boundary is the curve D described by the tip of the velocity vector **v** with magnitude $v_R(\theta)$ as θ varies from 0 to 360° (Fig. 4.2.5).

In particular, if Eq. 4.2.1, $Q(\theta) = k(\theta)\, v^2\,(\theta)$ holds, where $Q(\theta)$ is the wind effect induced by the wind speed $v(\theta)$, then

$$v_R(\theta) = \left[\frac{R}{k(\theta)}\right]^{1/2} \tag{4.2.5}$$

The wind velocity vector **v** varies continuously, that is, its tip describes a curve that evolves in time. If that vector were statistically stationary (see Sect. 2.3.1.1 for the definition of this term), it would be possible by using the theory

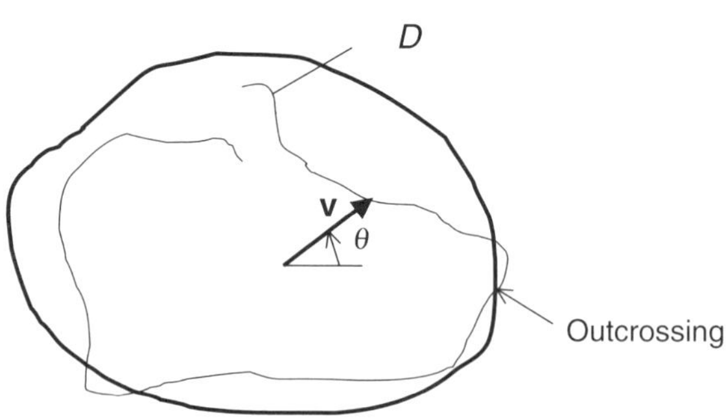

Figure 4.2.5. Limit-state boundary of the wind effect, and plot of the wind effect induced by the wind vector **V** as a function of the azimuth angle θ. Outcrossing of the boundary occurs twice in the figure.

of random processes to estimate the mean time interval T between successive *outcrossings* of the curve D by the vector $\mathbf{v}$. (The term *outcrossing* denotes crossing from inside the boundary, as indicated in Fig. 4.2.5.) The structure should be designed so that T is sufficiently large. For example, if R were the allowable stress in the member of interest, for nonhurricane regions T should in accordance with common practice be 50 years.

4.2.2.3.2 Mean Time Interval T Between Successive Outcrossings. The procedure for estimating the mean time interval T between outcrossings is similar for the vector process $\mathbf{v}$ and for a scalar random process ([4.13, p. 640]; see also Sect. 4.2.1.2, item 6). The theory governing the estimation of T is mathematically straightforward. Elements of the theory are reproduced here only for the general information of the reader.

For the time-dependent vector $\mathbf{v}(t)$ with origin O, it can be shown that

$$T = \frac{1}{\int_D E_0^\infty \, (\dot{v}_n|\mathbf{v}_R) f_{\mathbf{v}}(\mathbf{v}) \, dD} \tag{4.2.6}$$

where v_n is the projection of the vector $\mathbf{v}$ on the normal to line D; $E_0^\infty(\dot{v}_n|\mathbf{v}_R)$ is the average of the positive values of the time derivative of v_n, conditional on the tip of the vector $\mathbf{v}$ being on line D; $f_{\mathbf{v}}$ ($\mathbf{v}$) is the probability density of the vector $\mathbf{v}$; and dD is the elemental length along line D.

If polar coordinates are used, Eq. 4.2.6 can be written as

$$T \approx \frac{1}{\Sigma_{\theta_i} E_0^\infty \, [\dot{v}_n(\theta_i)|v(\theta_i) = v_R(\theta_i)] f_{v(\theta_i)}[v_R(\theta_i)] \, \{v_R^2 + [dv(\theta)/d\theta|_{v(\theta_i)=v_R(\theta_i)}]^2\}^{1/2} \, \Delta\theta} \tag{4.2.7}$$

where v_R is the magnitude of the vector $\mathbf{v}$ extending from origin O to the line D, θ is the angle defining the direction of $\mathbf{v}$, and $[v_R^2(\theta) + (dv/d\theta|_{v=v_R})^2]^{1/2}$ $\Delta\theta$ is the elemental length dD that corresponds in polar coordinates to the angular increment $\Delta\theta$. Some wind tunnel operators assume that $\dot{v}_n(\theta)$ is independent of $v(\theta)$ (this would imply that $E_0^\infty[\dot{v}_n(\theta)|v(\theta) = v_R(\theta)] = E_0^\infty[\dot{v}_n(\theta)]$); however, given the anisotropy of the wind vector process, others view this assumption as unacceptable. The difficulty in using the limit-state-boundary outcrossing approach lies in the reliable, unbiased estimation of the averages $E_0^\infty[\dot{v}_n(\theta)|v(\theta) = v_R\,(\theta)]$ (see comments on the outcrossing approach at the beginning of Sect. 4.2.2).

4.2.2.4 Sector-by-Sector Approach. The sector-by-sector approach is applicable within the framework of the wind tunnel procedure and has two variants. The first variant, currently used by some wind tunnel operators, is referred to as the *single-sector design criterion.* This variant yields *unconservative* results (Sect. 4.2.2.4.1). The second variant, referred to as the *mul-*

tisector design criterion, accounts for contributions to the probability of exceedance of specified wind effects from all sectors and yields *conservative* results (Sect. 4.2.2.4.2).

It would appear that realistic estimates of wind effects could be obtained by averaging outputs of the single- and multisector criteria. However, this remains to be verified, possibly by checking against results based on one-dimensional samples incorporating directional aerodynamic and extreme wind effects (Sect. 4.2.1.2).

The sector-by-sector approach is based on separate estimates of the effects induced by wind blowing from each of the sectors of the compass, that is, by each of the *sectorial wind speeds*. We consider the case of q sectors (e.g., $q = 8$ sectors of 45° each, or 16 sectors of 22.5° each), and an $\bar{N}$-year MRI (e.g., $\bar{N} = 50$ years).

4.2.2.4.1 Single-Sector Design Criterion [4.14]. Let the sectorial wind effects $Q_i(\bar{N}) = k_i v_i^2(\bar{N})$ ($i = 1, 2, \ldots, q$, where $q = 8$ or 16, say) be calculated for each sector i ($\bar{N}$ = MRI specified for design),[4] and let the *largest* of those sectorial wind effects be denoted by $Q_m(\bar{N})$. The single-sector design criterion consists of designing the structure so that the wind effect $Q_m(\bar{N})$ induces in the structure the limit state R (e.g., the allowable stress). Some wind-tunnel operators use this criterion in the belief that it results in a design wherein the MRI of the limit state R is $\bar{N}$. We show that the MRI of the limit state R for a member designed in accordance with this criterion is in fact *less* than $\bar{N}$. This is so because the single-sector design criterion does not take into account the effects of the wind speeds from all the sectors $i \neq m$, where m denotes the sector for which the largest sectorial wind effect $Q_m(\bar{N})$ occurs.

Consider the probability that the largest yearly wind effect Q, regardless of wind direction, does not exceed R, and denote it by $F_Q(Q \leq R)$. Assume for specificity that $\bar{N} = 50$ years and that the number of sectors is $q = 8$. The probability that $Q \leq R$ is equal to the probability that $v_1 \leq v_1(\bar{N}_1)$ *and* $v_2 \leq v_2(\bar{N}_2), \ldots,$ *and* $v_8 \leq v_8(\bar{N}_8)$:

$$F_Q(Q \leq R) = \text{Prob}[v_1 \leq v_1(\bar{N}_1), v_2 \leq v_2(\bar{N}_2), \ldots, v_8 \leq v_8(\bar{N}_8)] \quad (4.2.8)$$

In Eq. 4.2.8 one of the sectorial wind speeds, denoted by $v_m(\bar{N}_m)$, is such that $\bar{N}_m = 50$ years, whereas for all $i \neq m$, $\bar{N}_i \geq 50$ years. [By virtue of applying the single-sector design criterion, the most unfavorable sectorial wind effect is induced by $v_m(\bar{N})$; hence, to induce the limit state R, wind speeds in sectors other than the sector m need to have MRIs larger than $\bar{N}$.]

If it is assumed that the speeds $v_1, v_2, \ldots, v_8$ are mutually *independent* (this assumption implies that their mutual correlations are zero), it follows from Eq. 4.2.8 that

[4]For simplicity we use the notations $Q_i \equiv Q(\theta_i)$, $k_i \equiv k(\theta_i)$, and $v_i \equiv v(\theta_i)$.

$$F_Q(Q \leq R) = \text{Prob}[v_1 \leq v_1(\overline{N}_1)]\ \text{Prob}[v_1 \leq v_2(\overline{N}_2)] \cdots$$
$$\text{Prob}[v_8 \leq v_8(\overline{N}_8)] \tag{4.2.9a}$$
$$< \frac{1}{1 - 1/50} = 0.98 \tag{4.2.9b}$$

(see Sect. 2.4.2, Eq. 2.4.1); that is, the mean recurrence interval of the event $Q \leq R$ is less than 50 years.[5] Consider, for example, the case in which the effects induced by wind blowing from one of the sectors are dominant; that is, the MRI of the event that winds from that sector would cause R to be exceeded is $\overline{N}_m = \overline{N} = 50$ years, in accordance with the single-sector design criterion, while for the other sectors (i.e., for all sectors $i \neq m$) the mean recurrence intervals of the wind speeds inducing the state R are $\overline{N}_i = 250$ years. Equation 4.2.9*a* then yields $F_Q(Q \leq R) = (1 - 1/50) \times (1 - 1/250)^7 = 0.98 \times 0.996^7 \approx 0.95$, corresponding to an MRI of the event $Q > R$ of $1/(1 - 0.95) = 20$ years. Under the independence assumption the single-sector design criterion is unconservative. The criterion remains unconservative even if the mutual correlations of the sectorial speeds do not vanish but are relatively small (either positive or negative), as is typically the case for extreme directional wind speeds. The criterion would be correct only if all sectorial wind speeds were mutually perfectly correlated, a case that does not occur in nature, however.

4.2.2.4.2 Multisector Design Criterion. Unlike the single-sector design criterion, the multisector criterion reflects the fact that all sectorial wind speeds contribute to the probability of exceedance of a specified wind effect. The design criterion is that the probability $F_Q(Q \leq R)$ be less than the specified probability of nonexceedance of the limit state of interest. Consider, as an illustration, the case in which the event that winds from one sector cause the limit state R to be exceeded has an MRI of 120 years, whereas for each of the other seven sectors the corresponding MRI is 600 years. Then, under the assumption of mutual independence of the sectorial speeds, $F_Q(Q \leq R) = (1 - 1/120) \times (1 - 1/600)^7 = 0.9917 \times 0.99833^7 \approx 0.98$, corresponding to an MRI of the event $Q > R$ of approximately $1/(1 - 0.98) = 50$ years. The fact that the mutual correlations of the sectorial wind speeds are not zero means that the actual MRI of that event is greater than 50 years. The multisector design criterion is therefore conservative.

[5] Equation 4.2.9*a* follows from Eq. 4.2.8 by virtue of the definition of independent events. For example, consider the throw of a die and the throw of a coin. It is easy to see that since the outcome of the throw of the die and the outcome of the throw of the coin are independent events, the probability $P(1, \text{H})$ that those outcomes are, respectively, "1" *and* "heads" is equal to the product of the probabilities $P(1)$ of the outcomes "1" and $P(\text{H})$ of the outcome "heads"; that is, $P(1, \text{H}) = P(1)\,P(\text{H}) = (1/6)(1/2) = 1/12$.

4.2.3 Uncertainties in the Estimation of Wind Effects: Wind Load Factors

The mean recurrence interval (MRI) of wind speeds used for allowable stress design is typically 50 years for nonhurricane regions. This choice of MRI is entirely conventional.

Strength design requires that a limit state such as the yield stress be reached under a wind load higher than the 50-year wind load. The higher wind load is obtained via multiplication of the 50-year load by a factor, called the *wind load factor,* whose magnitude exceeds unity by an amount based on professional consensus. The ASCE 7-05 Standard specifies a wind load factor of 1.6,[6] meaning that for nonhurricane regions the design wind load that induces a limit state associated with strength design is 1.6 times the 50-year wind load. Under the assumptions implicit in the ASCE 7 Standard (see ASCE 7-05 Commentary C6.5.4 that (a) the nondirectional extreme wind speeds have an EV Gumbel distribution (Eq. 2.4.3) and (b) wind load estimates are not affected by knowledge uncertainties, the design wind load for strength design has a nominal MRI of about 500 years for a nominal wind load factor $LF_{wASCE98} = 1.5$, and of about 700 years for a nominal wind load factor $LF_{wASCE02} = 1.6$.[7]

In reality the wind load factor has a *dual* function. As just indicated, the wind load factor must magnify the 50-year load so that a load with a longer MRI is obtained for strength design calculations. However, it must also account for knowledge uncertainties in the estimation of the wind loads. The uncertainties pertain to the relevant micrometeorological, aerodynamic, and extreme wind estimates used to define the wind loads or wind effects, and are associated with, among others, sampling errors in the estimation of extreme speeds due to the limited size of the wind speed records; hurricane modeling uncertainties; errors in wind speed observations; errors in the estimation of the terrain roughness; uncertainties in the ratio between speeds averaged over, for example, 3 s and 1 h; uncertainties in the relation between

[6] The ASCE 7 Standard prescribes the same *nominal* wind load factor (i.e., 1.6) for both hurricane and nonhurricane regions. However, this is done solely as a convenience to the user. For hurricane regions the *effective* wind load factors implicit in the ASCE 7 Standard are larger than 1.6. As indicated in the ASCE 7-05 Commentary, in hurricane-prone regions the Standard does not define basic wind speeds as speeds with a 50-year MRI. Rather, it defines basic speeds as speeds corresponding to strength design divided by the square root of the nominal wind load factor (the square of the basic speeds so defined times the nominal wind load factor is thus proportional to the wind effect for strength design). The MRIs for the ASCE 7-05 basic wind speeds in hurricane regions are therefore larger than 50 years by amounts that depend on geographical location.

[7] In an earlier version of ASCE 7 the wind load factor was 1.5. In the ASCE 7-02 and 7-05 Standards the nominal wind load factor is 1.6. Since the basic wind speed map in the ASCE 7-02 and 7-05 Standards was not modified accordingly, the nominal MRI of the wind speed used for strength design calculations (i.e., the basic wind speed times the factor $\sqrt{1.6}$) is about 700 years rather than 500 years, as intended originally.

wind speeds in different roughness regimes; errors in aerodynamic pressure measurements; and uncertainties in the estimation of peak coefficients. The larger these errors and uncertainties, the larger the wind load factor must be.

In this section we define load factors in light of our remarks concerning their dual function and discuss their estimation briefly. Section 4.2.3.1 is devoted to definitions of the wind load factor. In Section 4.2.3.2 we discuss the modeling of the peak wind loads. Section 4.2.3.3 is concerned with the probabilistic modeling of the wind loads corresponding to any specified mean recurrence interval (MRI). In Section 4.2.3.4 we describe results of simulations based on such modeling which shed light on the types of uncertainty that are most significant for wind engineering purposes.

4.2.3.1 Definitions of the Wind Load Factor. In the ASCE 7-05 Standard the wind load factor is defined as the ratio

$$\mathrm{LF}_w = \left(\frac{1.6}{1.5}\right)\left(\frac{\text{point estimate of 500-yr speed}}{\text{point estimate of 50-yr speed}}\right)^2 \tag{4.2.10}$$

(Point estimates are estimates based on the assumption that knowledge uncertainties are negligible.) Equation 4.2.10 also assumes that wind effects (e.g., bending moments) with an $\overline{N}$-year MRI are proportional to the square of the wind speeds with an $\overline{N}$-year MRI. This assumption is correct for rigid buildings whose members' behavior is linearly elastic.

Owing to the existence of knowledge uncertainties, the estimate of a wind force with specified MRIs is itself uncertain; that is, it is characterized by a probability distribution representing a measure of those uncertainties. Therefore, a wind effect is determined not only by the specified MRI $\overline{N}$ of the wind-induced wind effect but also by the specified probability that the error in the estimation of the wind effect is not exceeded. In this approach, which is more realistic than the ASCE 7 Standard approach, the wind load factor is defined as

$$\mathrm{LF}_w = \frac{F_{\mathrm{pk}}(\tau, \overline{N}_2)_{p_2}}{F_{\mathrm{pk}}(\tau, \overline{N}_1)_{p_1}} \tag{4.2.11}$$

where $F_{\mathrm{pk}}(\tau, \overline{N}_2)_{p_2}$ is the peak wind effect corresponding to the MRI $\overline{N}_2$ and a p_2 probability of nonexceedance of the error in the estimation of the peak wind effect, $F_{\mathrm{pk}}(\tau, \overline{N}_1)_{p_1}$ is similarly defined (Fig. 4.2.6), and the parameter τ denotes the duration of the storm (about 20 min to 1 h) during which the peak effect occurs. For example, if $\overline{N}_2 = 500$ years, $\overline{N}_1 = 50$ years, $p_2 = 90\%$, and $p_1 = 50\%$, the wind load factor is computed as follows:

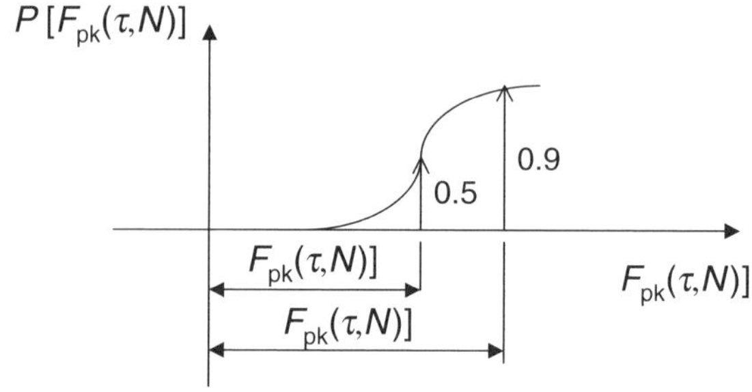

Figure 4.2.6. Cumulative distribution function of a peak wind effect $F_{pk}(\tau, \overline{N})$ showing values of the peak effect corresponding to the 50th and 90th percentage points.

$$LF_{\text{wind}} = \frac{F_{pk}(\tau, 500)_{90\%}}{F_{pk}(\tau, 50)_{50\%}} \tag{4.2.12}$$

The peak wind effect $F_{pk}(\tau, 50)_{50\%}$ may be assumed to be equal, approximately, to the point estimate of the wind effect. $F_{pk}(\tau, 500)_{90\%}$ is an estimate of the 500-year peak effect that has a 90% probability of not being exceeded.

The probabilistic distribution of the peak wind effects is based on the deterministic model of the wind effects developed in Sect. 4.2.3.2. This model is then modified (i.e., transformed into a probabilistic model) by inserting in it random factors reflecting the uncertainties of concern (Sect. 4.2.3.3). Monte Carlo simulations based on that probabilistic model yield the requisite estimated probability distribution of the peak wind effect of interest (Sect. 4.2.3.4).

4.2.3.2 Peak Wind Effect Model. In the absence of uncertainties, the peak wind load with an $\overline{N}$-year MRI may be written as

$$F_{pk}(\tau, \overline{N}) = c_{\text{str}} \frac{\rho}{2} C_{pk}(\tau) v^2(\tau, H_{\text{aero}}, z_0, \overline{N}) \tag{4.2.13}$$

In Eq. 4.2.13, $F_{pk}(\tau, \overline{N})$ is the peak wind effect during a storm of duration τ, c_{str} is a coefficient that converts a peak wind load to a peak wind effect (e.g., a bending moment), ρ is the specific mass of air, $C_{pk}(\tau)$ is a peak load coefficient, and $v(\tau, H_{\text{aero}}, z_0, \overline{N})$ is the wind speed with MRI $\overline{N}$, averaged over time τ, at the elevation H_{aero} (e.g., the eave height or the height of the building) in terrain with roughness length z_0 (e.g., τ = 1 h, H_{aero} = 15 m, z_0 = 0.3 m). If directionality is not taken into account, in Eq. 4.2.11 v denotes the wind speed regardless of its direction. If directionality is taken into account as in Sect. 4.2.2.2.2, the product $[C_{pk}(\tau)]^{1/2}v$ is replaced by the equivalent wind speed v_{eq}.

The wind speed $v(\tau, H_{\text{aero}}, z_0, \overline{N})$ upwind of the building depends on $v(T, H_{\text{met}}, z_{01}, \overline{N})$, the wind speed with MRI $\overline{N}$, averaged over time T, at standard elevation H_{met} in open terrain with roughness z_{01} (e.g., $T = 3$ s, $H_{met} = 33$ ft, $z_{01} = 0.10$ ft). With minor changes of notation, Eq. 2.3.6 yields

$$v(\tau, H_{areo}, z_0, \overline{N}) = r(\tau, T)\frac{(z_0/z_{01})^{\delta}[\ln(H_{aero}/z_0)]}{\ln(H_{met}/z_{01})}v(T, H_{met}, z_{01}, \overline{N}) \quad (4.2.14)$$

where r between the wind speed averaged over times τ and T (see Sect. 2.3.1.2). The interval τ is about 20 min to 1 h because it is typical of the duration of large-scale storms and of the prototype record length for wind tunnel measurements.[8]

4.2.3.3 Probabilistic Description of Peak Wind Effects as Functions of Knowledge Uncertainties. Owing to uncertainties with respect to the parameters in Eq. 4.2.14 (terrain roughness lengths, exponent δ, the ratio r between the wind speed averaged over times τ and T), these quantities are affected by random uncertainty factors with appropriate distributions, usually normal or truncated normal,[9] having unit mean and standard deviations based on experience and professional judgment. The estimated wind speed $v(T, H_{met}, z_{01}, \overline{N})$ is affected by sampling errors due to the finite size of the wind speed data set (Sect. 2.4.4). The peak coefficients C_{pk} estimated from a time history of duration τ are also affected by sampling errors, which for 1-h-long records are typically about 5%, and for 30- or 20-min records are about 7 and 10%, respectively [4.8, 4.15].

Aerodynamic errors inherent in wind tunnel simulations, errors due to the use of wind tunnel rather than full-scale measurements, errors in the wind tunnel estimation of the aerodynamic pressures, errors in the transformation of aerodynamic effects into stresses or other structural effects, and errors in the modeling of hurricane winds also need to be taken into account, again by using appropriate random uncertainty factors affecting the probabilistic wind load model.

Once a probabilistic wind load model is developed by incorporating in Eq. 4.2.13 the requisite random factors reflecting the various uncertainties, probabilistic descriptions of the wind effects are obtained by Monte Carlo simulation (see [4.16] for details).

4.2.3.4 Typical Results and Their Relation to ASCE 7 Standard Provisions and Current Practice. In this section we provide an example of numerical simulation results based on a set of assumptions on the uncertain-

[8] The relation between wind tunnel time and prototype time is inherent in the relation between nondimensional (reduced) wind speeds in the wind tunnel and the prototype, $[V/(Dt)]_{model} = [V/(Dt)]_{prototype}$, where V is the mean wind speed and D is a characteristic dimension of the building.

[9] The truncation is necessary so that negative arguments of the logarithmic function do not occur in the Monte Carlo simulations.

ties pertaining to the factors that determine peak wind effects [4.16]. Such assumptions need to be weighed by code-writing bodies with a view to reaching the consensus required for the incorporation of similar results in codes. For example, in [4.16] it was assumed that the errors in the estimation of the wind tunnel pressures are relatively small. This is not necessarily the case, especially for low buildings (see Sect. 4.2.1.4).

1. The type of extreme value (EV) distribution that best fits extreme wind speeds has a significant effect on the magnitude of wind load factors. For nonhurricane regions, estimated wind effects based on the assumption of the validity of the EV Gumbel distribution of nondirectional winds (defined as in Sect. 2.4.1) were found to be higher than those based on the EV reverse Weibull (EV RW) distribution with $c = -0.1$ by only 3% for a 50-year MRI but by almost 15% for a 500-year MRI. Wind load factors given by Eq. 4.1.10 are, tentatively, about 1.8 if based on the assumption that an EV Gumbel distribution is valid and 1.6 if the EV RW distribution with tail-length parameter $c = -0.1$ is assumed to hold. For hurricane regions, estimated load factors based on the EV RW distribution of nondirectional wind speeds are typically about 2.2. The EV RW distribution thus appears to yield wind load factors reasonably consistent with those specified in the ASCE Standard for nonhurricane regions, and implicit in the ASCE Standard for hurricane regions.

   According to results published in 1980 [4.17], the safety of engineered structures under nonhurricane wind loads, estimated by assuming the validity of the EV Gumbel distribution of the nondirectional wind speeds, is about one order of magnitude lower than under gravity loads. These results do not appear to be confirmed by experience. This suggests that the EV Gumbel distribution is conservative and lends additional credibility to the EV RW distribution with parameter $c \approx -0.1$ as a prudent and appropriate model of extreme wind speeds in nonhurricane regions.
2. The variability of the ratio that converts wind speeds with, for example, 3-s averaging time to speeds with 1-h averaging time can affect the estimated wind load factor significantly. The adoption of sustained wind speeds for codification purposes would reduce a source of significant errors and should be considered in the development of future versions of the ASCE 7 Standard.
3. An increase in the size of the extreme wind data set from, say, 35 years to 50 years, typically has a weak effect on the estimated values of the load factors.
4. For relatively low buildings in open terrain, wind load factors are typically affected weakly by even large uncertainties with respect to the roughness lengths of the open terrain over which the wind speed data

are measured and of open terrain upwind of the building. The effect of uncertainties pertaining to built-up terrain can be significant, however, and remains to be studied.

5. Whether the length of the wind tunnel records used to estimate response peaks is 30 min, say, or 60 min has a marginal effect on the design wind load estimation and the wind load factor. The relatively weak effect of uncertainties associated with typical lengths of wind tunnel records, and/or sizes of wind speed samples, is due to the fact that in the overall uncertainty estimation scheme these individual uncertainties are drowned by the aggregate effect of all the uncertainties involved in the estimates. This observation is obvious for structural reliability practitioners. Nevertheless, it is not always taken into account in wind tunnel practice, in which criteria based on very high quantiles of the distribution of the peak wind effect coefficient have been used without regard for the interaction among the variabilities of all the relevant factors affecting wind load effects (see Sect. 2.8). If confirmed, this result implies the possibility of reducing the size of database-assisted design records from the 1-h length currently deemed to be appropriate.
6. As was shown in Sect. 4.2.2.1, wind effects can be influenced significantly by the building orientation. For code development, building orientation can be viewed as an uncertainty parameter and can be incorporated as such in analyses that include the other relevant uncertainties in the estimation of the response. Such analyses entail Monte Carlo estimates of the peak wind effect distribution conditional on the angle of orientation of the building being contained in a specified sector. For the orientation to be included as a parameter, a probability distribution of the building orientation needs to be specified on the basis of approximate statistical data on actual building orientations at the location of concern. Calculations can then be performed that account for all relevant uncertainties, including building orientation [4.18].

4.2.4 Nonlinear Response: Ultimate Structural Capacity

Current design practice for wind loads consists of checking individual members and cross sections. Safety margins, primarily in the form of load and resistance factors, only provide insight into whether local limit states (e.g., the attainment of the yield stress in a member) are reached. However, such limit states can be exceeded in individual members with no penalty from an overall structural safety or serviceability viewpoint. The extent to which this is the case can be ascertained by performing a nonlinear analysis. Doing so for structures subjected to fluctuating wind loads has become possible only following the development in recent years of novel computational tools. Figure 4.2.7 shows results of nonlinear analyses from which it is possible to obtain quantitative results concerning the wind loads causing ultimate limit

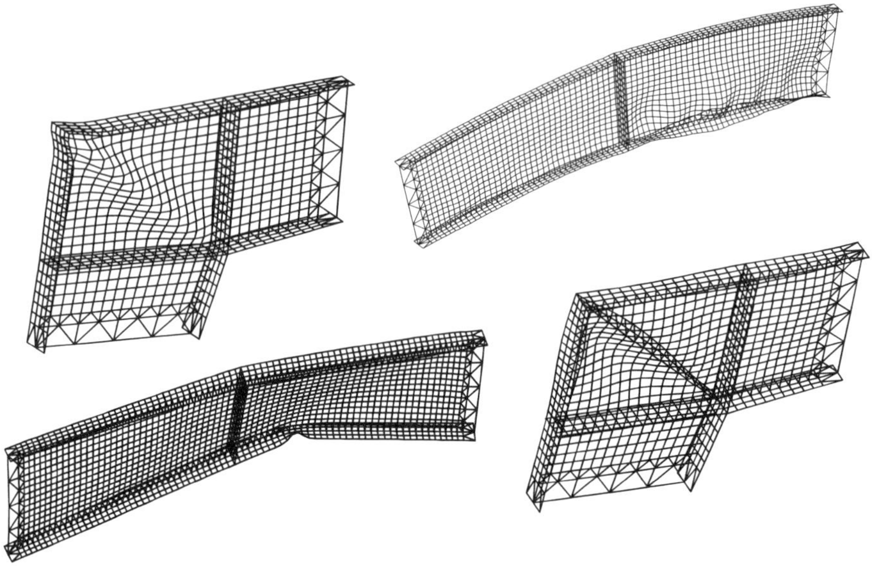

Figure 4.2.7. Local buckling in the frames of an industrial portal building.

states and their mean recurrence intervals (MRIs) [4.19]; see also [4.37]. Once the wind speeds that induce an ultimate limit state and their respective probabilities of exceedance are determined for each azimuth sector, the MRI of the ultimate limit state of the structure being considered can be estimated as in Sect. 4.2.2.4.2. By helping distribute material where it is needed and eliminating it where it is not, nonlinear analyses allow significant improvements in safety performance at no additional cost, and/or reductions in material consumption at no loss in safety performance.

4.3 FLEXIBLE BUILDINGS

Wind blowing from any given direction induces in flexible buildings fluctuating deflections in both the wind direction and the direction normal to the wind. Because the distribution of the fluctuating wind pressures is nonuniform over the building surface, the point of application of the wind load and the building's elastic center do not coincide. Therefore, the building experiences torsional deformations. Additional torsional deformations occur if the building's mass center and elastic center do not coincide.

Assume that a building has a rectangular shape in plan and that the wind blows in a direction normal to a building face. The response in the wind direction and the response normal to the wind direction are referred to as the *along-wind* and *across-wind response,* respectively (in Fig. 4.3.1, the along-

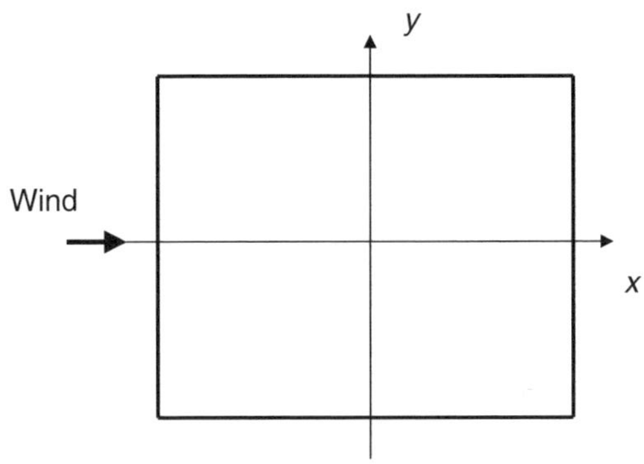

Figure 4.3.1. Notations.

wind response occurs in the x-direction, and the across-wind response occurs in the y direction). The along-wind response is due to the mean and the fluctuating drag force (Sect. 2.5.2), which in turn are due to positive pressures on the building's windward face, and negative pressures (suctions) on the building's leeward face. The across-wind force is due predominantly to the fluctuating lift force induced by vorticity shed in the building's wake (Sect. 2.5.3).[10] If the wind direction is not normal to one of the building faces, distinguishing between along-wind and across-wind response, or between the effects of drag and lift, is no longer meaningful in practice. For each wind direction it is necessary to determine the responses in two directions (which for a rectangular building are normal to the building faces) and in torsion. If the building's wind-induced motions are sufficiently large, the aerodynamics and the building motions interact, and the building can experience *aeroelastic effects*.

No analytical methods are available for calculating the response of flexible buildings to wind other than the along-wind response to flow normal to a rectangular building's face, over terrain with uniform upwind surface roughness (see, e.g., ASCE Sect. 6.5.8.2 and Sect. 4A.1 in this chapter). However, estimates of the along-wind response are not sufficient for design purposes, since the across-wind response is typically comparable to, or larger than, the along-wind response; for some types of building torsional effects are significant as well. Analytical methods for estimating across-wind response are available only for preliminary, order-of-magnitude calculations. The ASCE 7-05 Commentary C6.5.8 provides access to an interactive database on aerodynamic loads that may be useful in some cases for rough, preliminary

[10]The drag force includes a contribution due to the vorticity shed in the building's wake, but unlike the vorticity-induced lift, this contribution is negligible for practical purposes. The lift force also includes a small contribution due to pressures induced by the lateral turbulence fluctuations in the oncoming wind flow.

calculations. For wind skewed with respect to a building face, the wind loading cannot be defined analytically. If the building aerodynamics is influenced by neighboring buildings, even along-wind response calculations are unreliable. Appendix 4A contains methods for rough, generic estimates of along-wind, across-wind, and torsional response that can be useful in some cases for preliminary calculations.

For design purposes structural engineers resort to structural and serviceability response estimates based on ad hoc wind tunnel tests and extreme wind climate studies. Three approaches to the wind tunnel testing of flexible buildings are currently available: the database-assisted design (DAD) approach, the high-frequency force balance (HFFB) approach (both of which use rigid wind tunnel models), and the aeroelastic modeling approach. However, wind tunnel testing by itself is not sufficient for estimating wind effects. To obtain estimates for structural design and serviceability checking, wind tunnel test results must be used in conjunction with dynamic response estimates and extreme wind speed data, both of which are direction dependent.

Section 4.3.1 presents the DAD approach for the general case in which the mass and elastic centers do not coincide at all floors. (The approach is, clearly, also applicable in the particular case in which the mass and elastic centers always coincide.) The main advantages of the DAD approach are that (1) it accounts correctly for the distribution of the wind loading over the building surface, since that distribution is captured in the measurements performed and recorded simultaneously at the pressure taps, and (2) it allows the peak wind-induced moments, shear forces, and axial forces in the building's structural members to be calculated and combined simply, directly, and rigorously. The HFFB procedure does not have these advantages and can therefore be inaccurate, despite corrections based on generic pressure distributions, particularly for buildings with nonlinear modal shapes [4.20]. On the other hand, HFFB has the advantage of allowing quick testing of alternative geometries for preliminary aerodynamic assessments.

Owing to its clarity, the DAD approach allows response estimates provided for final design by the wind engineering consultant to be scrutinized effectively by the structural engineering user. In fact, just as specialized structural engineers can typically perform seismic response calculations without having to resort to earthquake engineering consultants, structural engineers are enabled by the DAD approach to perform wind response calculations on their own. To do so, they must be provided with:

- The requisite aerodynamic database (i.e., pressure coefficient time histories measured in the wind tunnel at a sufficient number of pressure taps for a sufficient number of wind directions), recorded electronically by the wind tunnel operator.
- Ratios between directional mean wind speeds at 10 m over open terrain on the one hand and at the top of the building on the other, which are

typically obtained in the wind tunnel, although in the particular case of horizontal terrain with uniform roughness, those ratios can be determined from micrometeorological models (see Sect. 2.3.2).

- Directional extreme wind speed data at or near the site of concern (Sect. 2.4).

Software for performing DAD estimates of the wind-induced response of flexible structures, based on the material presented subsequently, is provided on the site http://www.nist.gov/wind.

As was seen in Sect. 4.2.3, uncertainties inherent in estimates of wind effects must be accounted for in design by *wind load factors*. Unlike for rigid buildings, for flexible buildings it is necessary to take into account dynamic response parameters, that is, damping and natural frequencies of vibration. The damping, in particular, is poorly known and therefore has significant variability. In the United States fairly widely accepted values for damping ratios are 1% for steel buildings and 2% for reinforced concrete buildings. Smaller values are being considered in Japan, a fact possibly related to differences in construction types and techniques [4.21]. Actually, damping generally varies as a function of the magnitude of the response: It is higher for the larger responses considered in design for strength, and lower for the smaller responses considered in design for serviceability. However, the increase in damping with the magnitude of the response appears to stop at some point, and even to turn into a decrease [4.36]. The consequence of the uncertainties with respect to damping and, to a lesser extent, to the natural frequencies of vibration, is that wind load factors should be higher for tall buildings than for rigid buildings. Codes and standards do not reflect this requirement. The Monte Carlo simulation approach described in Sect. 4.2.3 can be adapted to obtain approximate estimates of the magnitude of the requisite wind load factors that account for uncertainties with respect to dynamic response parameters.

In view of their increasingly reliable performance, damping devices ([4.30]; see also Sect. 5.3.3.2) are now being used not only for the mitigation of accelerations and drift (Sect. 4.3.4), but also for the reduction of the response considered in design for safety [4.31].

The high-frequency force balance approach is described in Section 4.3.2. Section 4.3.3 is concerned with aeroelastic testing. Section 4.3.4 is devoted to serviceability issues.

4.3.1 Database-Assisted Design Approach to Estimation of Wind Effects

For flexible buildings, the DAD approach entails the following steps. First, obtain from the wind tunnel operator the aerodynamic input from simultaneous pressure measurements performed on a rigid model (Sect. 4.3.1.1).

Second, perform the dynamic modeling; that is, calculate the natural frequencies and modes of vibration, and develop and solve the equations of motion under excitation by winds within a sufficient range of speeds, blowing from each of a sufficient number of directions (Sect. 4.3.1.2). Third, for those wind directions, calculate the wind effects directly relevant to the design of structural members, as well as deflections and accelerations (Sect. 4.3.1.3). Fourth, obtain directional extreme wind speed data sets for a sufficient number of years or storm events (see Sect. 2.3). Fifth, for specified mean recurrence intervals, estimate wind effects, including deflections and accelerations, by accounting for the directional dependence of the dynamic response and the extreme wind speeds (Sect. 4.3.1.4) [4.41].

4.3.1.1 Aerodynamic Input. The aerodynamic input to the DAD approach consists of the simultaneously recorded aerodynamic pressure time histories at a large number of taps on the surface of a rigid model of the building (Fig. 4.3.2). Pressure time histories are recorded simultaneously at all the taps for each of a sufficiently large number of mean wind directions θ. The time

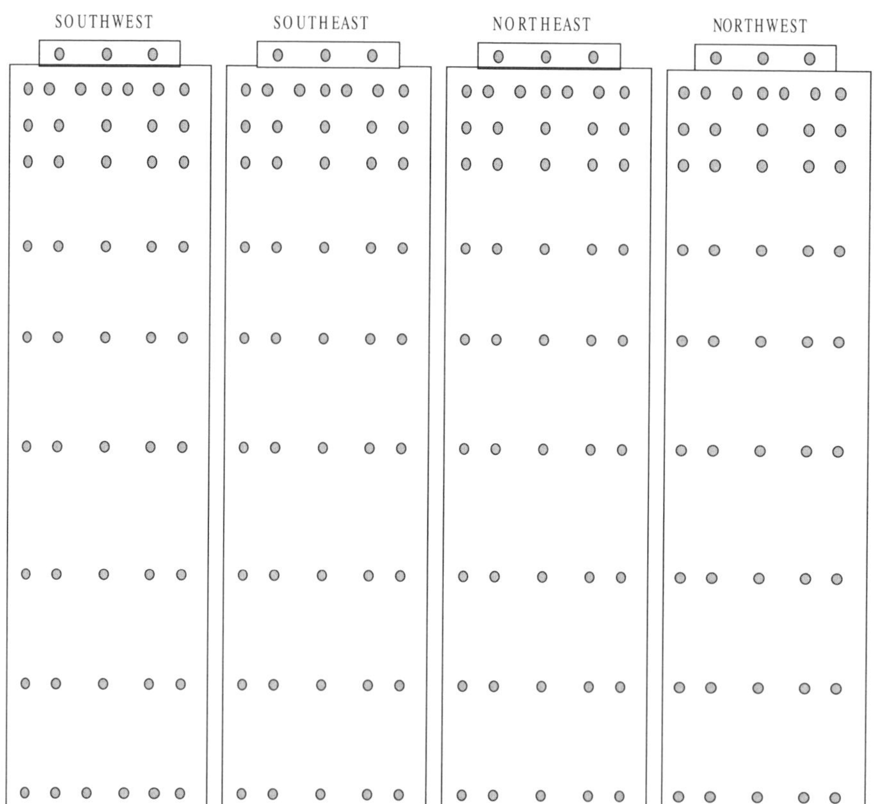

Figure 4.3.2. Arrangement of pressure taps on the facades of a tall building model.

histories are used to obtain generalized forces and torsional moments for the dynamic analysis of the structure.

Wind blowing from direction θ with mean hourly speed at height H above ground at the building site $\overline{V}_\theta(H)$ induces external pressures on each of the building faces. (Typically, H is the height of the building.) At a tap location denoted by l, the wind-induced pressure on a face parallel to a rectangular building's principal axis y acts in the x-direction (Fig. 4.3.1) and has the expression (in SI units)

$$p_{lx\theta}(t) = 0.62C_{plx\theta}(t)\overline{V}_\theta^2(H) \tag{4.3.1}$$

where for any specified wind speed $v_{\theta-\alpha}$ at 10 m over open terrain blowing from direction $\theta - \alpha$ (α is the veering angle, see Sect. 2.3.2.1.3), the corresponding mean speed $\overline{V}_\theta(H)$ must in general be obtained by measurement in the wind tunnel. $C_{plx\theta}$ denotes the pressure coefficient corresponding to the pressures $p_{lx\theta}$. An equation similar to Eq. 4.3.1 holds at a tap denoted by l for wind pressures $p_{ly\theta}$ acting in the y-direction on faces parallel to the x-axis. Wind forces in the x and y directions at the elevation z_l of a pressure tap can be obtained from the pressures $p_{lx\theta}(t)$ and $p_{ly\theta}(t)$, respectively, through multiplication by the respective areas A_{lx} and A_{ly} tributary to the tap. The wind-induced torsional moments are the moments *about the mass center* due to the wind forces at the elevations z_l. The pressures coefficients $C_{plx\theta}(t)$ and $C_{ply\theta}(t)$ are obtained in the wind tunnel and are made available to the structural engineer in electronic form. For a rectangular building the pressure components $p_{ly\theta}(t)$ of taps on faces normal to the direction x, and the pressure components $p_{lx\theta}(t)$ of taps on faces normal to the direction y, are equal to zero. The approach just described can be easily adapted for nonrectangular buildings.

The length of the pressure–time series measured in the wind tunnel usually corresponds to a prototype time of about 1 h. The wind tunnel sampling rate depends on the model scale and on the velocity scale (ratio of velocity in the wind tunnel to the velocity in full scale) and is on the order of hundreds of hertz (e.g., 400 Hz). The relation between model and prototype sampling rates is governed by the relation $(nD/V)_{\text{model}} = (nD/V)_{\text{prototype}}$. For example, if the model scale is $D_{\text{model}}/D_{\text{prototype}} = 1/500$, and if the velocity scale is $V_{\text{model}}/V_{\text{prototype}} = \frac{1}{5}$, the ratio of model to prototype sampling rates is $n_{\text{model}}/n_{\text{prototype}} = 100$. If $n_{\text{model}} = 400$ Hz, the corresponding prototype sampling rate $n_{\text{prototype}}$ is 4 Hz.

4.3.1.2 Dynamic Modeling

4.3.1.2.1 Natural Frequencies and Modes of Vibration [4.22]. We now describe a procedure for obtaining modal frequencies and shapes.[11] The total kinetic energy of the system is

[11] Estimates of natural frequencies and modes of vibration that account for P–δ effects are not covered herein.

$$T = \frac{1}{2} \sum_{n=1}^{nf} (m_n \dot{x}_n^2 + m_n \dot{y}_n^2 + I_n \dot{\varphi}_n^2) \tag{4.3.2}$$

where x_n and y_n are the displacements of the center of mass of the nth floor in the x and y directions, respectively, $\varphi(z_n)$ is the torsional rotation of the nth floor, and n_f is the total number of floors.

The total strain energy of the system is

$$V = \frac{1}{2} \{q\}^{\mathrm{T}}[k]\{q\} \tag{4.3.3}$$

where the superscript T denotes a transpose and

$$\{q\}^{\mathrm{T}} = \{x_1, x_2, \ldots, x_{nf}, y_1, y_2, \ldots, y_{nf}, \varphi_1, \varphi_2, \ldots, \varphi_{nf}\} \tag{4.3.4}$$

$$[k] = [a]^{-1} \tag{4.3.5}$$

$$[a] = \begin{bmatrix} [x_{i1,x} x_{i2,x}, \ldots, x_{in_f,x}][x_{i1,y} x_{i2,y}, \ldots, x_{in_f,y}][x_{i1,\varphi}, x_{i2,\varphi}, \ldots, x_{in_f,\varphi}] \\ [y_{i1,x} y_{i2,x}, \ldots, y_{in_f,x}][y_{i1,y} y_{i2,y}, \ldots, y_{in_f,y}][y_{i1,\varphi}, y_{i2,\varphi}, \ldots, y_{in_f,\varphi}] \\ [\varphi_{i1,x} \varphi_{i2,x}, \ldots, \varphi_{in_f,x}][\varphi_{i1,y}, \varphi_{i2,y}, \ldots, \varphi_{in_f,y}][\varphi_{i1,\alpha} \varphi_{i2,\alpha}, \ldots, \varphi_{in_f,\alpha}] \end{bmatrix} \tag{4.3.6}$$

$[k]$ is the system's stiffness matrix and $[a]$ is its inverse, that is, the flexibility matrix. The nine component matrices of matrix $[a]$ are represented on the right-hand side of Eq. 4.3.6 by their respective ith rows. The size of matrix $[a]$ is $3n_f \times 3n_f$. The terms of matrix $[a]$ are displacements in the x or y direction or torsional rotations at the center of mass of floor i ($i = 1, 2, \ldots, n_f$) due to a unit horizontal force in the x or y direction or unit torsional moment at the center of mass of floor j ($j = 1, 2, \ldots, n_f$) and can be obtained by using standard structural analysis programs. (For example, the term $y_{i2,x}$ is the y displacement of the center of mass of floor i due to a unit horizontal force acting at the center of mass of floor 2 in direction x.) In the stiffness matrix $[k]$ the restoring force $k_{ix,2y}$ represents the horizontal force in the x-direction at the center of mass of floor i induced by a unit horizontal displacement in the y-direction at the center of mass of floor 2.) The positions of the centers of mass and elastic centers can vary from floor to floor and are accounted for implicitly by the terms of the matrix $[k]$.

The displacements of the mass center and the torsional rotation at the elevation z_i of the ith floor for the freely vibrating structure form a vector $\{w(t)\}$ of dimension $3n_f$. Its terms are

$$\begin{aligned} w_1(t) &= x_1(t), w_2(t) = x_2(t), \ldots, w_{nf}(t) = x_{nf}(t), w_{nf+1}(t) \\ &= y_1(t), \ldots, w_{2nf}(t) = y_{nf}(t), w_{2nf+1}(t) = \varphi_1(t), \ldots, w_{3nf}(t) \\ &= \varphi_{nf}(t) \end{aligned} \tag{4.3.7}$$

The equations of motion of the undamped, freely vibrating system are obtained from Lagrange's equations and are easily shown to be

$$[M]\{\ddot{w}(t)\} + [k]\{w\} = \{0\} \tag{4.3.8}$$

where $[M]$ is a diagonal matrix of the floor masses (polar mass moments of intertia). These equations are coupled owing to the cross-terms of the strain energy function. The natural frequencies of vibration ω_i ($i = 1, 2, \ldots, 3n_f$) are obtained by solving the secular equation numerically and are ordered so that $\omega_1 < \omega_2 < \cdots < \omega_{3nf}$. (The secular equation is obtained by substituting in Eq. 4.3.6 the vector $\{w \cos \omega t\}$ for the vector $\{w(t)\}$.) It is permissible in practice to assume that modal frequencies are always distinct. (This can be achieved by seeing to it that, for example, stiffnesses in the x and y directions differ from each other, albeit by very small amounts.) The corresponding modes of vibration are the $3n_f$ eigenvectors corresponding to the eigenvalues obtained by solving the secular equation and constitute a matrix $[\mu]$ of dimension $3n_f \times 3n_f$. The eigenvector $\{\mu_j\}$ corresponds to the eigenvalue ω_j.

4.3.1.2.2 Equations of Motion of Structure Under Excitation by Wind from Direction θ. As indicated in Sect. 4.3.1.1, wind with mean velocity $\overline{V}_\theta(H)$ blowing from direction θ induces aerodynamic forces in the x and y directions at the elevation z_l of pressure tap l equal to the pressure components $p_{lx\theta}(t)$ and $p_{ly\theta}(t)$, times the pressure tap tributary area A_l. The aerodynamic torsional moment about the mass center due to the tap l is denoted by $M_{l\varphi\theta}(t)$ and is equal to the pressures at that tap times the respective tributary area A_l times the respective horizontal distance from the tap to the mass center at the tap level.

The equations of motion of the forced system are

$$[M]\{\ddot{w}_\theta(t)\} + [k]\{w_\theta\} = \{F_\theta(t)\} \tag{4.3.9}$$

where $\{F_\theta(t)\}$ is the vector of the wind forces (torsional moments) $F_{x1\theta}(t)$, $F_{x2\theta}(t), \ldots, F_{xn_f\theta}(t)$, $F_{y1\theta}(t), \ldots, F_{yn_f\theta}(t)$, $M_{\varphi 1\theta}(t), \ldots, M_{\varphi n_f\theta}(t)$ acting on floors $1, 2, \ldots, n_f$. The subscript θ indicates that the force and response vectors are associated with the direction θ from which the wind blows. The torsional moments $M_{\varphi 1\theta}(t), \ldots, M_{\varphi n_f\theta}(t)$ are taken with respect to the mass center of the respective floor. The pressures used to calculate the right-hand side of Eq. 4.3.9 are positive or negative if their directions are the same as or opposite to the directions of the principal axes of the building. Substituting in Eq. 4.3.8 the transformed variables

$$\{w_\theta(t)\} = [\mu]\{\xi_\theta(t)\} \tag{4.3.10}$$

where $[\mu]$ is the matrix consisting of the $3n_f$ vectors $\{\mu_j\}$ ($j = 1, 2, \ldots, 3n_f$), yields

$$[M][\mu]\{\ddot{\xi}_\theta(t)\} + [k][\mu]\{\xi_\theta(t)\} = \{F_\theta(t)\} \tag{4.3.11}$$

Premultiplying Eq. 4.3.11 by $[\mu]^{\mathrm{T}}$ yields

$$[\mu]^{\mathrm{T}}[M][\mu]\ \{\ddot{\xi}_\theta(t)\} + [\mu]^{\mathrm{T}}[k][\mu]\{\xi_\theta(t)\} = [\mu]^{\mathrm{T}}\{F_\theta(t)\} \tag{4.3.12a}$$

Owing to the orthogonality of the eigenvectors, Eq. 4.3.12*a*, to which modal viscous damping terms proportional to the modal damping ratios ζ_m are added, can be written as

$$M_m\ddot{\xi}_{m\theta}(t) + 2M_m\omega_m\zeta_m\dot{\xi}_{m\theta}(t) + M_m\omega_m^2\xi_{m\theta}(t) = \{[\mu]^{\mathrm{T}}\{F_\theta(t)\}]\}_m$$

$$(m = 1, 2, \ldots, 3n_f) \tag{4.3.12b}$$

In structural dynamics the quantity M_m and the quantity on the right-hand side of Eq. 4.3.12*b* are commonly called *generalized mass* and *generalized force,* respectively. It follows from the unforced equation of motion of the system that $M_m\omega_m^2 = [[\mu]^{\mathrm{T}}[k][\mu]]_m$. Once Eqs. 4.3.12b are solved numerically, the physical coordinates $\{w_\theta(t)\}$ [i.e., the coordinates $x_1(t), x_2(t), \ldots, x_{nf}(t), y_1(t), \ldots, y_{nf}(t), \varphi_1(t), \ldots, \varphi_{nf}(t)$; see Eq. 4.3.7] are given by Eq. 4.3.10. Accelerations $\{\ddot{w}_\theta(t)\}$ are obtained by differentiating Eq. 4.3.10 twice, the second derivatives of the generalized coordinates being known once Eqs. 4.3.12*b* have been solved.

From the requirement that the reduced frequency $nD/\overline{V}$ be the same for model and prototype, it follows that the time interval Δt between successive discrete values of the forcing $\{F_\theta(t)\}$ is

$$\Delta t = \frac{D}{D_{\mathrm{mod}}}\frac{\overline{V}_{\mathrm{mod}}(H_{\mathrm{mod}})}{\overline{V}(H)}\Delta t_{\mathrm{mod}}, \tag{4.3.13}$$

where prototype quantities are unsubscripted, and the subsscript "mod" pertains to the model. Equation 4.3.13 is applied separately for each wind speed and direction being considered in the calculations.

4.3.1.3 Calculation of Wind Effects for Design of Structural Members and Serviceability Checks

4.3.1.3.1 Wind Effects on Structural Members. We assume in the following that the structural system consists of beams, columns, and trusses, but the procedure we describe can easily be adapted to other systems.

The results of the dynamic modeling are used to calculate bending moments and axial forces in member *j*. The structure is subjected to time-dependent aerodynamic forces at the pressure tap locations and to inertial forces at each floor's center of mass. The wind effects at a cross section of the member can therefore be calculated at each time *t* by summing up the products of the aerodynamic and inertial forces times the respective influence coefficients that convert the forces into the wind effects. The influence coefficients are obtained by using standard structural analysis programs.

Three categories of influence coefficients are involved in the calculations:

1. Influence coefficients for effects induced by wind forces acting at tap l. These coefficients are identified by the subscripts W and l. They are (we omit designating the member's cross section):
   a. m_{jXWl} and m_{jYWl}, representing the bending moments induced about the local X and Y axes of member j by a unit load perpendicular to the building face at tap l
   b. P_{jWl}, the axial force in member j due to a unit load perpendicular to the building face at tap l
2. Influence coefficients for effects induced by inertial forces acting at the center of mass of floor n in directions x and y, identified by the subscripts Ixn and Iyn, respectively. They are:
   a. m_{jXIxn} and m_{jYIxn}, the bending moments about axes X and Y in member j due to a unit horizontal force acting through the center of mass in the x-direction at floor n
   b. m_{jXIyn} and m_{jYIyn}, similar moments due to a unit horizontal force acting through the center of mass in the y-direction at floor n
   c. P_{jIxn} and P_{jIyn}, the axial forces due to a unit horizontal force acting through the center of mass in the x and y directions at floor k
3. Influence coefficients for effects induced by torsional moments about the center of mass of floor n. These coefficients are identified by the subscripts Tn. They are:
   a. m_{jXTn} and m_{jYTn}, the moments induced about axes X and Y by a unit horizontal torque about the mass center at floor n
   b. P_{jTn}, the axial force induced by a unit horizontal torque about the mass center at floor n

These sets of influence coefficients are combined with the wind forces based on the recorded pressures at the taps, and with the calculated inertial forces, to yield the wind-induced internal forces. The wind speed $\overline{V}_\theta(H)$ therefore induces at time t in member j the following axial load and moments about axes X and Y due to the wind pressures and inertial forces:

$$P_{j\theta}(t) = \sum_{l=1}^{n_l} P_{jWl} p_{l\theta}(t) A_l - \sum_{n=1}^{n_f} [P_{jIxn} m_n \ddot{x}_{n\theta}(t) + P_{jIyn} m_n \ddot{y}_{n\theta}(t) + P_{jTn} I_n \ddot{\varphi}_{n\theta}(t)] \quad (4.3.14a)$$

$$M_{jX\theta}(t) = \sum_{l=1}^{n_l}[m_{jXWl}p_{l\theta}(t)A_l$$

$$-\sum_{n=1}^{n_f}[m_{jXlxn}m_n\ddot{x}_{n\theta}(t) + m_{jXlyn}m_n\ddot{y}_{n\theta}(t) + m_{jXTn}I_n\ddot{\varphi}_{n\theta}(t)] \quad (4.3.14b)$$

$$M_{jY\theta}(t) = \sum_{l=1}^{n_l}m_{jYWl}p_{l\theta}(t)A_l$$

$$-\sum_{n=1}^{n_f}[m_{jYlxn}m_n\ddot{x}_{xn\theta}(t) + m_{jYlyn}m_n\ddot{y}_{n\theta}(t) + m_{jYTn}I_n\ddot{\varphi}_{n\theta}(t)] \quad (4.3.14c)$$

where m_n is the mass of the nth floor, $\ddot{x}_{n\theta}(t)$ and $\ddot{y}_{n\theta}(t)$ are the accelerations in the x and y directions at the mass center of floor n, and I_n and $\ddot{\varphi}_{n\theta}(t)$ are the mass moment of inertia and the rotational acceleration about the mass center of floor n, assumed to be infinitely rigid.

The first sum on the right-hand side of each of Eqs. 4.3.14 shows clearly the dependence of the internal forces in the structural system on the distribution of wind pressures along the building height. This distribution is known if the DAD approach is used and must be guessed if the HFFB approach is used (see Sect. 4.3.2). Guessing is difficult or impossible in the presence of a nearby building that affects the aerodynamic response of the building of interest.

For design purposes it is not the individual internal forces that need to be considered but rather, combinations of internal forces governed by interaction equations and specified load combinations. For example, for steel structures, the following design interaction equations [4.23] or variants thereof could be used for member j:

$$\frac{P^f_{j\theta}(t)}{\phi P_j} \geq 0.2: \qquad b^f_{j\theta}(t) = \frac{P^f_{j\theta}(t)}{\phi P_j} + \frac{8}{9}\left(\frac{M^f_{jX\theta}(t)}{\phi_b M_{jX}} + \frac{M^f_{jY\theta}(t)}{\phi_b M_{jY}}\right) \leq 1 \quad (4.3.15a)$$

$$\frac{P^f_{j\theta}(t)}{\phi P_j} < 0.2: \qquad b^f_{j\theta}(t) = \frac{P^f_{j\theta}(t)}{2\phi P_j} + \left(\frac{M^f_{jX\theta}(t)}{\phi_b M_{jX}} + \frac{M^f_{jY\theta}(t)}{\phi_b M_{jY}}\right) \leq 1 \quad (4.3.15b)$$

where P_j, M_{jx}, and M_{jy} are the nominal axial and flexural strengths of member j, ϕ and ϕ_b are axial and flexural resistance factors, and the quantities in the numerators are the total axial load and moments due to the specified factored combinations (hence the superscript f). For allowable stress design, similar equations hold with unfactored quantities. Applicable load combinations specified by the ASCE 7-05 Standard, Sect. 2.3.2 include

$$1.2D + 1.0L + 1.6W \tag{4.3.16a}$$

$$0.9D + 1.6W \tag{4.3.16b}$$

where D is the dead load, L is the live load, and W is the wind load.

Some wind engineering consultants use methodologies which require that designers consider as many as 20 separate combinations of wind effects due to motion in the x and y directions and in torsion. These combinations are obtained by using a variety of approximate combination factors or approximate methods for adding peak effects: for example, the "point-in-time" method. In fact, the combined wind effects depend on the influence coefficients applicable to the member of concern rather than being generic. This dependence could not be accounted for in the past, owing to limited computational capabilities and to the fact that the aerodynamic information obtained in the wind tunnel was incomplete and was not recorded for later use by the structural engineer.

Equations 4.3.15 and 4.3.16 yield the time histories $b^f_{j\theta}(t)$, the quantities of direct interest in design, through the exact superposition of the time-dependent wind effects induced by forces in the x and y directions and in torsion. If the generalized coordinates have Gaussian distributions, as is typically the case for tall structures (with the possible exception of upper floors), it is possible to estimate the expectation of the peak of $b^f_{j\theta}$ by using the approximate expression

$$[b^f_{j\theta}]_{\text{peak}} \approx \text{mean}(b^f_{j\theta}) + g\ \text{rms}(b^f_{j\theta}) \tag{4.3.17}$$

where rms denotes root mean square, and g is a peak factor whose value is typically about 3.75 [4.24]. Alternatively, the method of Sect. 4.2.1.2, item 6 may be used to obtain statistics of the peaks of interest. That method is incorporated in the software for calculating wind effects provided on the site http://www.nist.gov/wind. The approach just described can be adapted as needed for other interaction formulas and load combinations.

4.3.1.3.2 Deflections and Accelerations. Peak floor drift and measures of accelerations at various points within the building need to be calculated for serviceability design purposes. The equations of motion yield the displacements $x_\theta(z_n, t)$ and $y_\theta(z_n, t)$ and the rotations $\varphi_\theta(z_n, t)$ about the center of mass, as well as the respective accelerations, at any time t and floor elevation z_n. The resultant accelerations at elevation z_n and at any position on the floor are calculated by adding vectorially the relevant acceleration components at each time t. The peak acceleration is the peak of the time series so obtained. Similarly, rms values of resultant accelerations may be calculated.

4.3.1.4 Estimation of Wind Effects with Specified Mean Recurrence Intervals by Accounting for the Direction-Dependent Extreme Wind Speeds and Dynamic Response. In Sect. 4.2.2.3 we presented expressions for wind effects induced by a mean wind speed at the top of the building $\overline{V}_\theta(H)$ for fixed wind direction θ. For year p (or storm p; $p = 1, 2, \ldots, p_{\max}$) the peak wind effects in year (or storm) p, $[b^f_{j\theta p}]_{\text{peak}}$, can then be obtained for each direction $\theta = 1, 2, \ldots, 8$ (or $\theta = 1, 2, \ldots, 16$, or $\theta = 1, 2, \ldots, 36$). For design purposes only the largest value of the quantities $b^f_{j,\theta p}$, that is, $\max_\theta\{[b^f_{j,\theta p}(t)]_{\text{peak}}\}$, is of interest. This largest value—the largest wind effect in each year (or storm)—is denoted by b^f_{jp}. The set of quantities b^f_{jp} is obtained for each year (storm). This set is a one-dimensional (univariate) sample, of size $p_{\max}$, of the peak wind effect b^f_{jp} occurring in each year (storm).

Consider the case of p hurricanes and assume that $p_{\max} = 999$ and a rate of arrival of hurricanes at the site being considered of $\upsilon = 0.5$ per year. Rank-order the time series b^f_{jp}. The largest, the second largest, and the fourth largest of the p values of the rank-ordered series are estimators of the wind effect b^f_j with a mean recurrence interval (MRI) of $(999 + 1)/(1 \times 0.5) = 2000$ years, $1000/(2 \times 0.5) = 1000$ years, and 500 years, respectively. To obtain the requisite wind effects, the equations of motion need to be solved for a sufficient number of wind speeds for each of the directions of interest (since for flexible buildings the wind effect is no longer proportional to the square of the wind speed, as is the case for rigid buildings). The resulting generalized coordinates and their second derivatives may be obtained for other wind speeds by interpolation.

To perform the calculations, a directional hurricane wind speeds database is needed. For example, the NIST simulated hurricane wind speed database provides 999 extreme wind events at a large number of locations (mileposts) along the Gulf of Mexico and North Atlantic coast and is publicly available at http://www.nist.gov/wind. In nonhurricane regions or hurricane regions where significant nonhurricane winds can occur, directional nonhurricane wind speed data are needed. Larger sets of such data can be obtained from smaller samples of recorded data by simulation. The implementation of the procedure just described requires the availability of the ratios between directional wind speeds at 10 m above open terrain and their counterparts at the top of the building.

4.3.1.5 Remarks on the DAD Approach

1. Equations 4.3.14 show that wind effects of interest cannot be calculated by a generic formula or generic sets of load combination formulas handed down by the wind engineering consultant to the structural engineer. Rather, they depend on structure-specific quantities—the influence coefficients—and on specified load combinations and load factors. Database-assisted design as developed in this section creates an effective bridge between the aerodynamic, dynamic, and extreme wind speeds

information on the one hand and the information needed by the structural engineer for design purposes on the other. DAD eliminates the need for approximations and guesses that characterize other approaches used in wind engineering practice. Those approximations served the profession well at a time where computation capabilities were relatively weak, but are no longer needed today except for preliminary design purposes. In our view DAD is to other approaches currently used in wind engineering as modern structural design software is, say, to approximations of moment distribution methods.

2. Estimates of the dynamic response of flexible structures have typically been performed in the past by using spectral (frequency-domain) methods, which were first developed by Liepmann for application to the buffeting problem [4.25]. Spectral methods entail the transformation of time-domain equations of motion into algebraic equations in the frequency domain, thus obviating the need for solving ordinary differential equations such as Eq. 4.3.12*b*. This feature was advantageous given the absence in the past of fast ordinary differential equations (ODE) solvers. ODE solvers currently available allow the fast solution of the equations of motion in the time domain. One advantage of time-domain solutions is that they do not entail the loss of phase information that occurs in the spectral approach. Time series of wind effects associated with various modes of vibration and load combinations can therefore simply be added to yield time series and peaks of total, combined wind effects.
3. Once the shapes and frequencies of the modes of vibration are calculated as indicated in Sect. 4.3.1.2.1, the analysis proceeds in exactly the same manner regardless of whether or not the distance between elastic and mass centers is zero at all floors.
4. The number of taps must be sufficient to allow a reasonably accurate representation of the pressures acting on the exterior surfaces.

4.3.2 High-Frequency Force Balance Approach to Estimating Wind Effects

In the HFFB approach the models are rigid and are supported by a high-frequency force balance that allows measurements of the base shears and moments, but essentially, deformations are so small that no aeroelastic effects can occur (Fig. 4.3.3). The HFFB approach is designed primarily for buildings with linear fundamental modal shapes in motions along the principal axes of the building cross section. The wind tunnel response measurements on which the HFFB approach is based consist of base moments and shear forces in the x and y directions and base torsional moments.

The expression for the generalized force induced by pressures parallel to the x-direction is

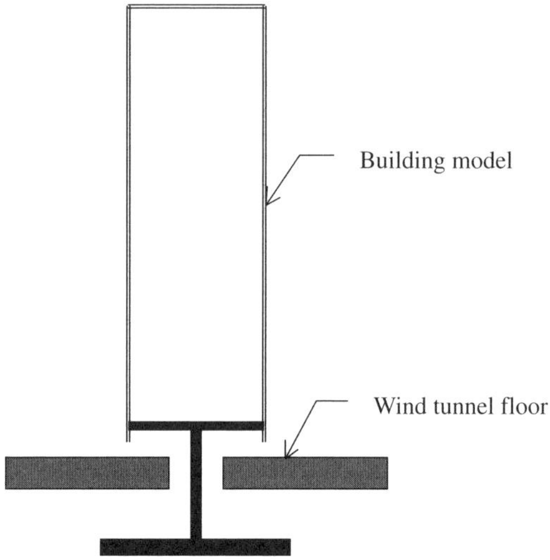

Figure 4.3.3. Schematic of a high-frequency force balance (HFFB) model.

$$Q_{x1}(t) = \int_0^H w_x(z, t)z \, dz \tag{4.3.18}$$

(see Sect. 2.6), where H is the building height, $w_x(z, t)$ is the wind loading parallel to the x direction per unit height, and z is the linear modal shape, taken to be equal to the height above ground. The HFFB approach takes advantage of the fact that this expression and its y-direction counterpart coincide with the wind-induced base moments measured on the wind tunnel model.

On the other hand, while the generalized aerodynamic torsional moment has the expression

$$Q_{\varphi 1}(t) = \int_0^H T(z, t)z \, dz \tag{4.3.19a}$$

where T is the aerodynamic torsional moment per unit height, the base aerodynamic torsional moment measured in the wind tunnel is

$$Q_{\varphi 1,\text{HFFB}}(t) = \int_0^H T(z, t) \, dz \tag{4.3.19b}$$

If the fundamental modal shapes are not linear, measured base moments may be subjected to corrections accounting for the modal shapes, so that, after correction, those moments represent approximations to the actual generalized forces. Similarly, it is possible to correct Eq. 4.3.19*b* so that it approximates Eq. 4.3.19*a*. However, the corrections depend on the distribution with height of the wind loading and on the aerodynamic torsion per unit height, which are in general unknown, especially for buildings affected by aerodynamic interference effects. The corrections and the corresponding approximations of the generalized torques and moments therefore depend on assumptions—guesses—on the wind load distribution that are more or less crude. Some wind tunnel operators assume that the distributions are the same as those that prevail for a building in terrain with uniform roughness, but aerodynamic interactions between buildings in urban settings can render this assumption inadequate; guessed-at pressure distributions appear to have been used even for the most recent estimates of wind effects on the World Trade Center twin towers.

4.3.2.1 Comparison Between the DAD and HFFB Approaches. As mentioned earlier, despite current correction methods, HFFB may be inaccurate, owing to the dependence of the internal forces in the structure on the unknown wind force distribution along the building height. Unlike in the HFFB approach, in the DAD approach the distribution of the wind loading and aerodynamic torsional moments with height is known, so it is possible to calculate wind effects accurately at all levels of the building as well as the response of buildings with any modal shapes and any number of modes of vibration. In addition, the DAD approach is capable of accounting for the effect of the distance at any floor between the floor's center of mass and the elastic center.

Higher wind tunnel testing costs required for DAD are offset at least in part by the fact that no separate testing is required to obtain pressures on cladding, as is the case if the HFFB approach is used. On the other hand, HFFB testing is less expensive (with the qualification just indicated). More important, it can also be performed faster, a useful feature in cases in which alternative building configuration designs need to be compared from the point of view of aerodynamic performance.

4.3.3 Aeroelastic Testing

Wind tunnel testing used in conjunction with the DAD or HFFB approach makes use of rigid models. Models used with the DAD approach experience no displacements, whereas models for the HFFB approach are supported at the base by a device allowing displacements that are too small to allow the development of aeroelastic effects.

Multidegree-of-freedom models for aeroelastic testing reproduce to the appropriate scale the distribution with height of the building's mass distribution and flexibility. If designed and constructed properly, such models replicate to a close approximation the dynamics of the prototype, including the modal shapes. However, they are expensive and their construction is time consuming (see Fig. 5.3.20 for an example of a similarly elaborate aeroelastic model of a suspension bridge).

A simpler type of aeroelastic model, called the *stick model,* is therefore used more commonly. The stick model is similar to the HFFB model except that its force balance device (1) has the flexibility required for the approximate simulation of the building motions, and (2) allows the simulation of damping. Various designs of the force balance, as well as analytical correction procedures, are employed to reduce the discrepancies between motions corresponding to linear fundamental modal shapes on the one hand and nonlinear modal shapes on the other [4.11, 4.26 4.27], and to account, to the extent possible, for the simplified reproduction in the wind tunnel of the prototype building's properties with respect to mass. Even simplified aeroelastic modeling of flexible buildings can provide useful approximate, or at least qualitative, representations of the extent to which aeroelastic effects are significant.

For most tall, even supertall buildings, those effects are minor or negligible. Tests have shown, for example, that this was true of the World Trade Center twin tower models. Moreover, aeroelastic tests indicate that the models commonly experience average positive aerodynamic damping. The explanation for such damping lies in the change of the relative velocity of the oncoming flow with respect to the building as the latter experiences fluctuating motions. It is prudent, however, not to count on positive aerodynamic damping in design. Given the typical absence of significant aeroelastic effects, DAD, or HFFB—to within the approximations of which it is capable—are typically sufficient for structural and serviceability design purposes.

However, there are as yet no definitive quantitative criteria on whether aeroelastic tests are needed or not. Tentative criteria applicable to untapered buildings with a square shape in plan suggest that aeroelastic tests may be necessary if, for wind normal to a building face, the calculated root mean square of the across-wind response is $\sigma_y > \alpha B$ where B denotes the width of the building and α is a factor that depends on the upwind terrain roughness. For open, suburban, and urban terrain, $\alpha \approx 0.015$, 0.025, and 0.045, respectively [4.28]. According to results reported in [4.29], for 500-m-tall buildings with ratio of height to width $H/D = 10$, square plan cross section, damping ratio $\zeta \approx 1.5\%$, and fundamental natural frequency in the x and y motions $n_1 = 0.1$ Hz, these criteria correspond to mean reduced velocities at the top of the building $V_r \approx 9$ and $V_r \approx 11$ in open and urban terrain, respectively ($V_r = V/n_1D$). Rough calculations based on the use of the Strouhal number suggest that significant vortex-induced vibrations do not in practice occur if the height of typical untapered buildings with a rectangular shape in plan is less than about, say, seven times the smaller dimension in plan.

There are several reasons why aeroelastic effects in tall buildings tend to be negligible for practical purposes. First, to a greater extent than for chimneys or stacks, vortex-induced forces on buildings tend to have broad frequency distributions, so that only a relatively small fraction of the total vortex-induced force is available for producing oscillations of the lock-in type. This diminishes the aeroelastic import of the vorticity. Second, unless the building is strongly tapered, three-dimensional effects associated with the flow near the top of the building are sufficiently strong to prevent the occurrence of self-excited forces in the upper portion of the building, where they would be most effective. Third, since the Strouhal number is constant, the nonuniform mean wind speed profile affects the theoretical vortex shedding frequencies along the building height and therefore reduces the extent to which the action of the vorticity is vertically coherent. A similar but a more pronounced favorable effect can be achieved by designing the building to be strongly tapered. Fourth, galloping and flutter oscillations typically occur at speeds greater than those associated with vortex-induced lock-in, and therefore are too large (correspond to MRIs that are too long) to be of interest in practice. Last but not least, the building configuration can play an important role in minimizing or suppressing vortex excitation and other aeroelastic effects. Aerodynamic testing was used for the design of the Taipei 101 building (Figs. 4.3.4, 4.3.5) with a view to reducing wind effects [4.31]. Chamfering of the corners (see Fig. 4.3.6) was determined to be effective. Other types of configuration suppress or greatly reduce wind-induced vorticity. One example is shown in Fig. 4.3.7.

For buildings with geometric configurations and structural systems that could result in significant aeroelastic response, it is prudent to conduct aeroelastic tests, even if the latter can yield only approximate or qualitative results. The role of aeroelastic effects should be checked in the tests not only for responses with 50-year or even 500-year MRIs, but for longer MRIs as well.

The 60-story John Hancock building in Boston, which has an elongated shape in plan, exhibited large torsional motions that needed to be reduced through the provision near the top of the building of two tuned mass dampers, one near each end of the building's long dimension. We understand that structural measures were required as well. Apparently, for legal or other reasons, details concerning the building's aeroelastic response were not made public. It may be hypothesized, however, that the torsional oscillations originally experienced by the building were due largely to the eccentricity of vortex-induced forces with respect to the building's elastic center and, possibly, to inadequate bracing against torsion.

4.3.4 Serviceability Requirements

Wind-related serviceability issues are of concern in two areas: (1) building envelope performance under wind-induced deformations, and (2) occupant discomfort due to building motion. For the performance of the building envelope to be adequate, the *peak interstory drift* must not exceed 1/300 to 1/400 of the story height under unfactored loads, although this criterion may vary depending on type of cladding or glazing (brittle cladding would require smaller ratios) and cladding attachment details. In absolute terms, interstory

Figure 4.3.4. Taipei 101 building. [Courtesy of ROC (Taiwan) Government Information Office.]

drift should not exceed 10 mm unless special details allow nonstructural partitions, cladding, or glazing to accommodate larger drift. However, this criterion must also be qualified, depending on specific building features. Total drift ratios with a 20-year MRI should, according to some practitioners, not exceed 1/600 to 1/400 of the building height (see Sect. CB1.2, Commentary to ASCE 7-02 Standard, p. 366). Some practitioners believe that larger values may be acceptable.

Occupant comfort is affected by the visual perception of building oscillations, especially torsional oscillations, which can be particularly unsettling to occupants, and by sufficiently strong and frequent accelerations of the upper floors, which can produce effects not unlike those associated, in more extreme cases, to seasickness. It has been hypothesized that occupant comfort is af-

Figure 4.3.5. High-frequency force balance model of the Taipei 101 building in a wind tunnel. (Courtesy of Rowan Williams Davies & Irwin Inc., Guelph, Ontario, Canada.)

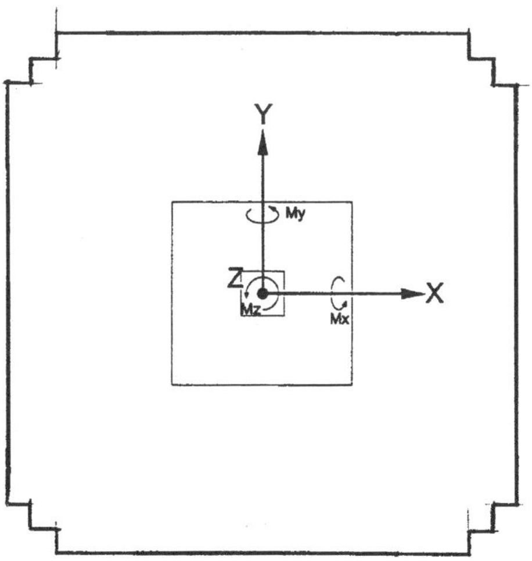

Figure 4.3.6. Cross section of the Taipei 101 building with chamfered corners (corner cuts). (Courtesy of Rowan Williams Davies & Irwin Inc., Guelph, Ontario, Canada.)

Figure 4.3.7. Endless column by Constantin Brancusi, Târgu-Jiu, Romania, erected in 1938 as a memorial to the Romanian soldiers fallen during World War I. (By permission of the World Monuments Fund. Copyright Kael Alford/World Monuments Fund, 2001.)

fected by rapid changes of acceleration (i.e., jerks), but to our knowledge, no criteria based on such changes have been developed so far. On the other hand, because occupant perception of accelerations is highly uncertain and has not been researched exhaustively, criteria on acceptable accelerations vary among codes and practitioners. For example, in North America proposed ranges of acceptable peak accelerations with 10-year MRIs at the top floors are 15 to 20 mg for offices and 10 to 15 mg for residential buildings. However, it has been determined that acceptable acceleration levels decrease as the oscillation frequency increases, so it has been suggested that these limits be reduced for higher frequencies of vibration, from the values just stated, which are assumed to be valid for frequencies of 0.1 Hz, to about half of those values for frequencies of 1 Hz. International Standards Organization (ISO) and Architectural Institute of Japan (AIJ) criteria similarly reflect a decrease in acceptable acceleration levels as the frequencies of vibration increase.

Compliance with comfort criteria can control structural design. To reduce the dynamic response, special damping systems are used, including various types of tuned mass dampers (an example is shown in Fig. 4.3.8), or liquid mass dampers, among others [4.30].

APPENDIX 4A: PRELIMINARY ESTIMATES OF FLEXIBLE BUILDING RESPONSE

The ASCE 7 Standard includes a procedure for calculating the along-wind response of flexible buildings in terrain with uniform roughness. However, it provides no indication on effects due to skew winds, or on across-wind effects, which typically exceed along-wind effects. Nor does it provide any indication of how to account for aerodynamic interference from nearby tall buildings. An alternative procedure for calculating along-wind response included in the National Building Code of Canada (NBC) [4.32] has the same limitations. However, unlike ASCE 7 Standard, [4.32] includes a procedure that yields estimates (albeit crude) of across-wind response.

Results yielded by ASCE 7, NBC, or similar generic procedures applied to flexible buildings, are only useful for preliminary design purposes. This is

Figure 4.3.8. Tuned mass damper used in the Taipei 101 building. (Courtesy of Motioneering, Inc., Guelph, Ontario, Canada.)

also true of generic estimates of tall building torsional response. Having stated these limitations clearly, we present in Sects. 4A.1, 4A.2, and 4A.3, respectively, simple procedures for generic estimates of along-wind, across-wind, and torsional response.

4A.1 Along-Wind Response

Procedures for calculating along-wind response incorporate a model of the wind profile upwind of the building and of the turbulence in the oncoming wind, a relation between the oncoming wind speeds and the along-wind fluctuating pressures on the building faces normal to the wind, and a method for the calculation of the dynamic response to these pressures. Internally, the procedures for calculating the fluctuating along-wind response make use of spectral methods. This implies (a) representing wind-induced fluctuating pressures at various points of the building faces by their *spectra* and *cross-spectra,* which are representations of large numbers of superposed sinusoidal components, each with a distinct frequency n; (b) relating these spectra and cross-spectra to those of the oncoming flow fluctuations; and (c) calculating the spectral densities, rms values, and peaks of the building responses by using the relation between the spectral densities of the pressures on the one hand and those of the response on the other. For additional material, see Sects. 2.3.3.1.3, 2.3.3.1.4, and 2.6.3. In all cases it is assumed that the wind direction is normal to a building face and that the terrain roughness upwind of the building is fairly uniform, with no significant obstacles near the building being designed.

Several procedures for calculating along-wind response have been proposed. For software that uses the logarithmic law description of the wind profile see http://www.nist.gov/wind. Section 4A.1.1 provides closed-form expressions for the along-wind response based on the logarithmic description of the wind profile and on the use of mean hourly wind speeds near the top of the building. The ASCE 7-02 procedure is a modified version of these expressions that uses the power law description of the wind profile specified in the Standard (based on the 3-s peak gust speed). According to [4.33], the use in the ASCE 7 version of a power law description in conjunction with 3-s peak gust speeds is a source of inconsistencies in the physical modeling of the wind flow that can introduce errors in the results. For this reason we do not recommend the use of the ASCE 7 procedure. Nor do we recommend the use of the National Building Code of Canada (NBC) procedure [4.32], which does not account for the imperfect correlation between windward and leeward pressures and for the dependence of wind spectra on height above ground.

4A.1.1 Closed-Form Expressions for the Along-Wind Response [4.34]. Consider a building with a rectangular shape in plan, in terrain with surface roughness length z_0 uniform over a sufficiently large distance upwind (see Sect. 3.2.2.3). The building has height H, width B (normal to the wind direction), and depth L (parallel to the wind direction), average pressure coefficient C_w on the windward face and C_l on the leeward face, average specific mass ρ_b, fundamental frequency n_1, and damping ratio ζ_1. The fundamental mode of vibration is assumed to be approximately linear. The mean hourly wind speed at elevation H is $\overline{V}(H)$. The following expressions may be used to calculate the peak along-wind deflection at elevation z:

$$x_{\text{pk}}(z) \approx \frac{0.5\rho u_*^2 C_D B z}{M_1(2\pi n_1)^2}\,[J + 3.75(\mathsf{B} + \mathsf{R})^{1/2}] \qquad (4\text{A}.1a)$$

where ρ is the specific mass of air and M_1 is the fundamental modal mass; if the building's average specific mass ρ_b is independent of elevation z, $M_1 = BLH\rho_b/3$),

$$u_* = \frac{\overline{V}(H)}{2.5 \ln (H/z_0)} \qquad (4\text{A}.1b)$$

$$J = 0.78Q^2 \qquad (4\text{A}.1c)$$

$$Q = 2 \ln \frac{H}{z_0} - 1 \qquad (4\text{A}.1d)$$

$$\mathsf{B} = \frac{6.71Q^2}{1 + 0.26(B/H)} \qquad (4\text{A}.1e)$$

$$\mathsf{R} = \frac{0.59Q^2 N_1^{-2/3}}{\zeta_1} \frac{C_{Df}^2}{C_D^2} \frac{\mathsf{C}(\eta_1)}{1 + 3.95N_1\,(B/H)} \qquad (4\text{A}.1f)$$

$$N_1 = \frac{n_1 H}{u_* Q} \qquad (4\text{A}.1g)$$

$$C_{Df}^2 = C_w^2 + C_\ell^2 + 2C_w C_\ell \mathsf{C}\,(\eta_2) \qquad C_D = C_w + C_\ell \qquad (4\text{A}.1h,i)$$

$$\mathsf{C}(\eta) = \frac{1}{\eta} - \frac{1 - e^{-2\eta}}{2\eta^2} \qquad (4\text{A}.1j)$$

$$\eta_1 = 3.55N_1 \qquad \eta_2 = \frac{12.32N_1\Delta}{H} \qquad (4\text{A}.1k)$$

Δ is the smallest of dimensions H, B, and L. The quantities J, B, and R are measures of the mean, quasi-static, and resonant response, respectively.

The peak acceleration at elevation z is

$$\ddot{x}_{\text{pk}}(z) \approx 4.0\,\frac{0.5\rho u_*^2 C_D B z}{M_1}\,\mathsf{R}^{1/2} \qquad (4\text{A}.2)$$

The ratio p between the along-wind response calculated by taking the dynamic amplification into account on the one hand and by neglecting it on the other can be written in the form

$$p \approx \frac{J + 3.75(\mathsf{B} + \mathsf{R})^{1/2}}{J + 3.75\mathsf{B}^{1/2}} \tag{4A.3}$$

If the fundamental modal shape is not a straight line, as was assumed in the development of Eqs. 4A.1, 4A.2, and 4A.3, corrections may be applied to the calculated deflections and accelerations to account for the nonlinearity of the modal shape [4.13, p. 340].

Numerical Example 4A.1. *Along-wind response estimated by Eqs. 4A.1 and 4A.2.* Consider a building with a square shape in plan and dimensions H = 656 ft and $B = L = 115$ ft, in terrain with uniform terrain roughness length $z_0 = 3.28$ ft (1 m) over an 8000 ft $\approx 12 \times 656$ ft distance upwind of the building. The fundamental modal shape is assumed to be linear, so Eqs. 4A.1 and 4A.2 are applicable. The fundamental frequency of vibration is assumed to be $n_1 = 0.175$ Hz and the damping ratio is $\zeta_1 = 0.01$. The average pressure coefficients are $C_w = 0.8$ and $C_l = 0.5$, so the drag coefficient is $C_D = C_w + C_l = 1.3$. The mean hourly wind speed is $\overline{V}(H) = 88.3$ mph $= 129.54$ ft/s. It is assumed that the specific mass of air and the average specific mass of the building are, respectively, $\rho = 0.00238$ slug/ft^3 (1.25 kg/m^3) and 0.381 slug/ft^3 (200 kg/m^3).

Equations 4A.1 yield $u_* = 9.78$ ft/s, $Q = 9.59$, $J = 71.83$, $\mathsf{B} = 590.9$, $N_1 = 1.22$, $\eta_1 = 4.35$, $\eta_2 = 2.63$, $\mathsf{C}(\eta_1) = 0.203$, $\mathsf{C}(\eta_2) = 0.308$, $C_{Df}^2 = 1.14$, $\mathsf{R} = 353$, and $x_{pk}(z) \approx 0.0024z$. For $z = H = 656$ ft, $x_{pk}(H) = 1.57$ ft, and $\ddot{x}_{pk}(H) \approx 0.76$ ft/s^2.

Numerical Example 4A.2. *Classification of a building as rigid or flexible.* For design purposes buildings are classified as rigid if their wind-induced resonant effects are negligible. For buildings with regular shapes in plan it is possible to determine whether those effects are negligible by performing an approximate dynamic analysis of the *along-wind response* (i.e., the building response in the wind direction induced by wind normal to a building face).

Consider a building with height $H = 150$ ft, flat roof, and a rectangular shape in plan (60 ft $\times$ 35 ft). Assume that the basic wind speed (i.e., the peak 3-s gust at 10 m above ground in terrain with exposure C) is $V = 100$ mph, that the building has exposure B in all directions, the fundamental natural frequency $n_1 = 0.9$ Hz, and the damping ratio $\zeta_1 = 0.02$. The aerodynamic coefficients are $C_D = 1.3$, $C_w = 0.8$, and $C_\ell = 0.5$. The wind is assumed to be normal to the long face of the building (i.e., to the dimension $B = 60$ ft).

First calculate the mean hourly wind speed at height H of the building. The mean hourly speed at 33 ft (10 m) above ground in exposure C (roughness length $z_{0\text{open}} = 0.066$ ft; see footnote b, Table 2.3.1) is $\overline{V}(z = 33 \text{ ft}) = 100/1.52 = 65.8$ mph $= 96.5$ ft/s (Sect. 2.3.1.2). The mean hourly wind speed at 150 ft over terrain with exposure B ($z_0 = 0.5$ ft; see footnote c, Table 2.3.1) is

$$\overline{V}(H = 150 \text{ ft}, z_0 = 0.5 \text{ ft}) = 96.5 \left(\frac{0.5}{0.066}\right)^{0.07} \frac{\ln(150/0.5)}{\ln(33/0.066)} = 102.0 \text{ ft/s}$$

(Eq. 2.3.6). We obtain from Eqs. 4A.1: $u_* = 6.77$ ft/s^2, $Q = 10.41$, $J = 84.49$, $\mathsf{B} = 658.65$, $N_1 = 1.92$, $\eta_1 = 6.80$, $\eta_2 = 5.51$, $\mathsf{C}(\eta_1) = 0.136$, $\mathsf{C}(\eta_2) = 0.165$, $C_{Df}^2 = 0.64 + 0.25 + 2 \times 0.8 \times 0.5 \times 0.165 = 1.02$, and $\mathsf{R} = 73.0$. From Eq. 4A.3, $p \approx 1.03$. It follows that if the building is designed as rigid, its response will be underestimated by 3%. Note that for this building $n_1 < 1$ Hz and the ratio $H/L = 150/35 = 4.39 > 4.0$.

4A.2 Across-Wind Response

Unlike the along-wind response, the across-wind response cannot be determined on the basis of theoretical representations of the wind-induced pressures. However, empirical expressions based on wind tunnel tests of generic building models have been proposed. The expressions are applicable provided that (a) no neighboring tall buildings strongly disturb the oncoming wind flow, and (b) for buildings with a square shape in plan the ratio between the calculated rms across-wind deflection σ_y at the top of the building is such that σ_y/D is less than about 0.015 for open terrain, 0.025 for suburban terrain, and 0.045 for centers of large cities, where D is the building depth. It is emphasized that the criteria pertaining to the ratio σ_y/D offer only a rough indication of tall building behavior.

The expressions for the peak deflections and accelerations at the elevation z of a building with height H are (dimensions in feet)

$$y_{\text{pk}}(z) = C\left[\frac{\overline{V}(H)}{n_1\sqrt{A}}\right]^p \frac{\sqrt{A}}{\zeta_1^{1/2}} \frac{\rho}{\rho_b} \frac{z}{H} \tag{4A.4}$$

$$\ddot{y}_{\text{pk}}(z) = (2\pi n_1)^2 y_{\text{pk}}(z) \tag{4A.5}$$

where A is the area of a horizontal section of the building, ρ is the specific mass of air, ρ_b is the mass of building per unit volume, and C and p are empirical constants. According to [4.32], $p = 3.3$. Although various authors suggest somewhat different values for the parameter C, $C = 0.00065$ is consistent with, for example, [4.32]. The rms responses may be assumed to be approximately four times smaller than the peak responses.

Numerical Example 4A.3. *Across-wind response estimation by Eqs. 4A.4 and 4A.5.* For the building, terrain, and wind speed of Numerical Example 4A.1, for $z = H$, Eqs. 4A.4 and 4A.5 yield

$$y_{\mathrm{pk}}(H) = 0.00065\left(\frac{129.54}{0.175 \times \sqrt{115 \times 115}}\right)^{3.3} \frac{115}{0.01^{1/2}} \frac{0.00238}{0.381} = 2.18 \text{ ft}$$

$$\ddot{y}_{\mathrm{pk}}(H) = 2.39 \text{ ft/s}^2$$

Note that $y_{\mathrm{pk}}/B = 2.18/115 = 0.019 < 0.025$, suggesting that Eqs. 4A.4 and 4A.5 are applicable and that the calculated across-wind response exceeds the calculated along-wind response.

Estimates of the across-wind response that may be more realistic than those yielded by Eqs. 4A.4 and 4A.5 may be obtained for a number of types of building in uniform terrain with various exposures by using methods described in [4.13, p. 345–351] and [4.35], which houses HFFB data publicly available, and a JAVA interface that automatically determines response quantities for a user-specified building. In general, such estimates should also be used only for preliminary design.

4A.3 Torsional Response

Like the across-wind response, the torsional response due to the eccentricity of the aerodynamic center with respect to the elastic center can only be estimated by using empirical expressions with parameters based on wind tunnel tests conducted on buildings with various shapes in terrain with uniform upwind roughness.[12] The peak base torsional moment may be written as

$$T_{\mathrm{pk}}[\overline{V}(H)] \approx 0.9\{\overline{T}[\overline{V}(H)] + 3.8T_{\mathrm{rms}}[\overline{V}(H)]\} \tag{4A.6}$$

where $\overline{T}$ is the mean base torsional moment and T_{rms} is the rms value of the fluctuating base torsional moment,

[12] This material is based on work performed at the University of Western Ontario kindly communicated by N. Isyumov, see [4.13], pp. 351–356 for details.

$$\overline{T}[\overline{V}(H)] = 0.038\rho L^4 H n_T^2 \left[\frac{\overline{V}(H)}{n_T L}\right]^2 \tag{4A.7}$$

$$T_{\text{rms}}[\overline{V}(H)] = 0.0017 \frac{1}{\zeta_T^{1/2}} \rho L^4 H n_T^2 \left[\frac{\overline{V}(H)}{n_T L}\right]^{2.68} \tag{4A.8}$$

Here n_T and ζ_T are the natural frequency and the damping ratio in the fundamental torsional mode of vibration and A is the area of the horizontal cross section of the building. The peak torsional moment is approximately

$$T_{\text{pk}} = 0.9(\overline{T} + 3.8T_{\text{rms}}) \tag{4A.9}$$

The peak horizontal acceleration induced by torsion at the top of the building at a distance v from the elastic center is

$$\ddot{\varphi}_{\text{pk}} v \approx \frac{7.6T_{\text{rms}} v}{\rho_b A H r_m^2} \tag{4A.10}$$

where $\ddot{\varphi}_{\text{pk}}$ is the peak angular acceleration and r_m is the radius of gyration of the horizontal cross section of the building.

For a building with a rectangular shape in plan (dimensions B and D),

$$L = \frac{2 \times 2 \times B^2/8}{\sqrt{BD}} + \frac{2 \times 2 \times D^2/8}{\sqrt{BD}} \tag{4A.11}$$

$$r_m = \frac{(B^2 + D^2)^{1/2}}{\sqrt{12}} \tag{4A.12}$$

Numerical Example 4A.4. *Torsional response of a tall building.* For the building, terrain, and wind speed of Numerical Examples 4A.1 and 4A.2, the natural frequency and the damping ratio in the fundamental torsional mode are assumed to be $n_T = 0.3$ Hz and $\zeta_T = 0.01$, respectively. Equations 4A.11 and 4A.12 yield $L = 115$ ft and $r_m = 46.95$ ft. We have $\overline{V}(H)/n_T L = 129.54/(0.3 \times 115) = 3.75$, $d = 115\sqrt{2}/2 = 81.3$ ft. Equations 4A.7 to 4A.10 yield

$$\begin{aligned} \overline{T}(129.54 \text{ ft/s}) &= 0.038 \times 0.00238 \times 115^4 \times 656 \times 0.3^2 \times 3.75^2 \\ &= 13{,}130 \text{ kip-ft} \end{aligned}$$

$$\begin{aligned} T_{\text{rms}}(129.54 \text{ ft/s}) &= 0.0017 \frac{1}{0.01^{1/2}} \times 0.00238 \times 115^4 \\ &\quad \times 656 \times 0.3^2 \times 3.75^2 = 14{,}400 \text{ kip-ft} \end{aligned}$$

$$T_{\text{pk}} = 0.9(13{,}130 + 3.8 \times 14{,}400) = 57{,}200 \text{ kip-ft}$$

$$\ddot{\varphi}_{\text{pk}} d \approx \frac{7.6 \times 14400 \times 10^3 \times 81.3}{0.381 \times 115 \times 115 \times 636 \times 46.95^2} = 1.26 \text{ ft/s}^2$$

REFERENCES

4.1. *Minimum Design Loads for Buildings and Other Structures,* ASCE Standard 7-05, American Society of Civil Engineers, Reston, VA, 2005.

4.2. *Eurocode 1: Actions on Structures—Part 1–4: General Actions—Wind Actions,* CEN TC 250, Draft 2002-06, 2002.

4.3. *Loading for Buildings,* Part 2, *Code of Practice for Wind Loads,* British Standard BS 6399, British Standards Institution, London, 1997.

4.4. *Structural Design Actions—Wind Actions—Commentary,* AS/NZ 1170.2 Supplement 1:2002, Standards Australia International, Sydney, and Standards New Zealand, Wellington, 2002.

4.5. E. Simiu, F. Sadek, T. M. Whalen, S. Jang, L.-W. Lu, S. M. C. Diniz, A. Grazini, and M. A. Riley, "Achieving safer and more economical buildings through database-assisted, reliability-based design for wind," *J. Wind Eng. Ind. Aerodyam.,* **91,** 1587–1611, 2003.

4.6. F. Sadek and E. Simiu, "Peak non-Gaussian wind effects for database-assisted design of low-rise buildings for wind loads," *J. Eng. Mech.,* **128,** 530–539, 2002.

4.7. T. M. Whalen, E. Simiu, G. Harris, J. Lin, and D. Surry, "The use of aerodynamic databases for the effective estimation of wind effects in main wind-force resisting systems: application to low buildings," *J. Wind Eng. Ind. Aerodynam.,* **77–78,** 685–692, 1998.

4.8. F. Sadek, S. Diniz, M. Kasperski, M. Gioffrè, and E. Simiu, "Sampling errors in the estimation of peak wind-induced internal forces on low-rise structures," *J. Eng. Mech.,* **130,** 235–239, 2004.

4.9. J. Main, "Interpolation issues in the estimation of peak internal forces using wind pressure databases," *Proceedings of the 6th Asia-Pacific Conference on Wind Engineering,* Seoul, Korea, C.-K., Choi, ed., 2005.

4.10. A. Rigato, P. Chang, and E. Simiu, "Database-assisted design, standardization, and wind direction effects," *J. Struct. Eng.,* **127,** 855–860, 2001.

4.11. *Wind Tunnel Studies of Buildings and Structures,* ASCE Manuals and Reports on Engineering Practice 67, American Society of Civil Engineers, Reston, VA, 1999.

4.12. N. Isyumov et al., "Predictions of wind loads and responses from simulated tropical storm passages," *Proceedings of the 11th International Conference on Wind Engineering,* Lubbock, TX, D. A. Smith and C. W. Letchford, eds., June 2–5, 2003.

4.13. E. Simiu and R. H. Scanlan, *Wind Effects on Structures,* 3rd ed., Wiley, New York, 1996.

4.14. E. Simiu and J. J. Filliben, "Wind tunnel testing and the sector-by-sector approach," *J. Struct. Eng.,* **131,** 1143–1145, 2005.

4.15. S. Coles and E. Simiu, "Estimating uncertainty in the extreme value analysis of data generated by a hurricane simulation model," *J. Eng. Mech.,* **129,** 1288–1294, Nov. 2003.

4.16. S. Diniz and E. Simiu, "Probabilistic descriptions of wind effects and wind load factors for database-assisted design," *J. Struct. Eng.,* **131,** 507–516, 2005.

4.17. B. Ellingwood et al., *Development of a Probability Based Criterion for American National Standard A58,* NBS Special Publication 577, National Bureau of Standards, Washington, DC, 1980.

4.18. P. Hanzlik, S. Diniz, A. Grazini, M. Grigoriu, and E. Simiu, "Building orientation and wind effects estimation," *J. Struct. Eng.,* **131,** 254–258, 2005.

4.19. S. Jang, L.-W. Lu, F. Sadek, and E. Simiu, "Database-assisted wind load capacity estimates for low-rise steel frames," *J. Struct. Eng.,* **128,** 1594–1603, 2002.

4.20. N. Lin, C. Letchford, Y. Tamura, B. Liang, and O. Nakamura, "Characteristics of wind forces acting on tall buildings," *J. Wind Eng. Ind. Aerodynam.,* **93,** 217–242, 2005.

4.21. Y. Tamura, K. Suda, and A. Sasaki, "Damping in buildings for wind-resistant designs," Invited Lecture, *Proceedings of the International Symposium on Wind and Structures for the 21st Century,* Cheju, Korea, 2005.

4.22. I. Venanzi, W. P. Fritz, and E. Simiu, "Structural design for wind of tall buildings with noncoincident mass and elastic centers: a database-assisted design approach," *Proceedings of the 6th Asia-Pacific Conference on Wind Engineering,* Seoul, Korea, C.-K. Choi, ed., 2005.

4.23. American Institute of Steel Construction, *Manual of Steel Construction: Load and Resistance Factor Design,* Vol. I, 3rd ed., AISC, New York, 2001.

4.24. A. G. Davenport, "Note on the distribution of the largest value of a random function with application to gust loading," *J. Inst. Civ. Eng.,* **24,** 187–196, 1964.

4.25. H. W. Liepmann, "On the application of statistical concepts to the buffeting problem," *J. Atmos. Sci.,* **19,** 793–800, 822, 1952.

4.26. *Wind Tunnel Testing: A General Outline,* Boundary Layer Wind Tunnel Laboratory, University of Western Ontario, London, Ontario, Canada, 1999.

4.27. Y. Zhou, and A. Kareem, "Aeroelastic balance," *J. Eng. Mech.,* **129,** 253–292, 2003.

4.28. C. S. Kwok and W. H. Melbourne, "Wind-induced lock-in excitation of tall structures," *J. Struct. Div., ASCE,* **107,** 57–72, 1981.

4.29. H. Kawai, "Vortex-induced vibration of tall buildings," *J. Wind Eng. Ind. Aerodynam.,* **41–44,** 117–128, 1992.

4.30. A. Kareem, T. Kijewski, and Y. Tamura, "Mitigation of motions of tall buildings with specific examples of recent applications," *Wind Struct.,* **2,** 201–251, 1995; http://www.nd.edu/~nathaz/journals/(1999) Mitigation_of_Motion_of_Tall_Buildings_with_recent_Applications.pdf.

4.31. P. A. Irwin, personal communication, 2005.

4.32. *NRC-CNRC User's Guide NBC 1995,* Structural Commentaries, Part 4, Institute for Research in Construction, Ottawa, 1995.

4.33. E. Simiu, F. Sadek, and M. Riley, "Discussion of 'Definition of wind profiles in ASCE 7,'" *J. Struct. Eng.,* **129,** 1564, 2003.

4.34. G. Solari, "Along-wind response estimation: closed form solution," *J. Struct. Div., ASCE,* **108,** 225–244, 1982.

4.35. Y. Zhou, T. Kijewski, and A. Kareem, "Aerodynamic loads on tall buildings: interactive database," *J. Struct. Eng.,* **129,** 394–404, 2003; see also http://aerodata.ce.nd.edu/.

4.36. Y. Tamura, personal communication, 2005.

4.37. D. Duthinh, "Nonlinear database-assisted design of low-rise buildings for wind," *Proceedings of the 6th Asia-Pacific Conference on Wind Engineering,* Seoul, Korea, C.-K. Choi, ed., 2005.

4.38. Y. Chen, G. A. Kopp, and D. Surry, "Interpolation of Pressure Time Series in an Aerodynamic Database for Low Buildings," *J. Wind Eng. Ind. Aerodynam.* **91,** 737–765, 2003.

4.39. F. Masters and K. Gurley, "Multivariate Stochastic Simulation of Wind Pressure over Low-rise Structures through Linear Model Interpolation," *Proceedings, 10th Americas Conference on Wind Engineering* (M. Levitan et al., eds.), Baton Rouge, LA, June 1–4, 2005.

4.40. T. Stathopoulos, personal communication, 2005.

4.41 W. Fritz, M. Iancovici, and E. Simiu, "Applicability of database-assisted design to tall buildings," *Proceedings of the 8th International Conference on Flow-Induced Vibrations,* Paris, 6–9, July 2004.

CHAPTER 5

BRIDGES

5.1 INTRODUCTION

Certain types of medium- and long-span bridges are subjected to aerodynamic forces generated by structural motions. These motions, called *self-excited,* are in turn affected by the aerodynamic forces they generate. Behavior associated with self-excited motions is referred to as *aeroelastic* (see Sect. 2.7).

Aeroelastic phenomena of interest in bridge engineering differ from those studied in aeronautical engineering in three ways [5.1–5.3].[1] First, bridge structures are typically *bluff,* although in modern suspension and cable-stayed bridges streamlined boxlike deck shapes are being used increasingly. Second, bridge decks have *appurtenances* such as handrails, curbs, parapets, center barriers, grillages, and so on, that can significantly affect the flow and the aeroelastic forces. Third, unlike flows typically considered in aeronautical engineering, the flows in which civil engineering structures are immersed are in most cases *turbulent.* Atmospheric turbulence depends on the thermal stratification of the flow. At very high wind speeds mechanical turbulence is dominant and the airflow may therefore be assumed to be neutrally stratified. Standard atmospheric models commonly used in wind engineering are applicable in this case. However, atmospheric flows are not necessarily neutrally stratified, even at relatively high wind speeds. The actual flow turbulence may therefore differ substantially from, and in some cases be considerably weaker than, the turbulence inherent in standard models. This may affect bridge be-

[1] Additional references that could not be included in this chapter owing to space limitations are available in [5.1–5.3].

havior unfavorably, particularly under vortex-induced excitation, which is typically stronger in smoother flow.

Given the limitations of analytical and numerical procedures, the aeroelastic characterization of bridge structures relies largely on laboratory testing and empirical modeling. Such testing is not always without its problems, however. Wind tunnels that achieve Reynolds numbers comparable to those typical of most types of civil engineering structures (e.g., high-pressure wind tunnels) are currently not capable of simulating atmospheric turbulence, which can significantly affect bridge aerodynamic and aeroelastic behavior. On the other hand, wind tunnels that attempt to simulate the features of atmospheric turbulence usually violate Reynolds number similarity requirements by factors on the order of 100 to 1000. Nevertheless, for most structures with sharp edges at which flow separation must occur both in the prototype and the model, and for properly modeled bodies with rounded shapes, ribbed surfaces, and corners with steplike shapes, it is commonly assumed that the violation of Reynolds number similarity is relatively inconsequential.

Short-span bridges usually consist mostly of simply supported or continuous girders, or trusses, for which wind effects are minor. Arches, prestressed concrete girders, and cable-stayed spans are commonly used for medium-span bridges. For long spans, suspension bridges are typical. A useful, though rough indicator of bridge flexibility as a function of span L (meters) is given by the expression

$$n_{1v} = \frac{100}{L} \tag{5.1.1}$$

where n_{1v} (Hz) is the fundamental frequency of vibration in vertical bending (heaving) [5.1].

Following this introduction, Sect. 5.2 is concerned with the description of aerodynamic and aeroelastic phenomena affecting bridge and cable behavior. To the extent allowed by the state of the art, aeroelastic behavior (vortex-induced oscillations, galloping, and flutter) is explained qualitatively by the nature of the flows around bluff bodies. Methods of aeroelastic analysis are presented in some detail. In Sect. 5.3 we briefly review a number of design rules and code provisions on the design of short-, medium-, and long-span bridges and provide information on the behavior, design, and testing of actual bridges. We present details on the design of the Akashi Kaikyo Bridge and on the dampers required for its main towers; discuss in a broader context aerodynamic and mechanical approaches to reducing wind-induced vibrations of towers and decks; address the topic of special aerodynamic effects, including parallel-deck and proximity effects, built environment effects, and topographic effects on bridge aerodynamics; discuss wind shielding; present material on suppressing cable vibrations; provide details on wind tunnel testing, with special reference to the Akashi Kaikyo Bridge; and comment briefly on computational fluid dynamics applications to bridges. In Appendix 5A we provide details on specifications governing the design of cable-stayed and suspension bridges in Japan, Appendix 5B is concerned with the three-dimensional analysis of coupled flutter, and Appendix 5C covers a procedure for estimating flutter velocity as well as buffeting effects.

5.2 WIND EFFECTS ON BRIDGES: BASIC FLUID MECHANICS AND AEROELASTIC FORCES AND MOTIONS

In this section we descripe aerodynamic and aeroelastic phenomena affecting bridge and cable behavior. To the extent allowed by the state of the art, aeroelastic behavior (vortex-induced oscillations, galloping, and flutter) is explained qualitatively by the nature of the flows around bluff bodies, in particular bluff bridge decks with various afterbody configurations. Methods of aeroelastic analysis are presented, and wind effects on cables are discussed in some detail.

5.2.1 Separated Flows Around Bridge Decks: Reynolds and Strouhal Numbers

Unlike streamlined aircraft wings, bridge structures typically have bluff cross sections around which the flow exhibits separation (Sect. 2.5.2) and vorticity shed in the body wake (Sect. 2.5.3). A schematic representation of separated flow and vorticity around a bridge is shown in Fig. 5.2.1.

The flow around a bluff body depends on the Reynolds number ($\text{Re} = VD/\nu$; V is the wind speed, D is the characteristic dimension of the body, and ν is the kinematic viscosity; see Sect. 2.5.1.2, Eq. 2.5.4*b*). Typical orders of magnitude of Reynolds numbers for bridges are for a stay cable with circular cross section and diameter $D = 0.15$ m, $V = 10$ m/s, $\text{Re} \approx 10^5$; for a deck with height $D = 1.5$ m, $V = 10$ m/s, $\text{Re} \approx 10^6$; and for a deck height $D = 5$ m, $V = 30$ m/s, $\text{Re} \approx 10^7$. For large prototype values of Re, the violation of the Reynolds number similarity criterion in the wind tunnel is unavoidable unless pressurized air tunnels are used.

Separation of flow over bodies with curved shapes depends on Re. For bodies with sharp corners, separation occurs at fixed points regardless of the value of Re (see, e.g., the windward bottom corner of the bridge in Fig. 5.2.1). For this reason it is commonly assumed that the flow in bodies with sharp corners is largely independent of Re. However, a certain distortion of the flow around the body can arise owing to viscous effects in the range of Reynolds numbers of about 500 to 2000 or less, as defined by the local scale. Examples are separated flows around handrails, parapets, or towers whose corners are chamfered in a steplike shape (Fig. 5.2.2) around which interferences of locally separated vortices can occur [5.4–5.6]. In such cases, to avoid serious discrepancies between the prototype and the wind tunnel model, local dimen-

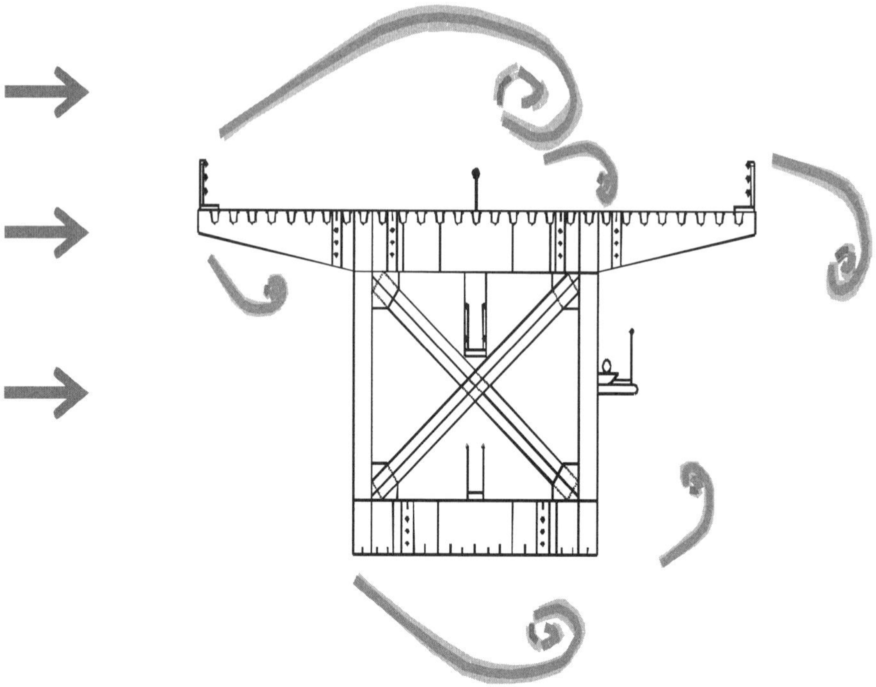

Figure 5.2.1. Vortex separation from sharp corners and protuberances of a typical bridge box deck.

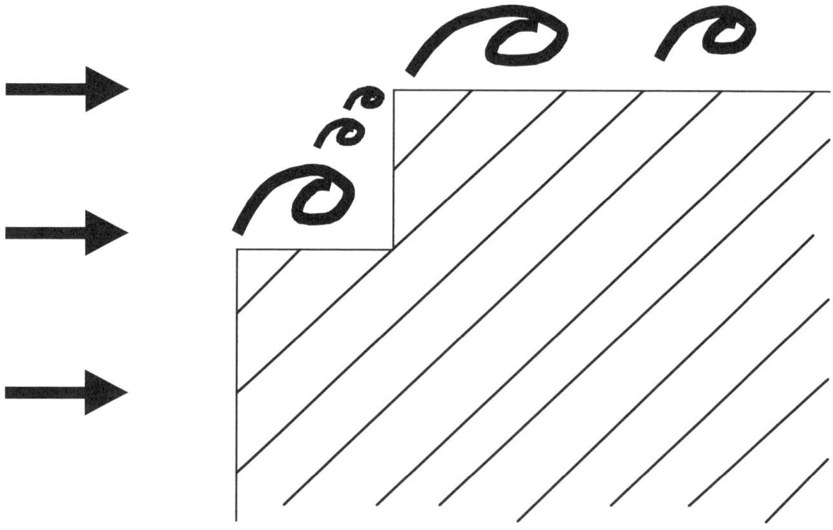

Figure 5.2.2. Interference of separated vortices in a sharp corner-cut section with stepwise configuration.

sions may have to be exaggerated in the model. An additional problem is that the small-scale components of the turbulence in the oncoming flow are affected by the Reynolds number. Since small-scale turbulence plays a role in determining where flow reattachment occurs (Sect. 2.5.2), the violation of the Reynolds number can affect bridge aerodynamics. Finally, large-scale (low-frequency) turbulent fluctuations are poorly simulated in the wind tunnel. This also can affect some simulations unfavorably (Sect. 5.2.6.4.1).

Unless Re is small, the flow around a bluff body experiences separation and forms an alternating von Kármán vortex street in the wake (Figs. 2.5.5 and 2.5.6). Vortex excitations and vortex-induced vibrations are due to alternating von Kármán vortices. For sufficiently large Re, flows in separation zones undergo strong turbulent mixing rather than being nearly laminar (see, e.g., Fig. 2.5.4*a*). The vorticity is therefore not clear and neat as in Figs. 2.5.5 and 2.5.6. Nevertheless, a dominant vortex shedding frequency can be defined.

It is an experimental fact that the vortex-shedding frequency n for a fixed body (i.e., a body at rest, with no significant oscillations) is proportional to the wind speed V and inversely proportional to the characteristic body dimension D. The proportionality constant St is called the *Strouhal number* (Sect. 2.5.3.1) [5.7, 5.8]. Strouhal numbers are shown in Fig. 5.2.3 for rectangular cylinders at rest with dimensions D and B normal and parallel to the wind direction, respectively. The jumps of the Strouhal number at $B/D \approx 2.8$ and $B/D \approx 6.0$ correspond to the reattachment of separated vortices to the surface of the cylinders. In the region $2.8 < B/D < 6$ the reattachment is

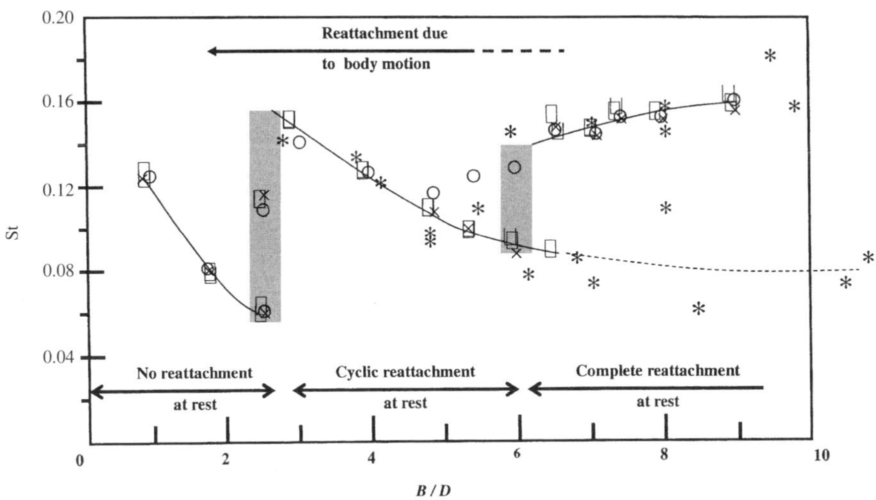

Figure 5.2.3. Strouhal number St for rectangular cylinders as a function of width-to-height ratio B/D and for bridge decks (based on wind speed inducing peak vortex excitation). Asterisks denote bridge decks. All other symbols denote rectangular cylinders. Indicatios on reattachment apply only to rectangular cylinders. (From [5.5].)

periodic. For $B/D > 6$ the flow is at all times fully reattached. Figure 5.2.3 also shows for a number of typical bridge prototype and model decks, Strouhal numbers obtained from observations of wind speeds inducing peak vortex-induced response. It is noteworthy that the Strouhal numbers of the bridge decks are in most cases similar to those of the rectangular cylinders.

If a body vibrates at its natural frequency under excitation by vorticity, the actual vortex-shedding frequency, which would normally be governed by the Strouhal number, may change within some relatively narrow limits, to become equal to the body's natural frequency. This synchronization effect, which is common to many nonlinear phenomena, is called *lock-in* (Sect. 2.7.1.1). Also, the along-span coherence of the shed vortices is greater in the presence of vortex-induced oscillations than for a fixed body (Sect. 2.7.1.2). Both phenomena are of an aeroelastic nature: The body motions influence the flow, and conversely.

5.2.2 Aerodynamics of Fixed Bodies

In this section we consider the case of fixed bodies (bodies at rest) and discuss pressure distributions and aerodynamic forces and coefficients.

5.2.2.1 Pressure Distributions. The flow around a bluff body is characterized by the presence and downwind motion of separated vortices and depends on the body configuration and the nature of the oncoming flow. Consequently, so do the pressure distribution and the resulting wind forces on the body. Consideration of basic pressure distributions for bodies at rest affords useful insights into the aeroelastic behavior of oscillating bodies.

Figure 5.2.4 shows mean pressure coefficients for two rectangular cylinders with ratios $B/D = 2$ and $B/D = 3$ and for a typical box bridge deck with overhangs on both sides of the deck. The cylinders and the deck are at rest, and the angle of attack (i.e., the angle of the mean flow velocity with respect to the horizontal) is $\alpha = 0°$. The pressure on the windward surface is *positive,* and its peak is equal to the dynamic pressure $\frac{1}{2}\rho V^2$ at the center. The pressures on the upper and lower surfaces and on the leeward surfaces are *negative*. The positive pressures on the windward side and the negative pressures on the leeward side result in a drag force. The pressures fluctuate in time, owing to the unsteady motion of the vortices. If the vortex shedding is regular, the forces acting on the body are periodic.

For a rectangular cylinder with ratio $B/D = 2$ and for a box deck, in smooth flow the pressure distributions on the upper and lower surfaces are nearly uniformly distributed, whereas in flow with turbulence intensity $I_u \approx 10\%$, the negative pressures become weaker toward the leeward side of the body. This is ascribed to the effect of the turbulence on the separation layers, which are brought increasingly close to the upper and lower surfaces by the flow turbulence.

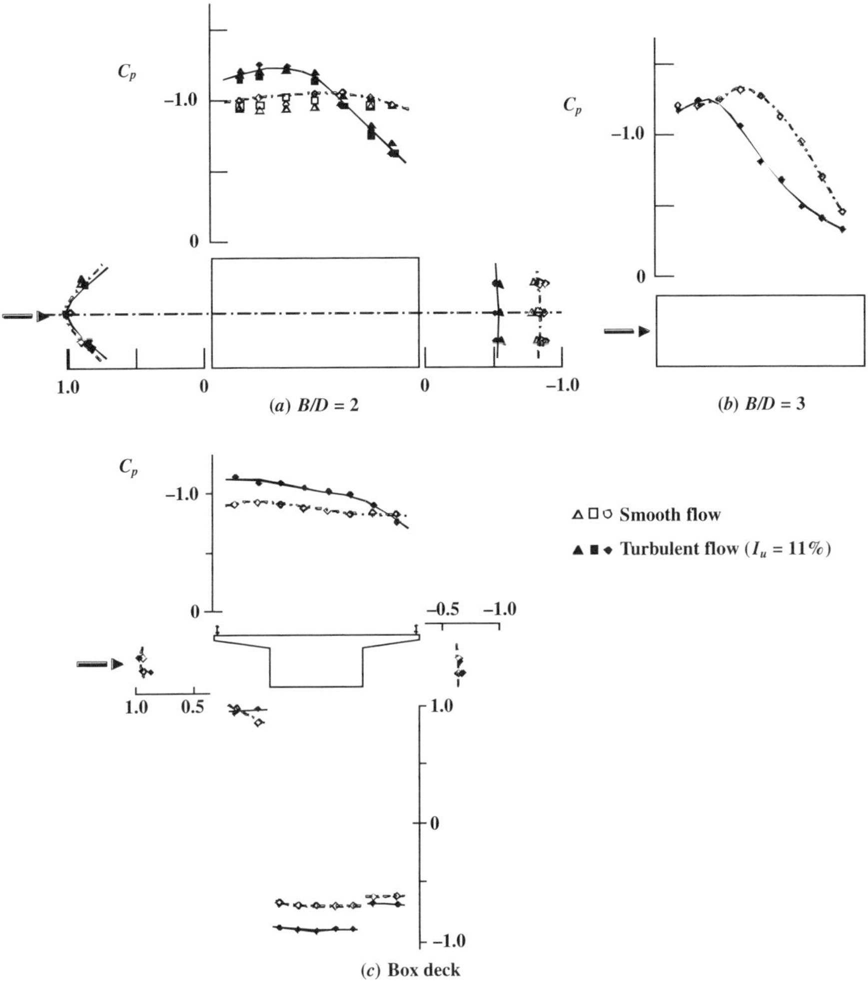

Figure 5.2.4. Mean pressure distribution for rectangular cylinders and box bridge decks at rest. (From [5.9].)

This example shows that turbulence can have a significant effect on deck aerodynamics. For a cylinder with ratio $B/D = 3$, that is, a cylinder with a longer along-wind dimension than one with $B/D \approx 2.8$ (the cutoff B/D ratio at which reattachment first occurs; see Fig. 5.2.3), a significant weakening of the negative pressures occurs toward the leeward side. This is the case in both smooth and turbulent flow [5.11].

5.2.2.2 Aerodynamic Forces and Force Coefficients. The aerodynamic pressures on a two-dimensional rectangular cylinder produce a drag force D parallel to the wind speed, a lift force L normal to the drag, and a torsional (pitching) moment M about the elastic center. The resulting forces and moment are expressed in terms of the dynamic pressure and the projected area A per unit span of the body as follows:

$$D = \tfrac{1}{2}\rho V^2 A C_D \tag{5.2.1a}$$

$$L = \tfrac{1}{2}\rho V^2 B C_L \tag{5.2.1b}$$

$$M = \tfrac{1}{2}\rho V^2 B^2 C_M \tag{5.2.1c}$$

where C_D, C_L and C_M are the nondimensional drag, lift, and moment coefficient, respectively.

For a circular cylinder with a smooth surface, the variation of the drag coefficient C_D with Re is shown in Fig. 5.2.5*b*. Like the variation of S (Fig. 5.2.5*a*), the variation of C_D reflects the dependence of the pressures on Re. In particular, C_D changes drastically from 1.0 to 0.22 and St changes from 0.2 to 0.5 in the *critical Reynolds number range* (Re $= 2.8 \times 10^5$ to 3.5×10^5) [5.10]. That range moves toward lower values of Re as the surface roughness increases (Fig. 5.2.5*c*) [5.12]. This is an important consideration for bridge cables (Sect. 5.2.7).

The drag coefficient C_D is shown in Fig. 5.2.6*a* for rectangular and H-shaped cylinders with various ratios B/D in smooth flow, and in Fig. 5.2.6*b* for rectangular cylinders in smooth and turbulent flows with various turbulence intensities. The peak drag coefficient C_D in smooth flow occurs for $B/D = 0.66$, for which the vortices shed in the wake cause the most intense base pressure (i.e., the lowest pressure on the leeward side of the cylinder). For turbulent flow the peak drag shifts to $B/D \approx 0.57$ [5.7, 5.8, 5.13, 5.14]. From Fig. 5.2.6*b* it may be inferred that the drag force on shallow bodies is smaller in turbulent flow than in smooth flow. For deep bodies the opposite is true. For a suspension bridge truss girder, $C_D \approx 2$; for a shallow streamlined box deck, owing to the absence or the weakness of separation effects, $C_D \approx 1$. Usually, bridge deck sections have ratios $B/D > 3$.

Figure 5.2.7 shows the three components of the forces induced by wind normal to the bridge on basic rectangular and H-shaped cylinders with $B/D = 2$ and $B/D = 5$, and for three bridge decks, as functions of various angles of attack α (i.e., of the angle between the wind velocity and its horizontal projection on the upper surface of the deck) [5.1, 5.13, 5.14]. For $\alpha \approx 0°$, $dC_M/d\alpha < 0$ for all the shapes of Fig. 5.2.7*a–d*; $dC_L/d\alpha$ is negative for $B/D = 2$ and positive for $B/D = 5$. For the shallow bridge box (Fig. 5.2.7*e*) and the truss girder of the Akashi Kaikyo bridge (Fig. 5.2.7*g*), both $dC_M/d\alpha$ and $dC_L/d\alpha$ are positive. The signs of $dC_M/d\alpha$ and $dC_L/d\alpha$ are aerodynamic properties of interest from an aeroelastic behavior point of view, as will be seen subsequently.

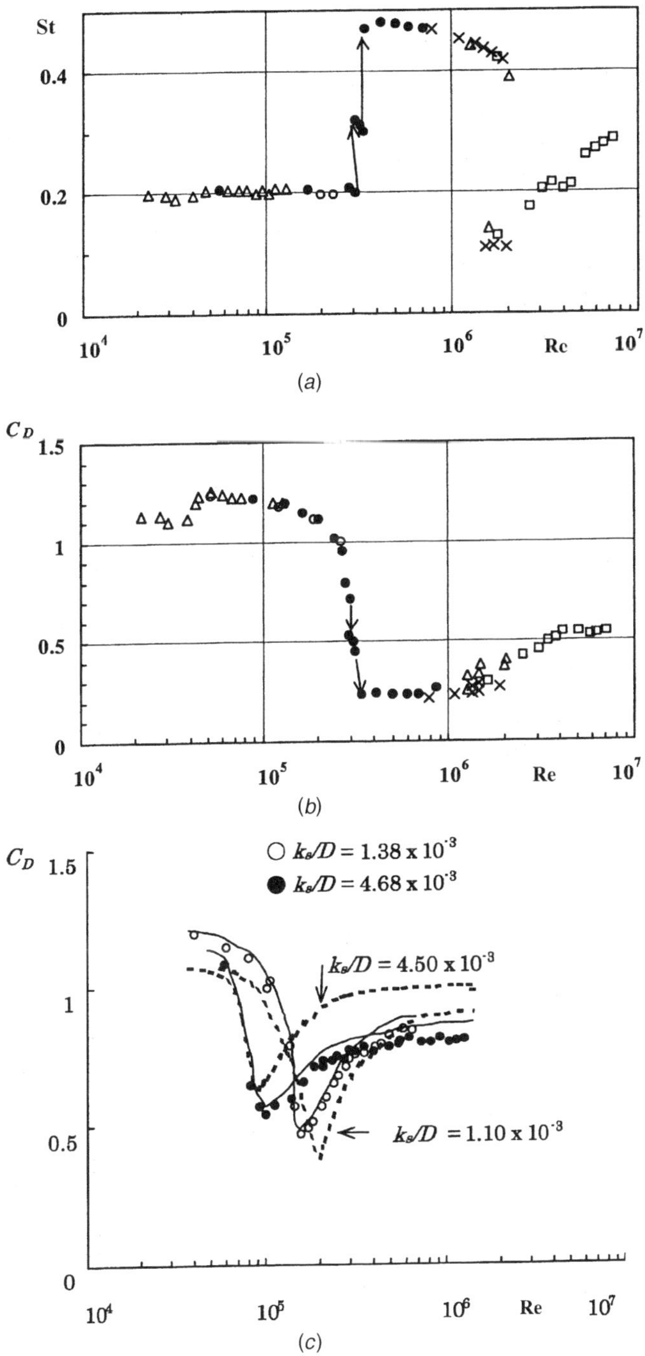

Figure 5.2.5. Strouhal number St and drag coefficient C_D for a circular cylinder vs. Reynolds number Re: (*a*) change of St at and near critical Reynolds number; (*b*) change of C_D at and near critical Reynolds number; (*c*) effect of surface roughness on the critical Reynolds number. [(*a, b*) From [5.10], courtesy of Cambridge University Press; (*c*) from [5.12]).

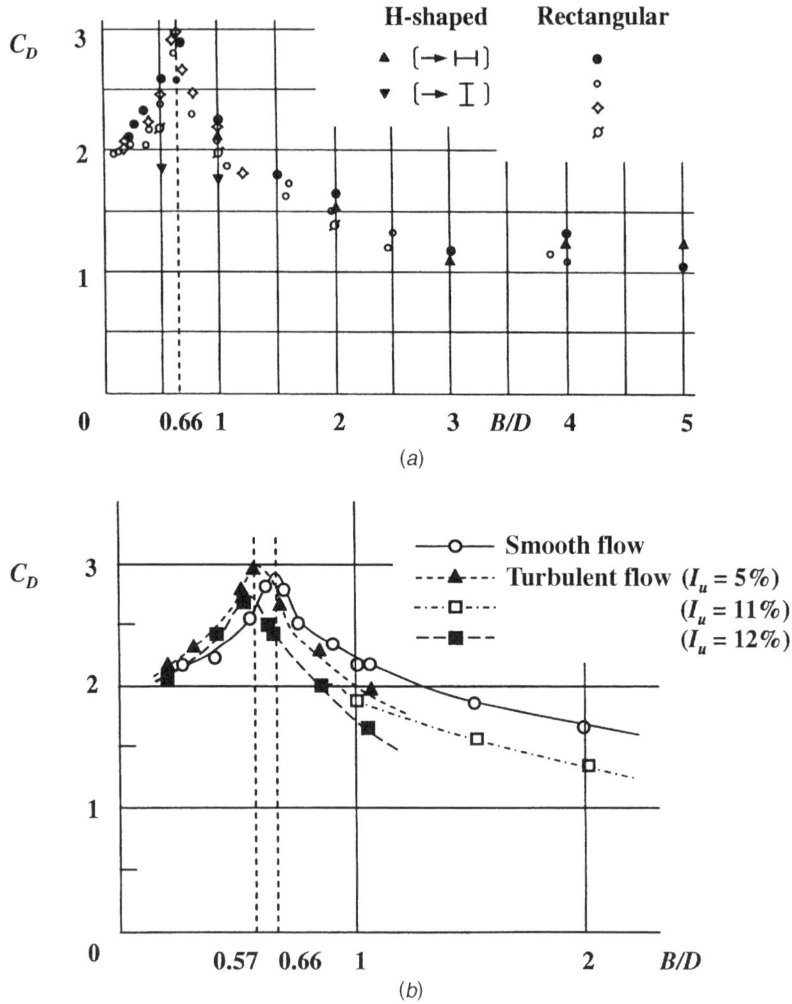

Figure 5.2.6. (*a*) Drag coefficient C_D of rectangular and H-shaped cylinders at $\alpha = 0°$ in smooth flow; (*b*) turbulence effects on drag of rectangular cylinders. (From [5.13, 5.14].)

5.2.3 Aerodynamics of Vibrating Bridges

5.2.3.1 Pressure Distributions. Flexible bridges experience wind-induced vibrations that cause the relative velocity of the flow with respect to the structure to change. Since pressures depend on the relative velocity of the flow with respect to the structure, the vibrations induce changes in the pressures as well. For every interval equal to one-eighth of the period of vibration, Figure 5.2.8*a* represents the distribution of pressure coefficients on the upper and lower surfaces of a structure experiencing harmonic vertical (heaving) vibrations. In Fig. 5.2.8*a* the structure is a rectangular cylinder with $B/D = 2$, the nondimensional vibration amplitude is $A/D = 0.1$, and the angle of

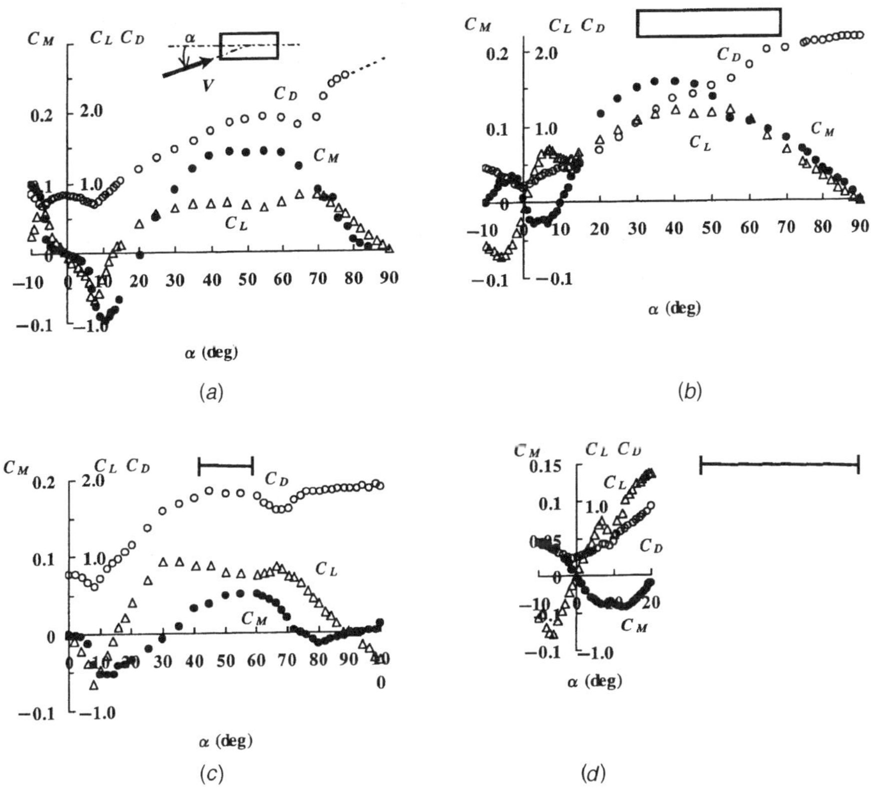

Figure 5.2.7. Static lift, drag, and moment: (*a*) rectangular cylinder, $B/D = 2$; (*b*) rectangular cylinder, $B/D = 5$; (*c*) H-shaped cylinder, $B/D = 2$; (*d*) H-shaped cylinder, $B/D = 5$; (*e*) shallow box bridge deck; (*f*) deep box bridge deck; (*g*) truss girder deck, Akashi Kaikyo Bridge. [(*a*)–(*d*) From [5.13, 5.14]; (*e*)–(*g*) courtesy of the Honshu–Shikoku Bridge Authority.]

attack is $\alpha = 0°$. Pressure distributions are shown for six values of the nondimensional ratio

$$V_r = \frac{V}{nD} \tag{5.2.2a}$$

where V denotes the wind velocity, D is a characteristic dimension, taken in this case to be the vertical dimension of the cylinder or deck, and n is the frequency of vibration. Equation 5.2.2*a* defines the *reduced velocity* V_r. In Fig. 5.2.8*b* the structure is a box deck with overhangs on both sides of the deck, the angle of attack is $\alpha = 5°$, and the pressure distributions are shown for two reduced wind velocities. Negative pressure coefficients designate upward pressures on the upper surface of the deck and downward pressures on the lower surface.

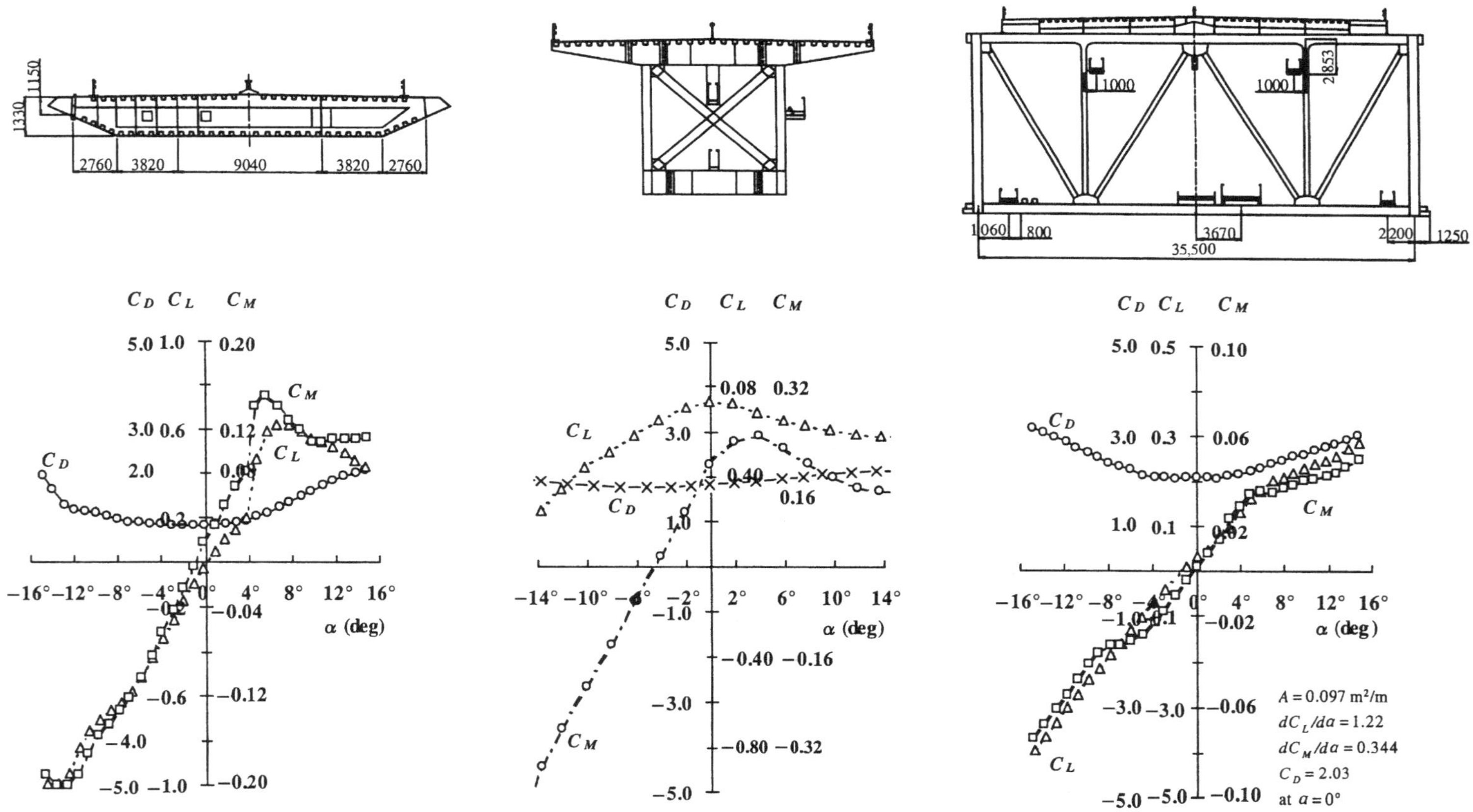

(e) (f) (g)

Figure 5.2.7. *(Continued)*

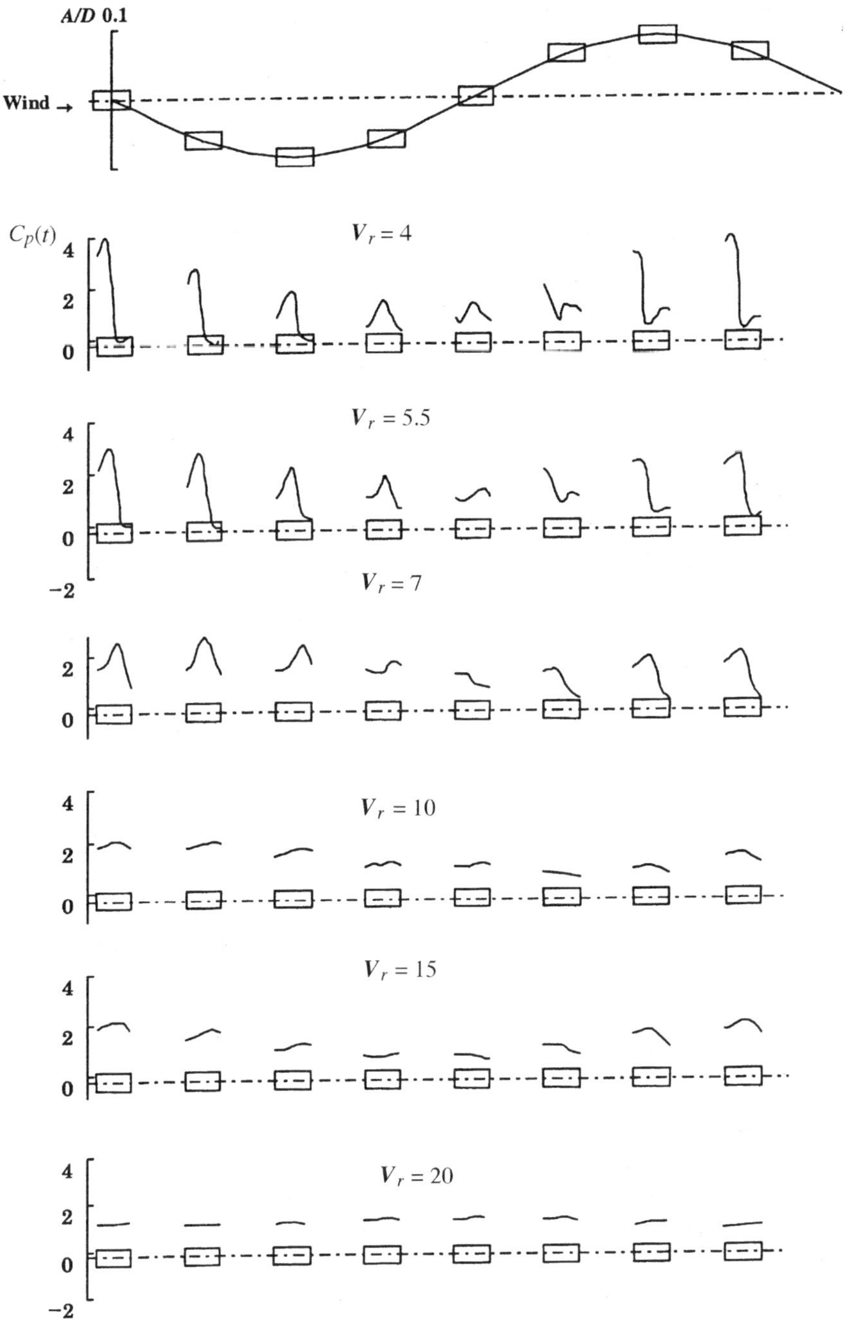

(*a*)

Figure 5.2.8. Instantaneous pressure distribution on upper and lower surfaces of bluff bodies in harmonic motion: (*a*) rectangular cylinder, $B/D = 2$; (*b*) box bridge deck. (From [5.9].)

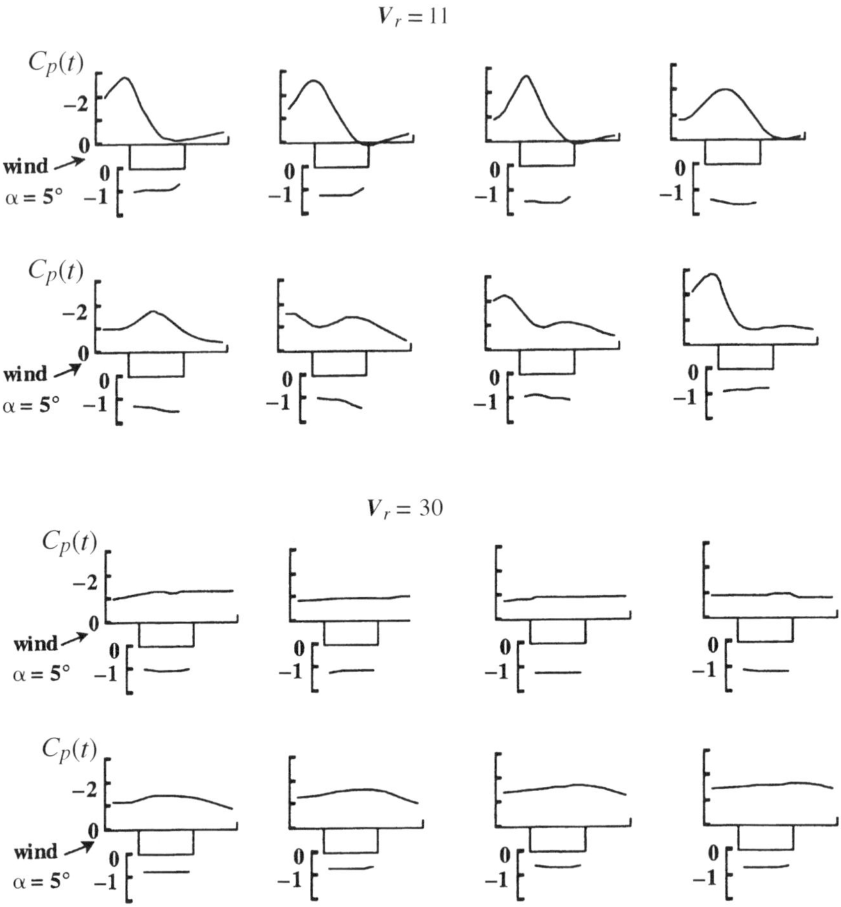

(b)

Figure 5.2.8. (*Continued*)

An interesting physical interpretation of the reduced wind speed is possible if Eq. 5.2.2*a* is rewritten as follows:

$$V_r = \frac{V}{nD} = \frac{1/n}{B/V}\frac{B}{D} = \frac{T_s}{T_v}\frac{B}{D} \tag{5.2.2b}$$

Equation 5.2.2*b* involves the period of vibration of the deck T_s and the time T_v required for the separated vortex to travel over the width (the along-wind dimension) of the bridge structure. Equation 5.2.2*b* is therefore a relation between the behavior of the separated flow and the vibration response. Small reduced velocities (large T_v) correspond to motions of the separated flow that are slow in relation to the period of vibration and to pressures that are nonuniform both in time and space (Fig. 5.2.8). Large reduced velocities correspond to fast motions of the separated flow (small T_v), which tend to produce distributions of instantaneous pressures over both surfaces of the vibrating deck that are relatively uniform in time and space.

For the cylinder with ratio $B/D = 2$ (Fig. 5.2.8*a*), a negative pressure peak occurs at smaller V_r. This is caused by the whirlpool of the vortex that forms downwind of the front edge. Owing to flow reattachment, a positive pressure region is present behind the negative peak. The flow reattachment occurs at points that move cyclically toward and away from the leeward edge. As V_r increases, the pressures tend to be more uniform, with weaker peaks. The reattachment point retreats toward the leeward edge and then disappears altogether; the flow over the surfaces becomes completely separated. As V_r is getting much larger, the pressure distribution on the surfaces is nearly uniform and is similar to the mean pressure distribution for the body at rest (Fig. 5.2.4*a*). The relation between pressure distribution and V_r reflects the physical significance of the reduced wind speed.

We now compare, for various values of V_r, pressure distributions for a box deck with overhangs on both sides (Fig. 5.2.8*b*). For $V_r = 11$ the instantaneous pressure distribution is similar to that for a fixed cylinder with ratio $B/D = 2$ at rest (Fig. 5.2.4*a*) or for low reduced wind velocities (Fig. 5.2.8*a*), indicating that reattachment occurs on the floor deck. On the other hand, there is no reattachment at $V_r = 30$, and the pressure distribution becomes relatively uniform, similar to the pressure distribution for the fixed deck (Fig. 5.2.4*c*). Intense pressure fluctuations such as those observed for bodies with relatively low V_r values also occur in torsional motions.

The pressures differ according to deck configuration, angle of attack, degree to which the body is two- or three-dimensional, reduced wind speed, type of vibration (whether across-wind, along-wind, or torsional), and vibration amplitude. Turbulence can also affect the pressures, as it intensifies the whirlpool in the separated flow and promotes reattachment closer to the windward side.

Under the action of aerodynamic forces, the bridge structure, including the towers, hanger cables, and stay cables, generally experiences mean and fluctuating deformations in the along-wind (horizontal) direction, the across-wind (vertical) direction, and in torsion. These motions can occur singly or be coupled. Conversely, the structure's or the members' motions produce aerodynamic forces, to which we turn in Sect. 5.2.3.2.

Figure 5.2.9 represents schematically regions of occurrence of across-wind vibrations of typical rectangular cylinders mounted elastically. Vortex-induced

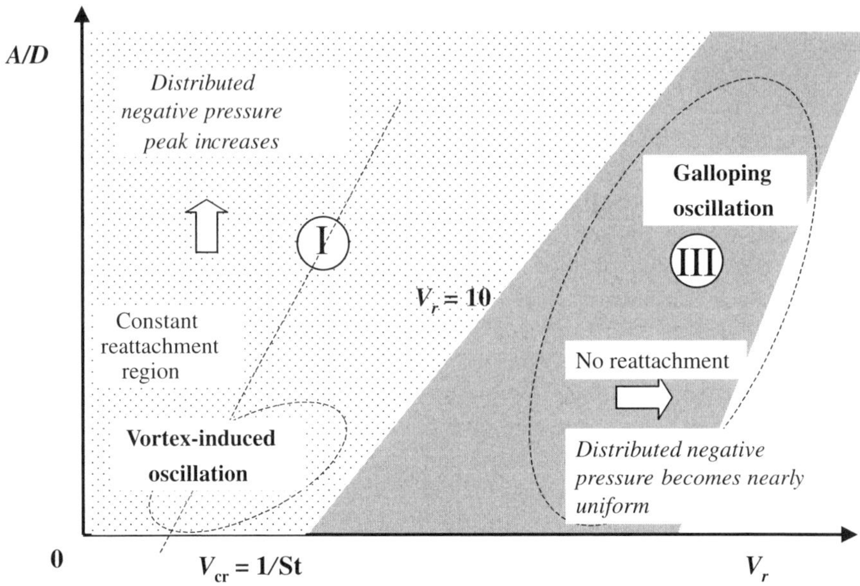

Figure 5.2.9. Conceptual diagram of vertical (heaving) response and flow reattachment for rectangular cylinders in smooth and turbulent flows.

vibrations (i.e., limited amplitude response at or near the wind speed defined by the Strouhal number) occur in region I. Galloping (i.e., vibration that increases as the wind speed increases) occurs in region III. The regions are defined by the nondimensional vibration amplitude A/D and the reduced wind speed V_r. In region I, reattachment of the separated flow can occur either once per period of vibration (this is particularly the case for shallow bodies with longer afterbody), or less than once per period. In region III the separation is complete (i.e., no reattachment occurs) and the mean pressure distribution nearly coincides with the distribution for stationary bodies. This allows the use for region III of a quasi-steady expression for the aerodynamic forces. In region II, which is present between regions I and III, the flow is completely separated, but no vibration occurs. The presence of turbulence in the flow tends to shift the boundaries between regions toward higher reduced velocities, but those boundaries become fuzzy and correspond to a more complicated variety of vibration patterns.

5.2.3.2 Motion-Dependent Aerodynamic Forces. Motion-dependent aerodynamic forces are termed *aeroelastic,* or *self-excited,* and produce aeroelastic, or self-excited, motions. Historically, the formulation of the motion-dependent aerodynamic forces problem has been based on results of vibrating aircraft wing theory [5.2, 5.3, 5.15–5.17], and has entailed the use of complex quantities. Imaginary parts in the expressions for the aeroelastic forces and the deformations associated with them reflect phase lags between force (moment) and displacement (rotation). Empirical adaptations of aircraft wing aeroelastic theory to suspension bridges were reported by various authors, including Pugsley, Scruton, Hirai and Okauchi, Ukeguchi, and Ito (see references in [5.18]); for the influential work by Scanlan, see references in [5.2, 5.3].

Denote the matrix of the motion-dependent complex aerodynamic forces and moments by $[F]$, the corresponding complex displacement vector by $\{u\}$, and the time rate of change of the displacement vector by $\{\dot{u}\}$. For small motions it may be assumed that the motion-dependent forces depend linearly on the displacements and velocities, as follows:

$$[F] = [F_D]\{u\} + [F_V]\{\dot{u}\} \tag{5.2.3}$$

It is convenient in applications to deal with real displacements and real motion-dependent forces and moments. With some changes of notation it is a simple exercise to recast Eq. 5.2.3 in the form shown in Sect. 2.7.3.1 (Eq. 2.7.3). As indicated in Sect. 2.7.3, the terms proportional to time rates of change of the displacements are called *aerodynamic damping terms.*

It is clear from the equations of motion of the body (Eqs. 2.7.4) and the expressions for the aeroelastic forces (Eqs. 2.7.5) that if the coefficients of the terms proportional to the rate of change of the displacements (rotations) are negative, the effect of those terms is equivalent to *positive* mechanical (viscous) damping, since when transferred to the left-hand side of Eq. 2.7.4 they augment the effect of the mechanical damping terms. For this reason, aeroelastic force (moment) terms with negative aerodynamic coefficients are said to constitute positive aerodynamic damping. Conversely, if the coefficients of the terms proportional to the rate of change of the displacements (rotations) are positive, the effect of the aerodynamic damping terms is equivalent to *negative* mechanical damping.

Positive mechanical damping dissipates energy through internal friction within the structure and therefore has a beneficial effect in that it limits the amplitude of the oscillatory motion. Positive aerodynamic damping has a similar effect, which is achieved through the transfer of energy from the body to the flow. Conversely, negative aerodynamic damping has a detrimental effect in that it increases the amplitude of the oscillatory motion through the transfer of energy from the flow to the body. For energy considerations in aeroelastic problems, see [5.16].

5.2.4 Vortex Excitation

For typical bridges with deep or shallow closed box decks with or without overhangs, or open girder decks with Π- or H-section, peak vortex-induced vibrations may occur for wind speeds on the order of 10 m/s. Basic structural bluff sections of tower or pylon structures, long members, hangers, and other substructures may be subjected to vortex excitation during construction or after completion of the bridge. In this section we discuss vortex excitation phenomena affecting bridge decks and towers. Those affecting cables are discussed in Sect. 5.2.7.

5.2.4.1 Vortex Excitation of Bridge Decks. We consider examples of vortex excitation observed in both prototype bridges and wind tunnel tests. Figure 5.2.10*a* is a view of the Trans-Tokyo Bay Bridge, a continuous steel girder bridge with 10 spans, of which the two longest spans are 240 m long. Figure 5.2.10*b* shows responses to vortex excitation of up to more than 50 cm, recorded under wind speeds of about 14 to 18 m/s before the installation of tuned mass dampers. Figure 5.2.10*b* also shows wind tunnel predictions of the peak response for two values of the logarithmic decrement of the damping. Sixteen sets of 10-ton tuned mass dampers installed inside the deck in the central four spans were found to control the vibration effectively. The Strouhal number, based on the wind speed inducing the peak amplitude, was close to the value applicable to rectangular cylinders.

For the old Tacoma Narrows Bridge, vertical bending vibrations were recorded before the bridge's collapse due to violent torsional (stall) flutter on November 7, 1940. Figure 5.2.11 shows wind tunnel test results obtained on

(*a*)

Figure 5.2.10. Trans-Tokyo Bay Bridge: (*a*) general view; (*b*) typical cross-section dimensions; (*c*) observed amplitude of full-scale vortex-induced response and comparison with wind tunnel test results for two values of the log decrement damping δ. [(*a*) From [5.22], courtesy of the Japan Highway Public Corporation.]

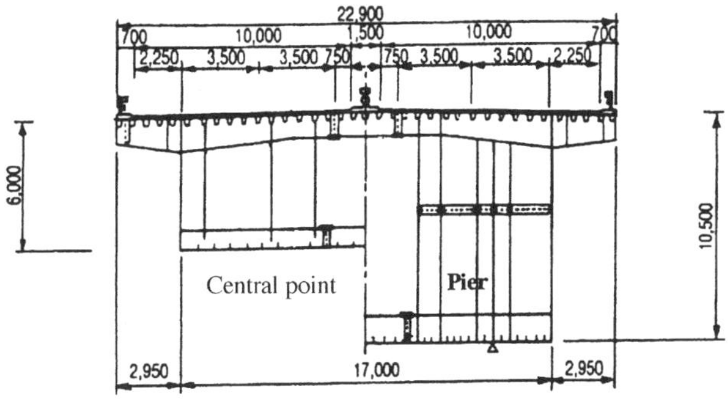

(*b*)

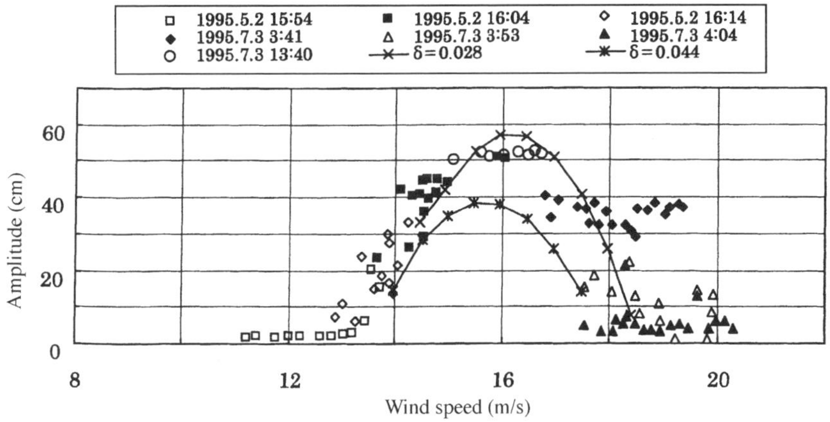

(*c*)

Figure 5.2.10. (*Continued*)

an aeroelastic model of the full bridge. As the wind speed was increased, several vibration peaks corresponding to successively higher modes were observed. An inset in Fig. 5.2.11 shows that the wind speeds and the corresponding frequencies for the peak vibrations can be plotted on a straight line, from which it follows that the Strouhal number was 0.094. This is close to the Strouhal number applicable to rectangular cylinders (Fig. 5.2.3).

For various short- and long-span bridges of various types and configurations that experienced vortex excitation (e.g., Kessock Bridge [5.20, 5.21], Great Belt East Bridge [5.23, 5.24], Erskine Bridge [5.25], Wye Bridge, Long Creek Bridge [5.26]), the dependence of the Strouhal number on the height/width ratio was close to the dependence shown in Fig. 5.2.3.

Figure 5.2.12, where A is the peak amplitude and δ is the logarithmic decrement, shows response plots obtained in wind tunnel tests for rectangular cylinders with ratios B/D = 1, 2, 3, 4, an H-shaped deck with ratio B/D =

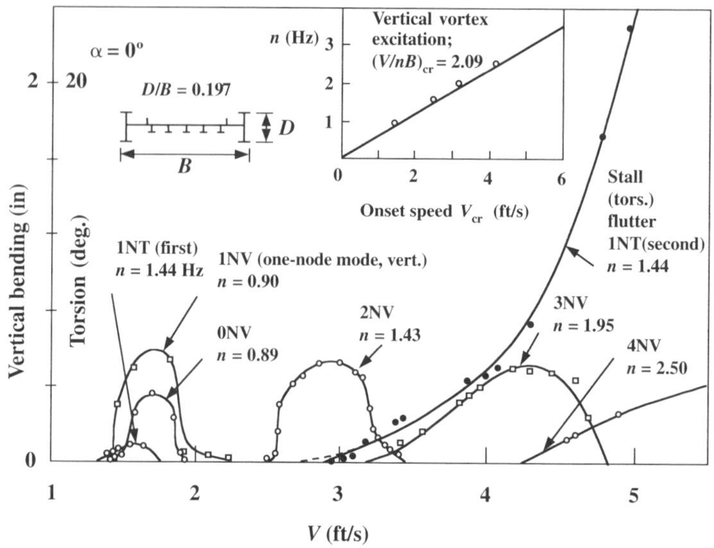

Figure 5.2.11. Aeroelastic full model wind tunnel test for the original Tacoma Narrows Bridge (vortex excitation and torsional flutter). (From [5.19].)

5 similar to the Tacoma Narrows Bridge, and ordinary box girder models with vertical and slanted webs [5.9, 5.27]. Data in Fig. 5.2.12 corresponding to *galloping vibrations,* which increase monotonically with wind speed, are denoted by **G** and are discussed later in this chapter. *Vortex-induced vibrations* experience a maximum at or near the critical speed V_{cr} and are denoted by **V**. For smooth flow (S.F.) and/or turbulent flow (T.F.), Fig. 5.2.12*a, c, d,* and *e* show the peak vortex-induced response of rectangular cylinders with ratios $B/D = 1$, 2, 3, and 4. Figure 5.2.12*f* and *g* show the peak torsional response of a rectangular cylinder with ratio $B/D = 2$ and an H-shaped deck with ratio $B/D = 5$. Figure 5.2.12*b* shows the vertical response of box girder bridge decks with vertical and with slanted webs. Figure 5.2.12 also contains information on the Scruton number Sc, defined as

$$\mathrm{Sc} = \frac{2\delta_h m}{\rho D^2} \tag{5.2.4a}$$

where δ_h is the logarithmic decrement for vertical oscillations and m is the body mass per unit length. For torsional vibration the Scruton number is similarly defined as

$$\mathrm{Sc} = \frac{2\delta_\alpha I}{\rho D^2 B^2} \tag{5.2.4b}$$

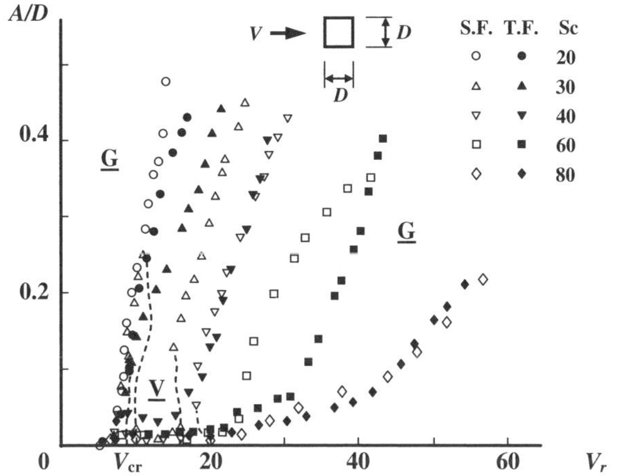

(a)

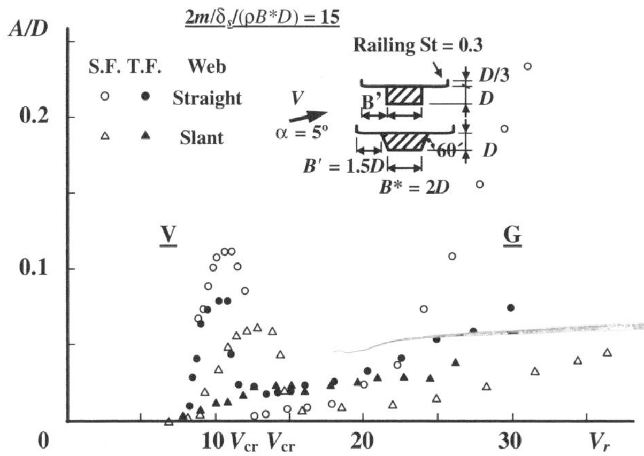

(b)

Figure 5.2.12. Wind-induced responses of rectangular and H-shaped cylinders and of typical bridge decks in smooth and turbulent flows: (*a*) vertical motions, rectangular cylinder, $B/D = 1$; (*b*) vertical motions, box decks; (*c*)–(*e*) vertical motions, rectangular cylinders, $B/D = 2$, 3, and 4; (*f*) torsional motions, rectangular cylinders, $B/D = 2$; (*g*) torsional and vertical motions, H-shaped cylinders, $B/D = 5$. [(*a*)–(*f*) From [5.9]; (*g*) from [5.27].)

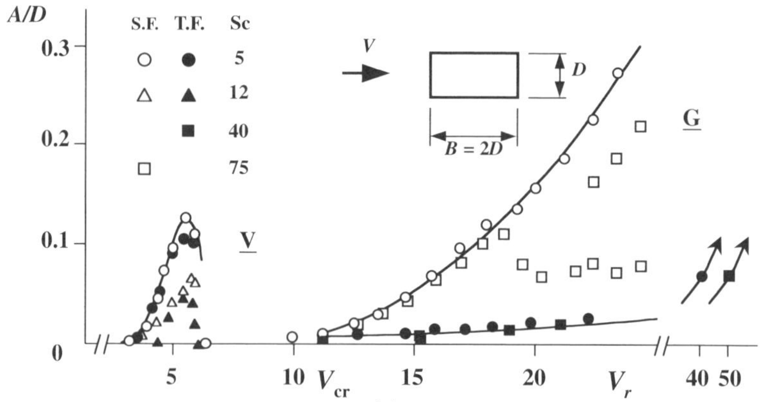

(c)

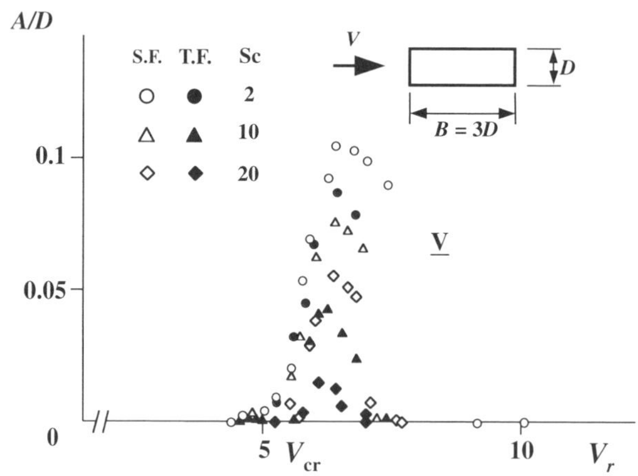

(d)

Figure 5.2.12. (*Continued*)

where δ_α is the logarithmic decrement for torsional oscillations and I is the polar moment of inertia of the body per unit length. The Scruton number is a measure of the combined roles of body inertia and mechanical (as opposed to aerodynamic) damping. The larger these two parameters are, the smaller is the peak vortex-induced response.

We discuss oscillations that may be assumed to be vortex-excited and are denoted in Fig. 5.2.12 by $\underline{\mathbf{V}}$ for vertical oscillations and $\underline{\mathbf{T}}$ for torsional oscillations. Consider first the square cylinder ($B/D = 1$; Fig. 5.2.12*a*). For smaller values of Sc, the response that occurs near the critical reduced velocity V_{cr}, defined as the inverse of Strouhal number St, increases monotonically as V_r increases. This is typical of galloping oscillations. If Sc is increased, the motion may turn to a vortex-induced oscillation. Further increasing Sc leads to the disappearance of the peak vortex-induced response. For a cylinder with ratio $B/D = 2$, the peak response occurs at $V_{cr}/2$, possibly due to intermittent

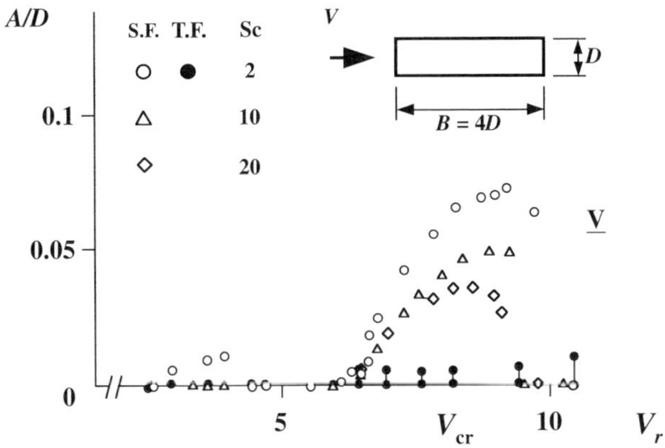

(e)

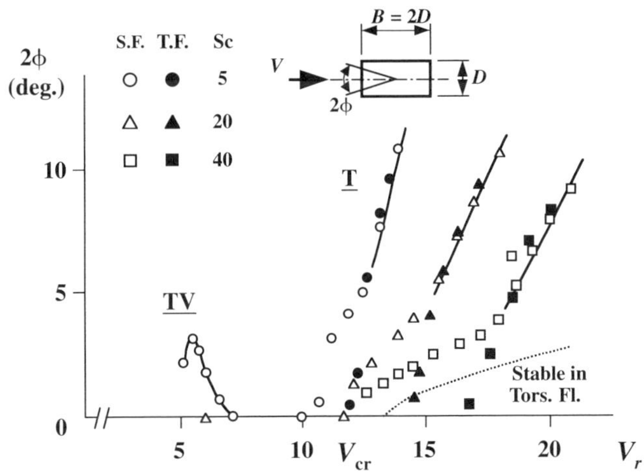

(f)

Figure 5.2.12. (*Continued*)

flow reattachment due to the body's cyclic motion. For $B/D = 3$ and 4 it occurs at V_{cr}. The box deck models have vortex-induced peak responses at the respective values of V_{cr}, which depend on whether the webs are vertical or slanted. The peak amplitudes decrease as the Scruton number increases. The presence of turbulence in the flow reduces the peak amplitudes except in the case of the square cylinder. No clear peaks can be seen in the vertical motion of a cylinder with $B/D = 4$ and the box deck with slanted webs, or in the torsion of a cylinder with $B/D = 2$. In the H-shaped deck with ratio $B/D = 5$ (Fig. 5.2.12*g*), the double amplitude of the vortex-induced torsional vibration, denoted by 2φ, is reduced considerably by turbulence with intensity 15% and is suppressed altogether by turbulence with an intensity of about 20%.

From data such as those of Fig. 5.2.12, it may be concluded that:

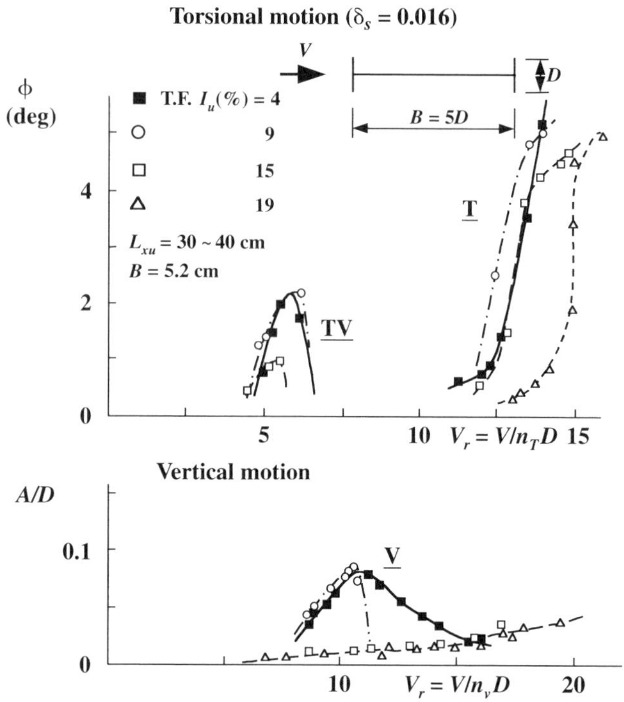

(g)

Figure 5.2.12. (*Continued*)

1. The onset of vortex-induced vibrations typically occurs at or near the critical reduced wind speed V_{cr} corresponding to the inverse of Strouhal number St, as shown for the fixed bodies in Fig. 5.2.3. However, for a cylinder with $B/D = 2$, the peak occurs at a reduced speed $V_{cr}/2$, as indicated earlier. Whether this may also be the case for certain types of bridge decks is not clear.
2. The peak vortex-induced response decreases as the Scruton number increases.
3. Turbulence typically reduces the peak vortex-induced response—hence the need, noted earlier, to consider effects of atmospheric flows that may not be turbulent.

Peak vortex-induced response can cause severe serviceability problems or even some damage, although it typically does not lead to collapse. Countermeasures are discussed in Sect. 5.3.3. They consist of improving the bridge configuration at the design stage, adding aerodynamic devices such as flaps or installing passive or active dampers.

5.2.4.2 Vortex Excitation of Tower Structures and Bridge Members. Vortex excitation may adversely affect towers or pylons of the single, gate, A-shape, or other type in cable-stayed bridges, main towers of suspension bridges, circular pipes, H-shaped suspended hangers and truss bridge members, and so on. Wind tunnel test results for an aeroelastic model of the Severn suspension bridge tower (136 m high, portal frame type) during erection are shown in Fig. 5.2.13.

The main 150-m-high tower of the Forth Road suspension bridge experienced vibrations with recorded peak amplitude of 1.05 m at the tip under 9- to 11-m/s winds normal to the bridge axis ($V_r \approx 6.75$, as referenced to the width D of the tower base). The vibrations prevented the requisite bolted connections from being performed until remedial measures were taken. For the 297-m-high main towers of the longest (1991-m) span of the Akashi Kaikyo Bridge, Fig. 5.2.14 shows test results for the vortex excitation in out-of-plane bending and torsion for the tower with main cables connected at the tip. The corner-cut section shown in the figure was adopted following numerous studies. It resulted in a reduction of the drag force and of large vibrations that could occur at relatively low wind speeds if the corners were

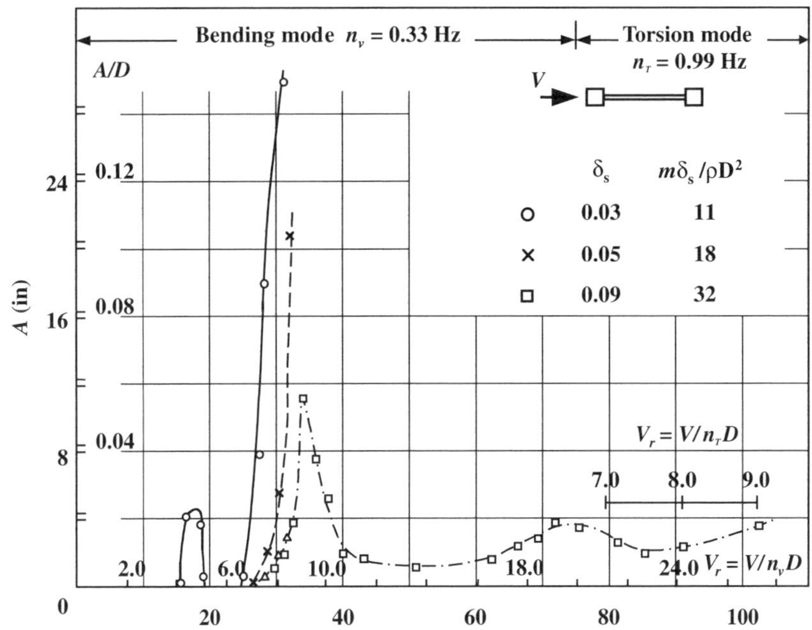

Figure 5.2.13. Effects of damping on amplitude of vibration at the top of a completed free-standing tower model for the Severn River Suspension Bridge. (From [5.28]; courtesy of Thomas Telford Ltd.)

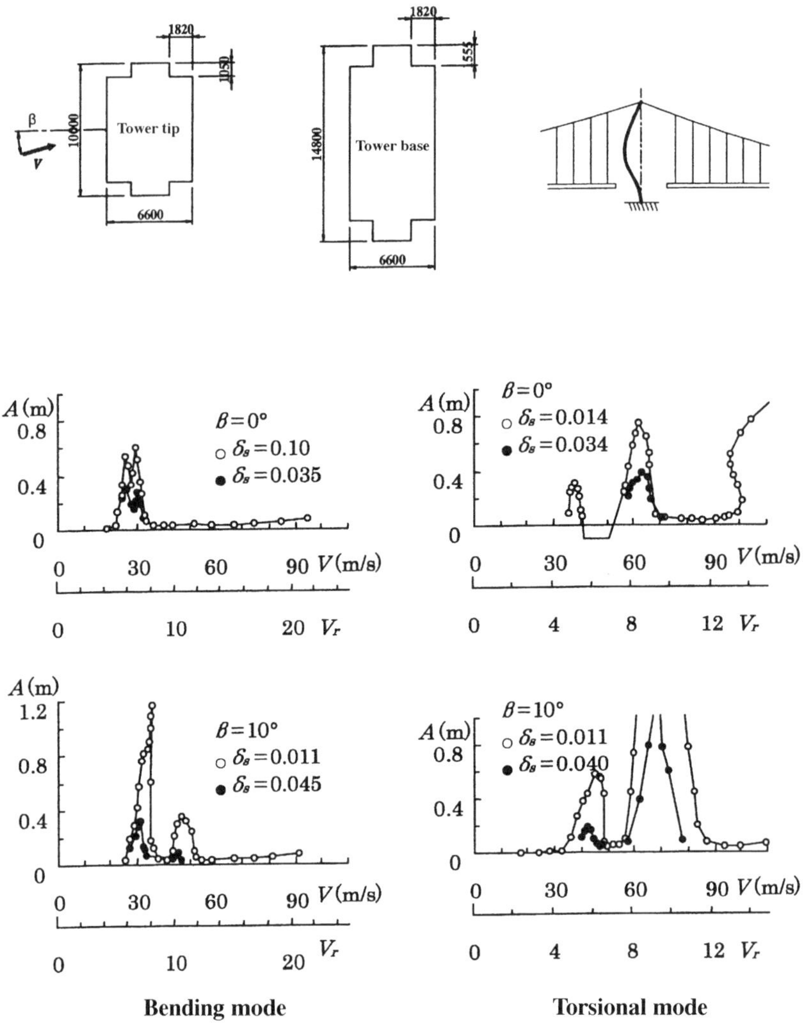

Figure 5.2.14. Effect of damping and wind incidence on the maximum amplitude of vibration in bending and torsion of a completed tower model connected with main cables at the tip for the Akashi Kaikyo Suspension Bridge (From [5.29]; courtesy of the Honshu–Shikoku Bridge Authority.)

not cut. The first mode of vibration was dominant for the tower with main cables at the tip, while higher modes were dominant for the single freestanding tower during erection. The gate-type tower consisted of two shafts. Owing to interference effects, the vortex-induced response was stronger for wind blowing at an angle $\beta = 10°$ (Fig. 5.2.14) than for $\beta = 0°$. The countermeasures for this case are discussed in Sect. 5.3.3.

Understanding of the effects of vortex excitations on individual members and hangers has improved to the point where undesirable effects can now be prevented routinely. Such effects have included fatigue cracks in vertical H-shaped tension members of the Commodore Barry Bridge, a large truss bridge in Pennsylvania, which occurred after the bridge's completion in 1974. Bending and torsional vibrations in H-shaped members of truss bridges and hangers of arch bridges have occurred in a number of cases. For circular steel hangers in arch bridges, countermeasures include strengthening of the gusset plates and attachment of tripping devices that break the vorticity, although such devices increase drag.

5.2.5 Flutter Instability: Galloping, Torsional Flutter, and Classical Flutter

As noted earlier, for bridges, large vortex-induced vibrations occur at relatively low wind speeds. At higher wind speeds bridges may experience three types of aeroelastic motion: (1) *galloping,* sometimes called *bending flutter;* (2) *torsional flutter,* also referred to as *flutter;* and (3) *classical flutter,* also called *coupled flutter.*

A bridge may be modeled as an n-degree-of-freedom system under unsteady aerodynamic forces. The system's equation of motion is

$$[M]\{\ddot{u}\} + [C]\{\dot{u}\} + [K]\{u\} = [F_V]\{\dot{u}\} + [F_D]\{u\} \tag{5.2.5}$$

[5.30, 5.31], where $[M]$, $[C]$, and $[K]$ are the mass, damping, and stiffness matrices of the structure, and $\{\ddot{u}\}$, $\{\dot{u}\}$, and $\{u\}$ are the acceleration, velocity, and displacement vectors of the motion, respectively. The unsteady motion-dependent aerodynamic forces on the right-hand side of Eq. 5.2.5 are defined by the matrices $[F_V]$ and $[F_D]$, which are functions of reduced wind speed (or reduced frequency), and the velocity and displacement vectors (Sect. 5.2.3.2). Aerodynamic forces proportional to the acceleration of the motion are typically negligible compared with the inertial terms. Equation 5.2.5 can be rewritten in the form

$$[M]\{\ddot{u}\} + [D]\{\dot{u}\} + [S]\{u\} = [0] \tag{5.2.6}$$

where $[M]$ is a mass matrix and $[D]$ and $[S]$ are damping and stiffness matrices, respectively. The physical variables u may be expressed in terms of generalized coordinates q. Let the damping and stiffness terms being considered be only those associated with aerodynamic damping and aerodynamic stiffness, and let the matrices $[D]$ and $[S]$ be written as sums of symmetric and nonsymmetric matrices indexed by the subscripts 1 and 2, respectively; that is,

$$[D] = [D_1] + [D_2],\ [S] = [S_1] + [S_2] \tag{5.2.7}$$

It can then be shown that the increment of total energy ΔE of the bridge's harmonic vibration over a period T_0 can be written as

$$\Delta E = I_1 + I_2 = -\int_0^{T_0} [\{\dot{q}\}^{\mathrm{T}}[D_1]\{\dot{q}\} + \int_0^{T_0} \{\dot{q}\}^{\mathrm{T}}[S_2]\{q\}]\, dt \tag{5.2.8}$$

where I_1 and I_2 denote the first and second integrals of Eq. 5.2.8, respectively.

The term I_1 is associated with a diagonal aerodynamic damping matrix and expresses the energy transfer due to individual one-degree-of-freedom modal motions. The term I_2 is associated with the nondiagonal aerodynamic stiffness matrix and pertains to energy transfer due to coupled modal motions. If the term I_1 dominates, the aeroelastic vibrations are one-degree-of-freedom motions of the galloping (bending flutter) or torsional (stall) type. If the term I_2 dominates, the vibrations are coupled flutter motions (bending and torsion, accompanied in some cases by along-wind bending). One-degree-of-freedom motions are caused by negative aeroelastic damping effects. Coupled flutter motions are due to phase lag effects between different modes of vibration. Consideration of mechanical damping, which was assumed to be zero, does not alter this qualitative picture.

From the expression of the aeroelastic forces (Eq. 2.7.5) it follows that the matrix $[D_1]$ is proportional to the wind speed V and the matrix $[S_2]$ is proportional to V^2. Coupled flutter due to the predominance of the term I_2 therefore tends to occur in higher wind speeds than torsional flutter does. This observation is consistent with observed behavior of actual bridges and with the fact that mechanical damping is considerably more effective in preventing galloping or torsional flutter than in preventing coupled flutter motions.

5.2.5.1 *Galloping.* Galloping is a bending oscillation that may affect deep box bridge decks, in which case it is a vertical oscillation, or may affect towers during the erection of suspension and cable-stayed bridges, in which case it is an across-wind horizontal oscillation. Galloping has been reported to have caused damage of the upper single hexagonal shaft ($B/D = \frac{3}{2}$) of an inverted 60-m-high Y-shaped pylon under winds of about 40 m/s [5.32].

It was noted that the occurrence of vortex excitation is associated with reattachment of separated vortices, while galloping can occur only at the higher speeds associated with the absence of flow reattachment. Typically, for bridge decks peak vortex-induced response occurs at reduced velocities $V_r \approx V_{\mathrm{cr}}$, while galloping occurs for $V_r >> V_{\mathrm{cr}}$ (see, e.g., Fig. 5.2.12*b* and *c*). Figure 5.2.12*d* and *e* show that galloping does not occur, even at the high speeds represented in the plots, for rectangular cylinders with ratios $B/D = 3$ or 4. For such cylinders the separated flow reattaches during any single oscillation period, whereas no reattachment occurs for lower B/D ratios.

For a box deck with vertical webs of Fig. 5.2.12*b*, and for a rectangular cylinder with $B/D = 2$, galloping that is possible in smooth flow for a certain range of speeds will occur in turbulent flow only at considerably higher speeds. This is due to the promotion by the turbulence of flow reattachment and the suppression of galloping that the reattachment entails.

A square cylinder (Fig. 5.2.12*a*) shows significant galloping in both smooth and turbulent flow. A large Scruton number tends to inhibit galloping, just as it tends to inhibit vortex-induced vibration. For higher reduced velocities, where no reattachment of separated vortices occurs, it was shown earlier that the mean pressure distribution on the upper and lower surfaces tends to be nearly the same in the vibrating body and in the body at rest. This allows the application of quasi-steady theory; that is, the self-excited forces may be described by forces acting on the body at rest. Therein lies one basic difference between galloping and torsional or classical flutter, for which the self-excited forces must be described as functions of reduced frequency.

The necessary condition for galloping is that in the neighborhood of $\alpha = 0$,

$$\frac{dC_L}{d\alpha} + C_D < 0 \tag{5.2.9}$$

(see Sect. 2.7.2). Figure 5.2.7*a* shows that for a rectangular cylinder with ratio $B/D = 2$, Eq. 5.2.9 holds. We emphasize again that quasi-steady theory is applicable to bodies that do not experience flow reattachment, such as rectangular cylinders with ratios $B/D < 2$ or so in flows with relatively high reduced velocities.

While one possible measure for preventing the strong vortex-induced excitation of bridge decks is the provision of flaps, which tend to suppress flow reattachment, suppression of galloping motion can be achieved by devices that promote reattachment (see Sect. 5.3.3.1). The respective countermeasures have opposite effects, entailing trade-offs that need to be considered in design.

5.2.5.2 *Torsional Flutter.* The collapse of the first Tacoma Narrows Bridge on November 7, 1940, was caused by torsional flutter. Figure 5.2.11, based on wind tunnel tests, shows that the amplitude of the torsional response increases nonlinearly with reduced velocity. Figure 5.2.15 shows the patterns of the wind tunnel flow around an H-shaped section similar to the shape of the deck of the first Tacoma Narrows Bridge. The patterns correspond to three instants during one cycle of a periodic torsional motion. Also shown are pictures of the torsional displacement of the bridge at each of those three instants. Note the vortices generated at the first of the three instants at the front edge of the deck, which subsequently travel toward the deck's rear edge. At the first instant, a large vortex separated from the front edge is associated with strong negative pressures on the windward half of the lower deck surface,

Figure 5.2.15. Torsional motion at time of collapse by torsional (stall) flutter, original Tacoma Narrows Bridge (from 16-mm Paramount News Film, 1940), and smoke visualization of separated vortex flow. (From [5.18].)

while a somewhat flatter flow is produced on the upper surface. At the second instant, the vortex started at the earlier instant, which has traveled along the lower surface of the deck, is broken. The flow over the upper surface is flatter than at the earlier instant. At the third instant the flow pattern is approximately symmetric in relation to the pattern occurring at the first instant. The flow patterns of Fig. 5.2.15 produce pressures consistent with the moments that cause the torsional motion.

The separated vortices around the bluff body are similar to von Kármán vortex sheets. Qualitatively, the patterns typical of flows during torsional flutter are enhanced counterparts of the corresponding patterns around bodies at rest.

Torsional responses in Fig. 5.2.12*g* are roughly similar for smoother flows with turbulence intensities of up to about 15%. The turbulence generally intensifies the separated vortex's whirlpool flow. This explains why for relatively high turbulence intensities torsional flutter may occur even in a square cylinder.

Torsional flutter can also occur in suspension bridge truss girders, particularly when parapets, fascia plates, handrails, and other appurtenances render the aerodynamics relatively similar to the aerodynamics around an H shape. Wind tunnel studies aimed at achieving control of torsional flutter in bridges with truss girders have been conducted in the UK (e.g., for the 1964 Forth Road Bridge), the United States (e.g., for the 1958 Mackinac Bridge, for which the susceptibility to torsional flutter was reduced by the provision of floor grating, and for the new Tacoma Narrows Bridge), Portugal (for the 1966 Tagus River Bridge), and Japan, where numerous truss girder bridges have been built, beginning in the 1950s.

Figure 5.2.16 shows motion-dependent aerodynamic forces on various bridge girders measured in wind tunnel tests. Figure 5.2.16*a* depicts the aerodynamic derivatives A_i^* (defined as in Sect. 2.7.3) measured in wind tunnel tests for the Golden Gate Bridge for zero mean angle of attack. Positive values over a certain reduced wind speed range result in negative aerodynamic damping. The propensity to flutter is due in large measure to the fact that the aerodynamic derivative A_2^*, which governs the torsional moment associated with torsional deformation, is positive for all but relatively small reduced velocities. Figure 5.2.16*b* shows that for an H-shaped deck with ratio B/D = 5, $M_{\varphi I}$ is positive over nearly the entire range of reduced wind speeds, a feature associated with a high propensity for torsional flutter [5.33].[2] In contrast, A_2^* for shallow box deck sections or wing sections is negative, which means that for such sections, torsional flutter is not expected to occur (Fig.

[2] Some practitioners consider that it is more practical to approach the flutter problem in terms of real variables. However, others recognize that complex variables have the advantage of preserving phase information, which can offer useful insights into modal coupling phenomena. The quantity $M_{\varphi I}$ is the counterpart of the aerodynamic derivative A_2^* in the alternative expression for the aeroelastic forces (for notation, see Eq. 5.2.11).

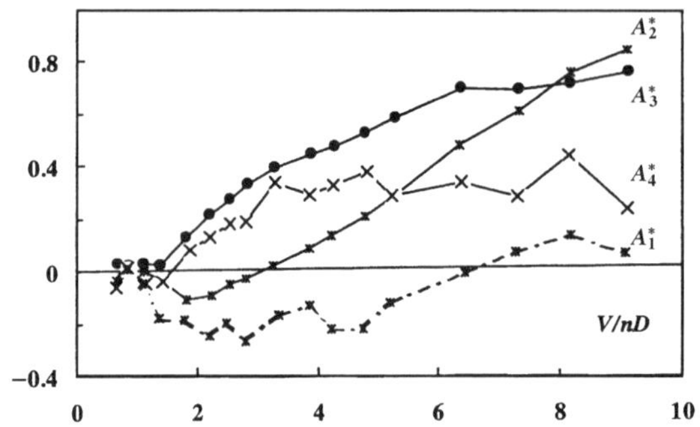

(*a*)

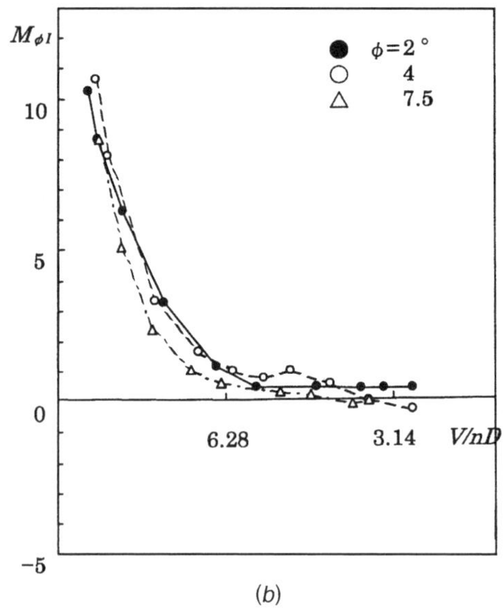

(*b*)

Figure 5.2.16. Unsteady aerodynamic force characterizations for bridge decks: (*a*) aerodynamic derivatives, Golden Gate Bridge (truss deck); (*b*) component $M_{\varphi I}$, H-shaped girder, $B/D = 5$; (*c*) set of eight aerodynamic derivatives, three streamlined box decks and airfoil; [5.2, 5.95]; (*d*) set of 12 real and imaginary aerodynamic force and moment components, Akashi Kaikyo Bridge, original design ($\alpha = 0°$). [(*a*)–(*c*) From [5.2, 5.95]; (*d*) courtesy of the Honshu–Shikoku Bridge Authority.)

5.2.16*c*) [5.95]; for the shapes of the deck sections see [5.2, p. 254]. This is also true of the Akashi Kaikyo bridge, for which $M_{\varphi I}$ is small (Fig. 5.2.16*d*).

By analogy with the galloping case (Eq. 5.2.9), it is tempting to use the inequality $dC_M/d\alpha + C_M|_{\alpha=0} < 0$ as a necessary condition for the occurrence of aeroelastic torsional vibration. There is indeed a correspondence between the negative sign of the sum $dC_M/d\alpha + C_\alpha$ at $\alpha = 0°$ and the onset of torsional

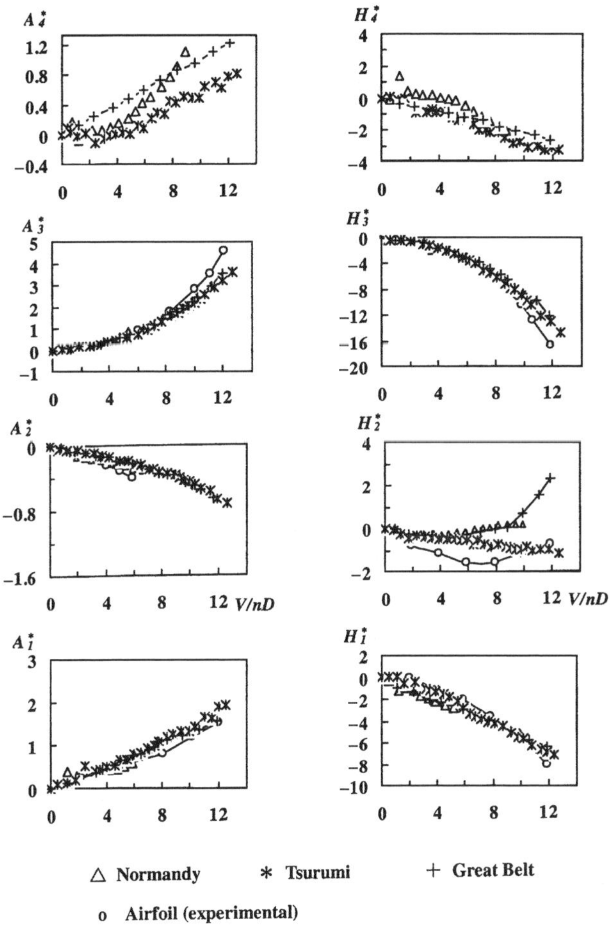

(c)

Figure 5.2.16. (*Continued*)

flutter (see, e.g., the rectangular and H-shaped cylinders with $B/D = 2$ and 5 in Fig. 5.2.7). However, unlike in the case of galloping, owing to the torsional motion, the relative velocity of the flow with respect to the body differs from point to point. This feature has prevented the success of attempts to develop a quasi-steady approach for torsional flutter that is similar to the approach applicable to galloping.

5.2.5.3 Classical Flutter

5.2.5.3.1 General Aspects of Coupled Flutter. The self-excited vibrations of bridge structures discussed in Sects. 5.2.5.1 and 5.2.5.2, galloping and

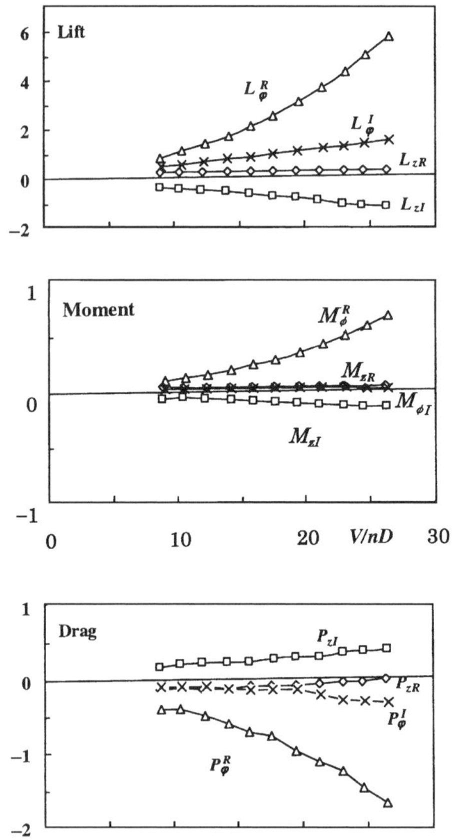

(*d*)

Figure 5.2.16. (*Continued*)

torsional flutter, could be related explicitly to the interaction between separated vortices and the structure oscillating in one degree of freedom (bending or torsion). On the other hand, classical flutter, or coupled flutter, is a type of self-excited motion that engages several degrees of freedom of the system. It may be argued that torsional flutter is a form of classical flutter in which the effect of the aerodynamic derivative A_2^* (or the coefficient $M_{\varphi I}$) is dominant.

In contrast, for the truss girders of the Akashi Kaikyo Bridge, the aerodynamic coefficients are such that galloping or torsional flutter cannot occur (Fig. 5.2.16*d*). However, coupled flutter motion can occur, and involves vertical bending, torsion around the longitudinal axis, and along-wind bending of the bridge deck.

The possibility of coupled flutter is studied by using (a) wind tunnel tests to observe the response of aeroelastic models of the full bridge (this can usually be done only for large projects), or (b) analyses applied to a three-dimensional computer model of the bridge, based on unsteady aerodynamic forces measured for rigid section models with elastic springs.

Coupled flutter analysis methods for long-span bridges include, among others, Bleich's method [5.34] and Selberg's simplified method [5.35]. In attempting to explain the collapse of the first Tacoma Narrows Bridge, Bleich assumed coupling between vertical bending and torsion (even though the observed response showed that torsion was dominant by far) and used similar vibration modes in vertical bending and torsion. In addition, the expressions used for the aeroelastic forces were those applicable to an airfoil section, which do not reflect correctly the aeroelastic behavior of H-sections. Selberg [5.35] introduced the following simplified empirical expression for the critical coupled flutter velocity (i.e., the lowest velocity for which coupled flutter motion occurs):

$$V_F = 0.44\omega_\alpha b\sqrt{1-\left(\frac{\omega_h}{\omega_\alpha}\right)^2}\sqrt{\frac{\sqrt{\nu}}{\mu}} \tag{5.2.10}$$

where $\mu = \pi\rho b^2/2m$ and $\nu = 8r^2/b^2$, ω_h and ω_α are natural circular frequencies of the vertical and torsional motion, respectively, b is the bridge's half width, m is the bridge mass per unit length, and r is the radius of gyration of the bridge cross section. Selberg's equation is independent of the bridge aerodynamics and can therefore produce incorrect results.

Aeroelastic lift, torsional moment, and drag acting on the deck are given by Eqs. 2.7.5. An alternative expression, based on traditional usage in aeronautics and specified in the Honshu–Shikoku Bridge code provisions [5.36], is

$$L_h = \pi\rho b^2\omega^2\left(L_{hR}h + L_{hI}\frac{\dot{h}}{\omega} + L_{pR}p + L_{pI}\frac{\dot{p}}{\omega} + L_{\alpha R}\alpha + L_{\alpha I}\frac{\dot{\alpha}}{\omega}\right) \tag{5.2.11a}$$

$$M_\alpha = \pi\rho b^4\omega^2\left(M_{hR}h + M_{hI}\frac{\dot{h}}{\omega} + M_{pR}p + M_{pI}\frac{\dot{p}}{\omega} + M_{\alpha R}\alpha + M_{\alpha I}\frac{\dot{\alpha}}{\omega}\right) \tag{5.2.11b}$$

$$D_p = \pi\rho b^2\omega^2\left(P_{hR}h + P_{hI}\frac{\dot{h}}{\omega} + P_{pR}p + P_{pI}\frac{\dot{p}}{\omega} + P_{\alpha R}\alpha + P_{\alpha I}\frac{\dot{\alpha}}{\omega}\right) \tag{5.2.11c}$$

in which the coefficients, $L_{\alpha R}$, . . . , $P_{\alpha I}$ are functions of the reduced velocity and b is half the deck width.

The coefficients in the expressions for the aeroelastic forces and moments are inferred from the decaying or growing free vibrations of an elastically mounted rigid section deck model in the wind tunnel. The free vibration method, first used by Scanlan in the early 1960s, is in our opinion the most useful approach developed so far. Identification of aerodynamic coefficients based on measurements of elastic forces obtained for harmonically forced models can also be performed. The aerodynamic derivatives are obtained from measurements of the reaction components in the mounting system, although eliminating the effect of the inertial forces poses some problems in still air. The reliability of the experimentally determined aeroelastic coefficients is an important concern. It should also be recalled that the linearity of the expressions for the aeroelastic forces is valid in principle for small deformations only.

The 3911-m-long Akashi Kaikyo Bridge was the first for which the horizontal motion of the deck was included in the analytical modeling. Depending upon bridge span, that inclusion can be essential. The acquisition of nine complex aeroelastic components requires the design of appropriate mounting devices. Some of the terms applicable to horizontal motion may be described by a quasi-static model (see Appendix 5C, Eqs. 5C.12). For the Akashi Kaikyo Bridge, following the experimental identification of the aeroelastic forces in section model tests, an analysis was performed of a large three-dimensional finite elements model (FEM) of the full bridge (Fig. 5.2.17). The model was used subsequently with the aeroelastic input based on those tests. A standard solution of the flutter motion equations was then applied to the system so obtained. The analysis accounted for the wind-induced inclination of the bridge deck with respect to the mean horizontal flow (i.e., the angle of attack). For an overview of the analysis, see Appendix 5B.

5.2.5.3.2 Performance with Respect to Flutter. The design of long-span suspension bridges to achieve satisfactory performance economically requires

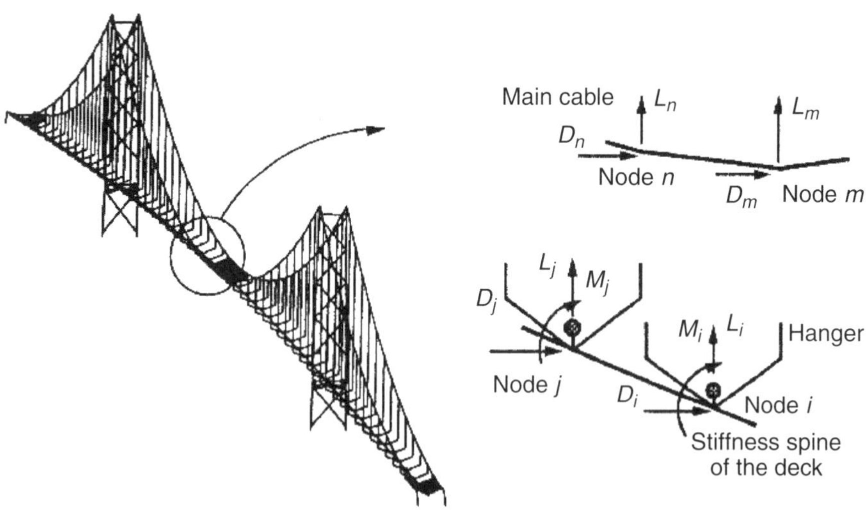

Figure 5.2.17. FEM and loading of a suspension bridge.

efficient aeroelastic design. This can be achieved through (1) the selection of cross-sectional deck shapes with favorable flow patterns, and (2) aeroelastically effective structural design, resulting in a reduction of modal and cross-modal integrals, particularly for torsional motions. Modal shapes consistent with such reduction can be achieved through (a) the selection of a favorable ratio of side spans to the main span, (b) cross sections that are stiffer near the supports and along the side spans than in the central part of the main span, (c) stiff pylon towers, and (d) stiff cables. For the Akashi Kaikyo Bridge the truss stiffening deck, selected from several alternatives (see Sect. 5.3.2) [5.71, 5.72], was of equal width throughout the bridge. However, a vertical plate for stabilization was installed beneath the deck floor only on the center span (see Fig. 5.3.2). An alternative proposed deck consisted of a boxlike shape that was shallow in the central part of the main span and was torsionally stronger and bluffer near the main span supports and along the side spans.

The longer the bridge span, the more the effect of the tower dynamics becomes significant from an aeroelastic point of view. For example, according to calculations for a three-span suspension bridge with a 3000-m main span and 1500-m side spans, replacing flexible steel pylons by stiffer, reinforced concrete pylons resulted in an increase of the critical flutter speed of about 25%, whereas for a bridge with a 2000-m main span and 1000-m side spans, the increase was about 10% [5.48]. For the latter bridge, replacing cables with allowable stress of about 1000 MPa or less by cables with allowable stress of about 1200 MPa or more (and a correspondingly lower area) resulted in a decrease in the critical flutter speed, reported to be about 25%.

Improvements in performance with respect to flutter can be achieved elegantly by adopting the spindle-type suspension bridge concept depicted in Fig. 5.2.18. Analyses reveal that aeroelastic effects on the central part of the main span play a dominant role in the onset of flutter. To achieve better aeroelastic performance, the deck should be aerodynamically stable over that part of the main span, while a less stable deck is acceptable for the remaining parts of the bridge. The gap between the two parallel bridge decks that make up the central part of the main span helps to improve aeroelastic performance substantially. This is also true of gaps between box girders in the Straits of Messina project (see Sect. 5.2.6.4.1).

5.2.6 Buffeting Response

5.2.6.1 Turbulence Effects on Bridge Vibration. The response of a bridge to turbulent wind is illustrated in Fig. 5.2.19, which contains records of the wind direction and speed, and of the bridge's vertical and horizontal buffeting response, taken during the passage of a typhoon over the site of the Akashi Kaikyo Bridge. The wind direction varied within an interval of about $\pm 30°$ from the direction normal to the bridge axis. The buffeting response increases with the wind speed much more weakly for the vertical motions (as well as for the torsional motions, not shown in Fig. 5.2.19) than for the horizontal motions. The bridge motions are due to randomly varying pressures

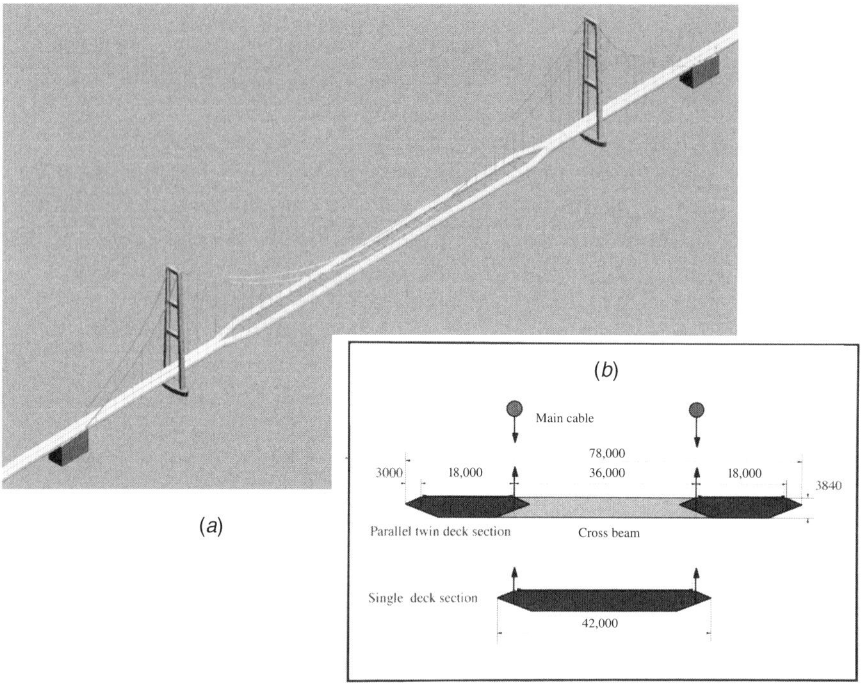

Figure 5.2.18. Proposed schematic of a spindle-type bridge for very long spans. (From [5.48, 5.49].)

induced by the gusty wind; no aeroelastic (vortex-induced, galloping, or flutter) motions occurred for the range of wind speeds shown.

This section is devoted to the modeling of the effect of wind gustiness on vortex-induced, galloping, and flutter motions. As mentioned earlier, the turbulence in the wind flow promotes flow reattachment, which tends to render the flow around the bluff body more similar to the flow around a streamlined body. Turbulence thus tends to mitigate peak vortex-induced oscillations. This is especially the case when reattachment occurs closer to the leading edge. Since galloping is associated with fully separated flows, turbulence also tends to increase the reduced velocity at which galloping oscillations set in, since it tends to promote reattachment. If the turbulence intensity is sufficiently high, torsional flutter occurs at higher reduced velocities than in smooth flow. Although the critical velocity at which coupled flutter sets in appears, typically, to be somewhat lower in turbulent flow than in smooth flow, this is not always the case. Tests in a wind tunnel with a simulated atmospheric boundary-layer flow can provide useful information on bridge behavior in turbulent flow, despite violation of the Reynolds number in the laboratory and other limitations on the faithful reproduction of natural winds.

If the turbulence intensities being considered are low, it is acceptable to represent the wind-induced forces on the bridge as a superposition of un-

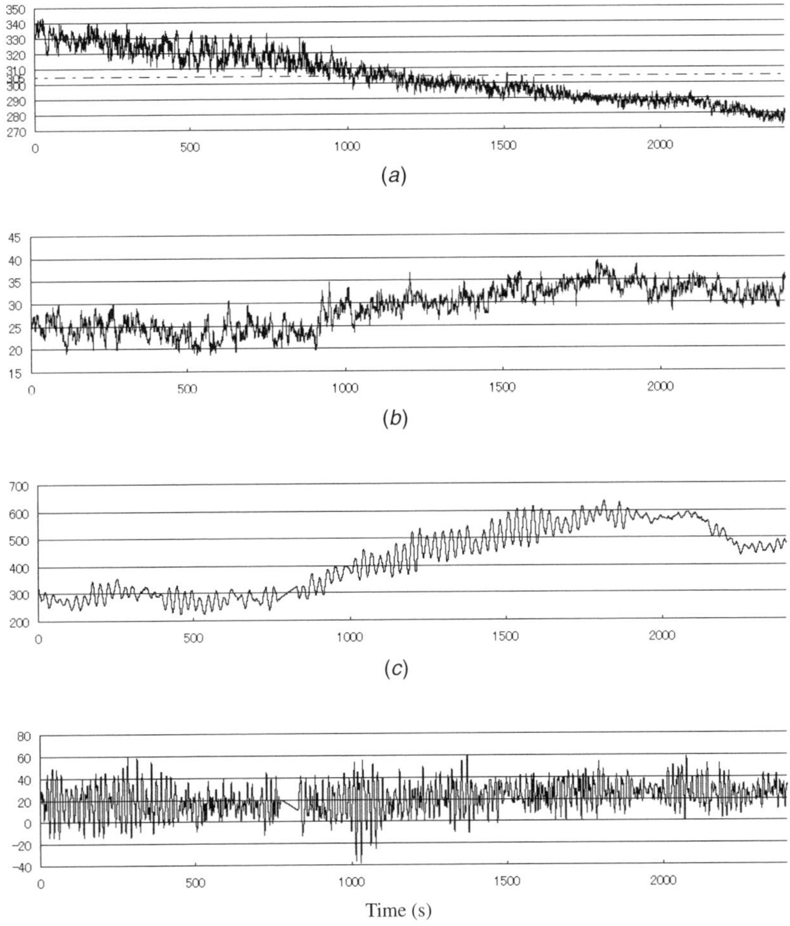

Figure 5.2.19. Field measurements, Akashi Kaikyo Bridge at typhoon passage: (*a*) wind direction, β (dashed line, normal to bridge axis); (*b*) wind speed, V (m/s); (*c*) lateral bending, p (cm); (*d*) vertical bending, h (cm). (From [5.18]; courtesy of the Honshu–Shikoku Bridge Authority.)

steady, motion-dependent forces and buffeting, time-dependent forces. The analysis must account for the role of the spatial correlation of the turbulence along the longitudinal axis of the bridge. Because the properties of the turbulent wind flow and of the unsteady aeroelastic forces are functions of frequency, bridge response analyses have so far been performed in the frequency domain. The three-dimensional nature of the bridge response problem, entailing as it does, coupling of vertical, torsional, and horizontal motion, needs to be accounted for in the analysis [5.37, 5.39, 5.40].

In the following sections we discuss a number of test results and analytical approaches concerning the response of long-span bridges to gusty winds.

5.2.6.2 Aeroelastic Full Bridge Model Tests in Boundary-Layer Wind Tunnels. Figure 5.2.20 shows buffeting responses of the 1/100 scale full-bridge aeroelastic model of the Akashi Kaikyo Bridge, observed at the midpoint of the center span for the original design of the deck. The response was observed in a boundary-layer wind tunnel. The 100-s-long time series correspond to about 1000 s for the prototype bridge and were recorded for several mean wind speeds. For the wind speed $V = 6.68$ m/s, the responses are

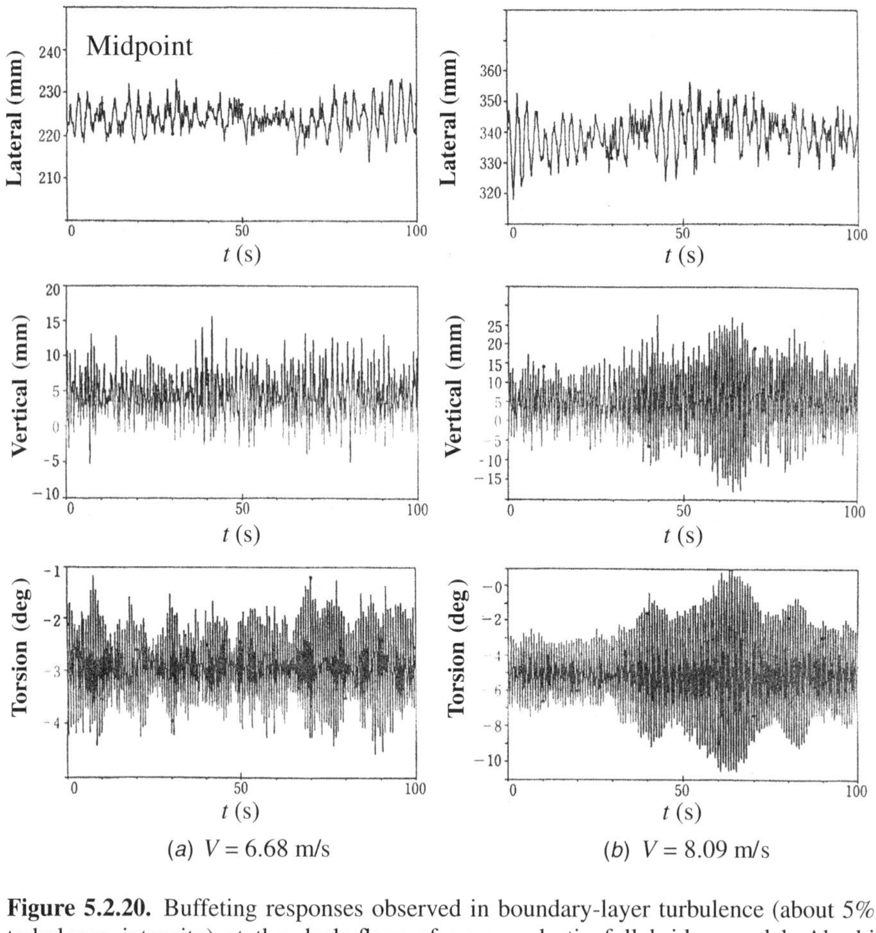

Figure 5.2.20. Buffeting responses observed in boundary-layer turbulence (about 5% turbulence intensity) at the deck floor of an aeroelastic full bridge model, Akashi Kaikyo Bridge. (From [5.37, 5.39]; courtesy of the Honshu–Shikoku Bridge Authority.)

typical of speeds for which no strong vortex-induced or flutter response occurs. For $V = 8.09$ m/s, the vertical and torsional responses are affected by flutter. For the same model in smooth flow, the onset of flutter occurred at a speed slightly higher than in turbulent flow; that is, for this model, turbulence had an unfavorable effect on flutter response.

Results of spectral analyses of wind tunnel model responses are shown in Fig. 5.2.21 for four wind speeds. For $V = 1.84$ m/s, the horizontal and torsional motions are coupled, as are the vertical and torsional motions. For $V = 4.43$ m/s, there is coupling between the vertical and horizontal motions

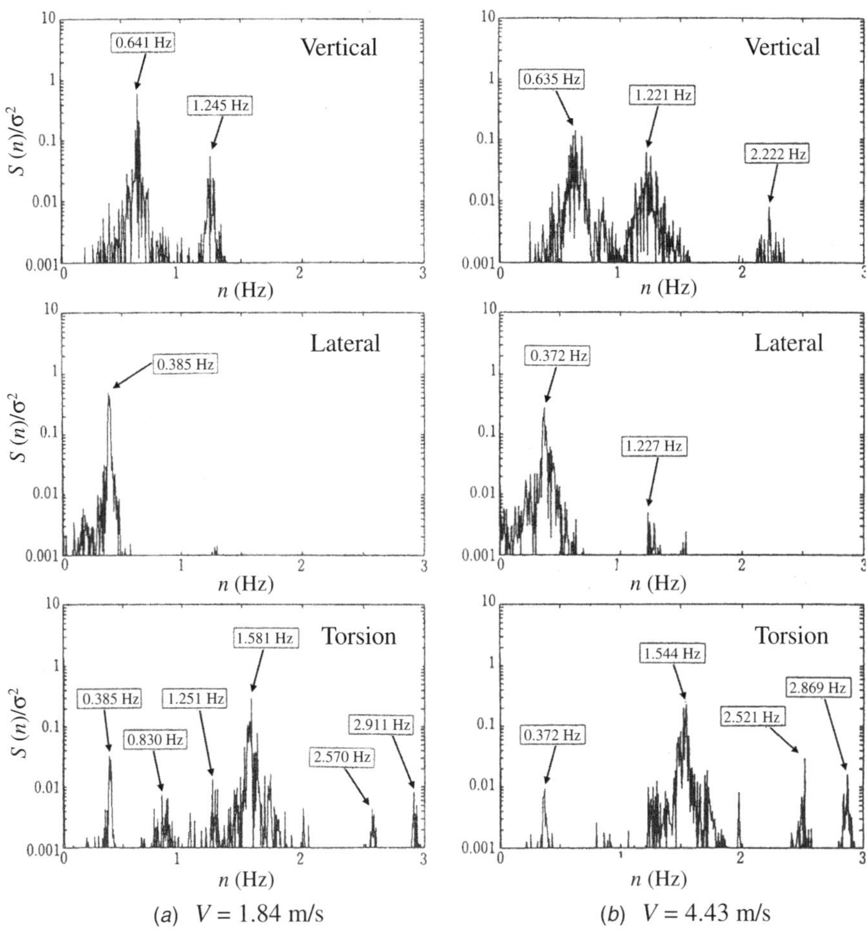

Figure 5.2.21. Power spectra of buffeting responses in boundary-layer turbulence (about 10% turbulence intensity) at the deck floor of an aeroelastic full bridge model, Akashi Kaikyo Bridge. (From [5.1]; courtesy of the Honshu–Shikoku Bridge Authority.)

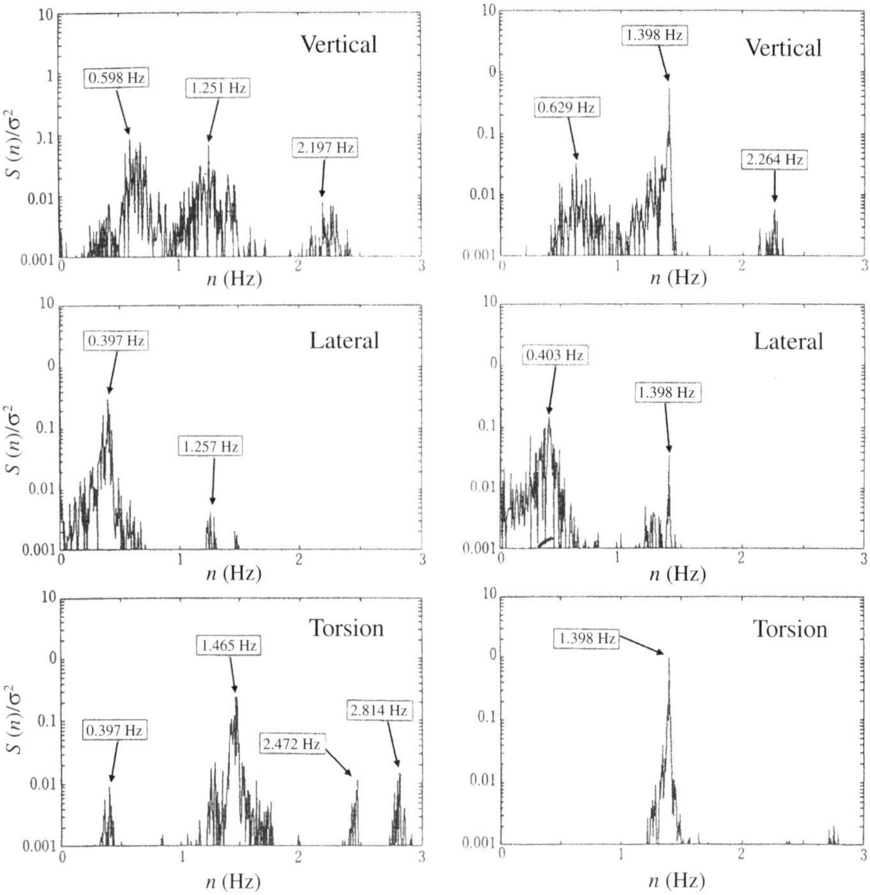

(*c*) $V = 6.27$ m/s (*d*) $V = 7.77$ m/s

Figure 5.2.21. (*Continued*)

and between the vertical and torsional motions. Note that frequencies of vibration are in some instances very close to each other but not identical (e.g., 1.245 Hz vs. 1.251 Hz for $V = 1.84$ m/s). This is ascribed to the effect of aerodynamic terms proportional to displacements. These terms may be interpreted as aerodynamic stiffness terms, which alter slightly the frequency of vibration, just as terms proportional to the time rate of change of the displacements are interpreted as aerodynamic damping terms. The vibrations are accompanied by coupled static deflections induced by wind [5.37, 5.39].

5.2.6.3 Expressions for the Aeroelastic and Buffeting Forces. Two force components (lift and drag) and a torsional moment are acting simultaneously on the bridge deck. They are

$$L = L_h + L_b \tag{5.2.12a}$$

$$M = M_\alpha + M_b \tag{5.2.12b}$$

$$D = D_p + D_b \tag{5.2.12c}$$

where the first term on the right-hand side of each of Eqs. 5.2.12 is the unsteady aeroelastic force (moment) given by Eqs. 2.7.5 or 5.2.11, and the second term is the fluctuating gust force (moment) caused by buffeting, given by Eqs. 2.7.9. There are 18 aeroelastic coefficients and, as mentioned earlier, they must be available as functions of angle of attack, given the fact that the bridge deforms vertically and torsionally and that the angle of wind incidence therefore varies in time. The coefficients can be obtained experimentally, although those coefficients associated with horizontal motion have so far been difficult to measure directly and have been assigned quasi-steady expressions.

The aeroelastic coefficients should in principle be measured in flow with the requisite turbulence features, although in some analyses coefficients measured in smooth flow are used instead. This is not acceptable, however, for bluff bodies with significant, lively flow separation zones, particularly at lower reduced velocities.

Uncertainties inherent in the use of Eqs. 5.2.12 may be due to, among other factors, possible interference effects between aeroelastic and buffeting forces. Also, the dependence of the aeroelastic forces on vibration amplitudes is likely nonlinear. However, measured unsteady aeroelastic forces (moments) may be assumed to be linear as long as they are used to determine the critical wind speed of flutter onset as a bifurcation point. These uncertainties and limitations should be kept in mind by the analyst and the bridge designer.

5.2.6.4 Buffeting Response of a Full-Span Bridge in the Presence of Aeroelastic Effects

5.2.6.4.1 Solution in the Frequency Domain. To perform and validate calculations pertaining to a full bridge model tested in the wind tunnel, it is necessary to use wind flow models based on the measured properties of the wind tunnel flow. These properties are usually not consistent with those of atmospheric flows, especially for low-frequency components of the turbulence. Large uncertainties associated with low-frequency turbulence result in uncertainties in the estimation of buffeting response, to which low-frequency vibration modes contribute significantly. An analytical approach to estimation of the aeroelastic behavior and dynamic behavior of a full bridge subjected to the action of turbulent wind is presented in some detail in Appendix 5C. Details of the flutter and buffeting analysis for the Akashi Kaikyo Bridge are reported in [5.43–5.45]. One conclusion of the Akashi Kaykio flutter analyses was that disregarding cross-modal effects (off-diagonal terms in the matrix **E**; see Eq. 5C.10) yielded incorrect estimates of the flutter velocity.

The Straits of Messina Bridge has a 3300-m-long main span that consists of a central box girder supporting two railway tracks, and two side box girders, each supporting two running lanes and an emergency lane. The three box girders are separated by gaps and are connected by transverse beams. Calculations and wind tunnel measurements for the yet-unbuilt bridge, conducted in Italy and independently in Denmark, are described briefly in [5.46].

Buffeting response calculations were reported to indicate that for the Tsing Ma bridge in Hong Kong (1377-m main span), the effect of including the cable and tower dynamics in the modeling was relatively small [5.47].

5.2.6.4.2 Analysis in the Time Domain. Time-domain analyses require the representation as time series of both the aeroelastic and the buffeting forces.

AEROELASTIC FORCES. These are in practice obtained experimentally as functions of reduced velocity, since measurements aimed at obtaining the indicial response function directly from experiment do not appear to have worked satisfactorily so far. One method for effecting a transformation from the frequency to the time domain is based on work performed in an aeronautical engineering context [5.42, 5.52] and is described, for example, in [5.43–5.45].

BUFFETING FORCES. The treatment of buffeting is illustrated by considering the lift force L_b. The lift induced by the longitudinal fluctuating velocity u may be described by using an indicial response function Φ'_{Lu} representing the elemental lift force induced at the nondimensional time $s = Vt/(B/2)$ by a longitudinal velocity equal to a Dirac δ-function. The first term on the right-hand side of Eq. 2.7.9*a* can be written in the time domain as

$$L_u = \frac{1}{2}\rho V^2 B C_L \int_{-\infty}^{s} \frac{2u(\sigma)}{V} \Phi'_{Lu}(s-\sigma)\, d\sigma \qquad (5.2.13)$$

However, Eq. 5.2.13 pertains only to the buffeting force L_u as applied at one point. The issue of spatial correlations is best treated by spectral approaches. Effecting time-domain simulations of buffeting effects due to horizontal and vertical components, u and w, based on spectral and cross-spectral representations of the pressures, is possible in principle but would require an effort that appears not yet to have been developed to the point where routine application is warranted.

5.2.7 Wind Effects on Cables

Serious vibration problems in cables that have occurred in cable-stayed bridges are due primarily to the cables' considerable unsupported lengths. The phenomena underlying these vibrations differ from those typical of vibrations due to vortex excitation, galloping of single cables, and wake-induced vibrations for parallel cables. Rain- and wind-induced vibrations in stay cables were observed by Hikami in 1984 [5.53] and have been the object of intense research efforts in the recent past. Dry inclined cable galloping [5.54] has been shown by tests to occur even in the absence of rainfall.

Three basic approaches have been proposed for the mitigation of stay cable vibrations. First, mechanical damping can be added by using specially designed dampers. Second, the aerodynamic behavior of the cables can be modified through the provision of helical wires, pattern indentations, or grooving strips or holes to control the generation of aerodynamic forces. Third, a method for interconnecting cables has been applied in a number of cases where cable damage has occurred.

Figure 5.2.22*a* shows the Scruton number Sc as a function of cable length for cable-stayed bridges constructed in Japan [5.55]. The symbol ● in Fig. 5.2.22*a* denotes circular cables with polyethylene coating commonly used for cable-stayed bridges. As the length of the cable increases, the Scruton number tends to reach a constant value of roughly 30. The mechanical damping of the cables can be as low as $\delta = 0.005$ (logarithmic decrement; this corresponds to a ratio of damping to critical damping ζ of less than 0.1%) (Fig. 5.2.22*b*) [5.56]. Conservative values assumed in design practice are as low as $\delta = 0.003$. This explains the susceptibility of stay cables to dangerous vibrations.

5.2.7.1 Vortex Excitation. At the critical value of Re for stay cables, typically the value of the Strouhal number is $\mathrm{St} = n_s D/V = 0.2$ (n_s is the vortex-shedding frequency and D is the cable diameter), although slightly lower values may in some cases be adequate. The cable is subjected to a fluctuating lift force with frequency n_s and a drag force with frequency $2n_s$. If n_s is close to or coincides with the fundamental natural frequency of vibration of the cable, n_1, a resonant large-amplitude vortex-induced vibration occurs. Because the wind speed that excites the fundamental mode is very low, resulting in small vortex-induced forces and oscillations, significant vibrations occur for higher modes of vibration. These higher-mode vibrations may have relatively small amplitudes, in which case they can easily be suppressed through a small addition of damping. Generally, an added logarithmic decrement value $\delta = 0.01$ is sufficient for this purpose.

5.2.7.2 Rain- and Wind-Induced Vibration. Rain- and wind-induced vibration is one of the more violent types of wind-induced vibration observed in stay cables. It was described in [5.53], following damage observed in 1984 in the Meiko-West Bridge (Fig. 5.2.23). Following the Meiko-West Bridge observations, rain- and wind-induced vibrations have been reported throughout the world. (Examples are the Farø Bridge in Denmark [5.57]; Hitsuishi-jima, Iwakuro-jima, Aratsu [5.59, 5.60], and Tenpozan Bridges in Japan; Nanpu and Yanpu Bridges in China; Erasmus Bridge in Holland; Fred Hartman and East Huntington Bridges in the United States [5.61]; and Normandie

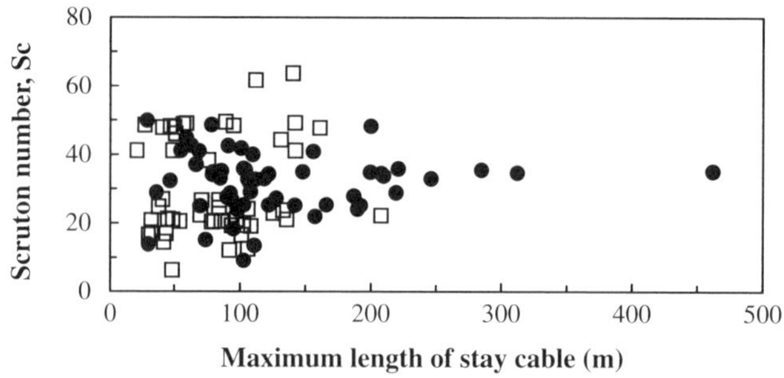

(*a*)

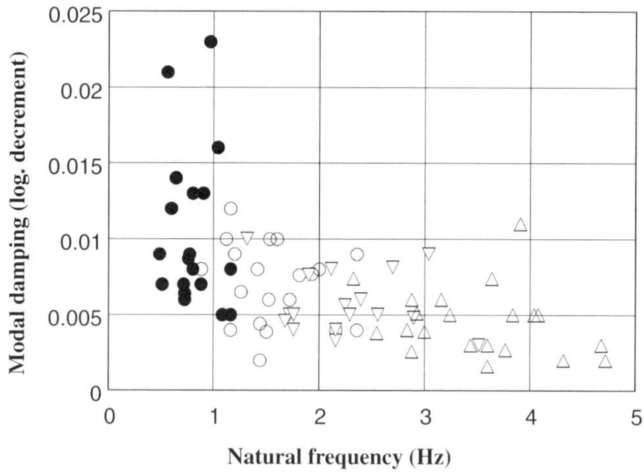

(*b*)

Figure 5.2.22. (*a*) Maximum length of stay cables vs. Scruton number; (*b*) modal damping vs. natural frequency in stay cables of cable-stayed bridges. Mode 1, ●; mode 2, ○; mode 3, ▽; mode 4, △; mode 5, □. [(*a*) From [5.55]; (*b*) from [5.56].]

Bridge in France [5.58].) A violent vibration of the stay cables has been observed in the 184-m center span Aratsu Bridge, resulting in the maximum double amplitude reaching 60 cm, about 3.5 times the cable diameter. The vibrations were suppressed by using oil dampers employed for automobiles [5.59, 5.60].

Rain- and wind-induced vibrations were observed to be most severe in winds whose directions deviate somewhat from the direction of a bridge's longitudinal axis in bridges located in river estuaries or at the sea coast. They occur in planes close to the plane of cables, are dominant in the lower four modes of vibration, usually have frequencies of vibration of about 3 Hz or less [5.1], and typically take place in winds with speeds $V > 6$ m/s or reduced

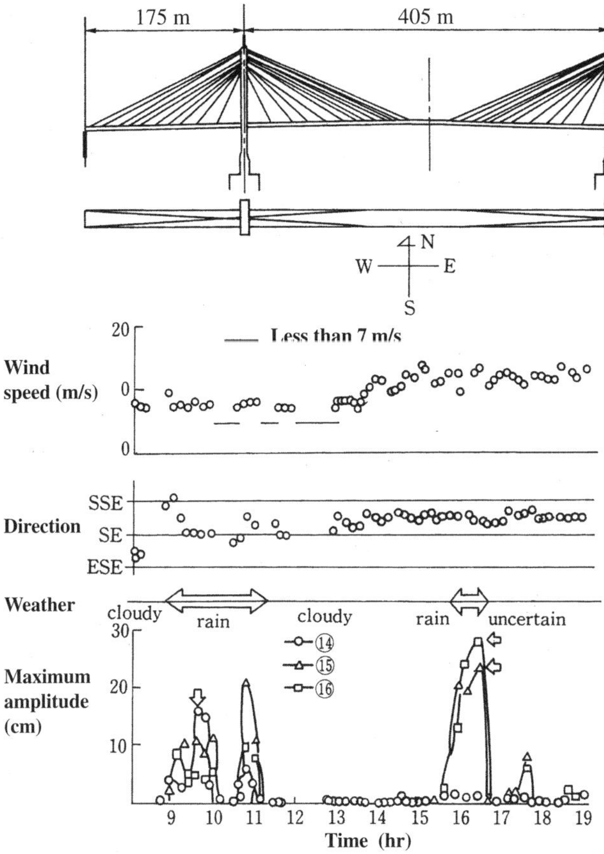

Maximum Amplitude

	Mode	δ
⑭	1	0.025
⑮	2	0.011
⑯	3	0.009

Figure 5.2.23. Rain- and wind-induced maximum-amplitude vibration observed in the Meiko-West Bridge vs. wind speed and direction, and time. (From [5.53]; courtesy of the Japan Highway Public Corporation.)

velocities $V_r > 20$. They tend to be stabilized by turbulence, but this tendency depends on turbulence levels and must be considered with caution.

Most investigations of rain- and wind-induced vibrations in stay cables have been conducted in the wind tunnel. However, some actual observations on bridges [5.61] and field studies using prototype cables [5.62] have been reported. It has been concluded that the onset of the vibrations is affected in a complex way by rivulets that form on the upper and lower surfaces of the cable, flows on the downwind side of the cable, and nonuniformity along the cable of Kármán vortex shedding, whose frequency is lower than the frequency based on the Strouhal number for the dry cable. In some instances the presence of rivulets on the lower side of the cable only is sufficient to generate the vibrations.

Some wind tunnel tests suggest the existence of dry inclined cable galloping in the absence of rainfall [5.54]. According to [5.63], rain- and wind-induced vibration of stay cables is initiated in the transition range around the critical Reynolds number and is consistent with the galloping criterion applied to drag and lift coefficients for that range.

Figure 5.2.24 shows nondimensional rain- and wind-induced vibration amplitudes observed in the wind tunnel vs. cable frequencies of vibration. Large vibrations occur in wind speed higher than 10 m/s, with no apparent dependence on vibration frequency. Vibrations induced at the lower speeds in Fig. 5.2.24*a* appear to be caused by a slightly different type of rivulet. It is inferred from these observations that vibrations are affected mostly by the formation of rivulets, meaning that the reduced velocity does not appear to be a useful parameter for plotting the experimental results.

Figure 5.2.25 is a plot of the Scruton number Sc vs. the reduced velocity at the onset of vibrations observed in the wind tunnel tests. The vanishing reduced velocity corresponds to the nonoccurrence of vibrations. Based on this plot, it was tentatively suggested that a Scruton number Sc $\geq$ 60 will definitely assure that rain- and wind-induced vibrations are under control. This is reasonably consistent with the requirement that $\delta > 0.02$, based on information on actual vibrations presented in [5.65, 5.66] (Fig. 5.2.22*a*).

5.2.7.3 Cable Vibrations in the Wake of a Parallel Cable. If two cables with diameter D are parallel, at the same elevation, and a short distance X away from each other, the downwind cable may experience vibrations forced by the wake flow of the cable upwind. For smaller X/D ratios, the vibrations that occur are called *wake galloping;* for larger ratios, the vibrations are called *wake flutter.*

Wake galloping of the downwind cable occurs in an elliptic path with a large vertical axis and is characterized by large amplitudes. In parallel cables of cable-stayed bridges, for ratios X/D from 2 to 5, wake galloping is significant. It can occur in the fundamental mode of vibration, or if spacers are installed to suppress the fundamental mode, subspan vibrations can occur. The reduced velocity at the onset of wake galloping is $V_r > 20$ or so, and the amplitude can be as much as three times the cable diameter [5.1, 5.2, 5.65, 5.66]. Wake galloping suppression can be achieved by increasing the damping and increasing the distance between the cables. The range $X/D =$ 6 to 7 appears to be desirable, as larger ratios might give rise to wake flutter.

Wake flutter has an elliptic path with a larger horizontal axis and occurs for $X/D > 10$ or so. It is typically a problem for transmission lines. However, violent vibrations of this type have been observed in parallel hanger cables ($X/D = 9$) of the Akashi Kaikyo Bridge. From wind tunnel tests it was concluded that the vibrations observed were consistent with a negative aerodynamic damping with logarithmic decrement $\delta = -0.2$ or stronger, so that just providing additional damping was not an effective means of vibration suppression. Smaller amplitudes might be achieved by increasing the cable's frequency of vibration [5.67].

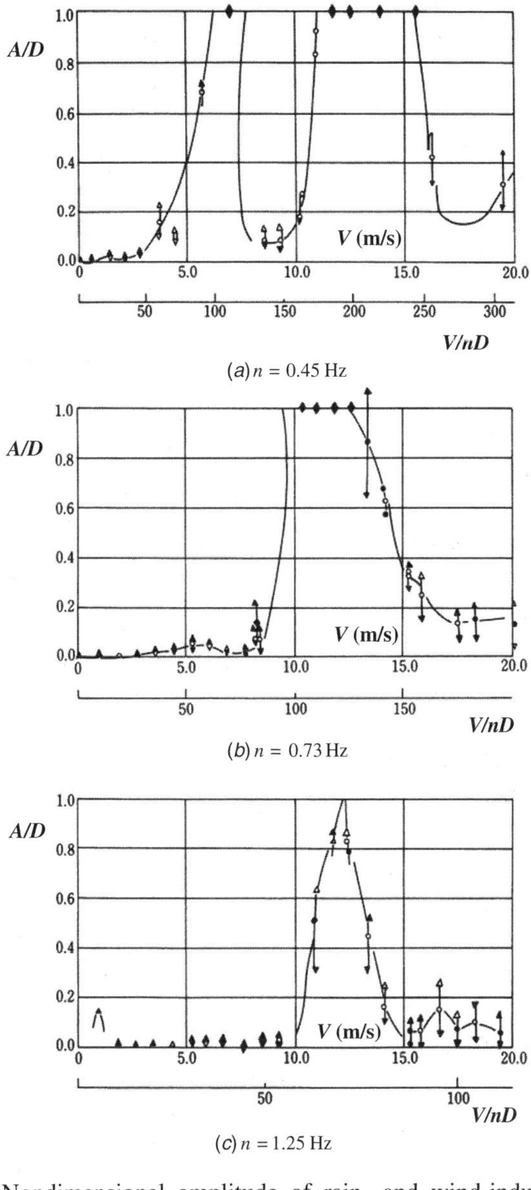

Figure 5.2.24. Nondimensional amplitude of rain- and wind-induced vibration vs. reduced wind speed for various natural frequencies. (From [5.64].)

5.3 BRIDGE DESIGN: SPECIFICATIONS, DESIGN DATA, AND CASE STUDIES

In this section we review briefly a number of design rules and code provisions on the design of short-, medium-, and long-span bridges and provide infor-

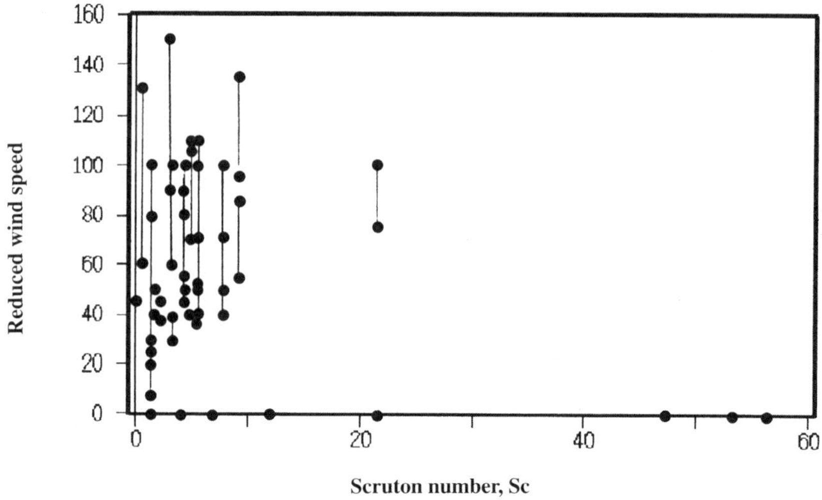

Figure 5.2.25. Reduced wind speed at the onset of rain- and wind-induced vibration vs. Scruton's number Sc. (From [5.65, 5.66].)

mation on the features of a number of actual bridges. We discuss details of the analysis and design of the Ahashi Kaikyo Bridge and on the dampers required for its main towers. Also discussed in this section are aerodynamic and mechanical approaches to reducing wind-induced vibrations of towers and decks, aerodynamic effects for twin-deck bridges or bridges built in close proximity to each other, wind shielding, the suppression or mitigation of wind-induced vibrations in cables, details on the wind tunnel testing of bridges with special reference to the Akashi Kaiyko bridge, and the application of computational fluid dynamics to bridge aerodynamics.

5.3.1 Wind Effects on Short- and Medium-Span Bridges: BD 49/01 Standard and Design Rules for Aerodynamic Effects on Bridges

The BD 49/01 Standard (Vol. 1, Sect. 3, Part 3) [5.68] specifies design requirements for bridges with respect to aerodynamic effects and includes provisions for wind tunnel testing. Additional guidance is available with respect to actual bridge configurations and archived test data available in the library of the Institution of Civil Engineers. The library also contains material on theoretical and computational fluid dynamics.

Bridge structures may be subjected to several types of aerodynamic excitation that produce motions that can result in damage to parts or in total collapse. The Standard considers:

1. Motions of limited amplitude, which can cause unacceptable stresses or fatigue damage. Such motions can be due to vortex-induced vibrations

or to the combined effects of vorticity and turbulence. The design must suppress or accommodate such motions so that unacceptable stresses or damage does not occur.

2. Large-amplitude motions, including galloping, torsional (stall) flutter, and classical flutter. The bridge must be designed so that these motions do not occur.
3. Nonoscillatory divergence due to static aerodynamic torsional instability (see Sect. 2.7), which the design should also prevent.

5.3.1.1 Aerodynamic Susceptibility Parameter. As part of the process of checking the suitability of a bridge design, the susceptibility parameter P_b is determined as follows:

$$P_b = 16\frac{\rho b^2}{m}\frac{V}{n_B b}\frac{V}{n_B L} \tag{5.3.1}$$

where m is the mass per unit length, b is the overall deck width, V is the hourly mean wind speed, L is the length of the longest span, and n_B is the natural frequency in bending (vertical motion). If:

1. $P_b < 0.04$, the bridge may be considered to be subjected to insignificant effects with respect to all forms of aerodynamic excitation.
2. $0.04 \le P_b \le 1.00$, the bridge may be considered to be within the scope of the design rules specified by the Standard.
3. $1.00 > P_b$, the bridge is considered to be very susceptible to aerodynamic excitation.

Normal highway bridges of less than 25-m span are generally in category 1. Bridges with spans greater than 250 m are likely to be in category 2. This classification superseded an earlier classification in the 1993 version [5.68], which was based on span length only (see also [5.69]). The parameter P_b is intended to test the sensitivity of the bridge deck to vortex excitation, galloping, torsional flutter, coupled flutter, and gust responses. The criteria to be satisfied also pertain to safety with respect to fatigue as well as to subjective perception of the motion, or comfort. The validation of the parameter P_b relies significantly on calibration. The numerical factor 16 in Eq. 5.3.1 is an arbitrary scaling factor. The inclusion of the length L in Eq. 5.3.1 is related to the along-span correlation of the flow turbulence and of the vortex excitation.

The fatigue aspect of vortex excitation is addressed through the presence in the expression for P_b of the square of the ratio V to the frequency n_B. As follows from the definition of the Strouhal number, the critical velocity V_{cr} that causes the peak vortex-induced oscillation amplitude is proportional to n_B. Thus, when V/n_B is high, V/V_{cr} is also high, reflecting the relatively frequent occurrence of peak vortex-induced oscillations and therefore a high cycle count. Equation 5.3.1 shows that the value of P_b is then high as well.

In addition to assessing the bridge design in light of the parameter P_b, it is appropriate to assess the bridge sensitivity in light of past experience. Table 5.3.1 lists calculated values of P_b for various bridges. The selection is not consistent and includes bridges with spans longer than 250 m. The sensitivity parameter P_b indicates that bridges with spans exceeding 200 to 250 m can have aerodynamic properties that do not require that wind tunnel tests be conducted. However, unless more data are obtained for calibration purposes, it is not advisable to rely uncritically on Eq. 5.3.1 as a definitive criterion.

5.3.1.2 Outline of Primary Clauses in Standard BD 49/01. The Standard states that bridges belonging to category 2 are to be considered adequate if they satisfy the relevant criteria with respect to (1) response under vortex excitation, (2) response under excitation by flow turbulence, and (3) galloping, torsional flutter, and classical flutter. On the other hand, for bridges in category 3, stability is outside the scope of the rules and must be verified by means of further studies using archived test data for the configurations being designed or for similar configurations, proven theoretical or computational fluid dynamics (CFD) procedures, or wind tunnel tests on scaled models.

The clauses for vortex excitation pertain to (1) critical wind speeds at the peak amplitude as functions of ratio of width to height of the deck and other limiting criteria, and (2) maximum amplitudes for fatigue strength assessments. However, there are no provisions concerning design improvements (change of deck configuration or installation of devices such as tuned mass dampers).

Estimates of galloping, torsional flutter, or classical flutter response are to be derived from formulas on the respective critical wind speeds (for classical flutter a formula similar to Selberg's formula is used). Alternatively, values of critical wind speeds may be determined by wind tunnel tests. Limiting criteria must be satisfied to assure that no galloping, torsional flutter, or classical flutter occurs for speeds less than the specified wind speed. Coefficients of 1.25 for the UK and 1.4 for tropical cyclone-prone locations are applied as safety margins.

Dynamic magnification effects may be disregarded if the fundamental frequency of vibration is greater than 1 Hz. Otherwise, they are to be considered by performing a dynamic analysis.

Other requirements are specified with respect to fatigue, proximity effects (see Sect. 5.3.4.1), and wind tunnel testing (see Sect. 5.3.5). No provisions concerning cables and towers are included in the Standard.

5.3.2 Wind Effects on a Long-Span Bridge: Code of Practice and Case Study

5.3.2.1 Code of Practice. Committee activities on guidelines and manuals for design to withstand wind effects on the Honshu–Shikoku Bridges in Japan started in 1964. Documentation was subjected to revisions in 1967 and 1976,

TABLE 5.3.1. Aerodynamic Susceptibility Parameter P_b for Various Bridge Structures[a]

Bridge Name (Deck Type, Year Completed)	Long Creek (Π-Shape, 1967)	Ishikari Estuary (Shallow Box, 1972)	Kessock (Box, 1982)	Trans-Tokyo Bay (Box, 1995)	Tacoma Narrows (H-Shape, 1940)	Great Belt East (Shallow Box, 1998)	Osteroey (Box, 2000)
Longest span, L (m)	217	288	240	240	853	1624	595
Deck width, B (m)	10	13.3	21.92	22.9	11.89	31	13.6
Deck height, D (m)	2.4	2.2	3.25	6.0	2.44	4.4	2.5
B/D	4.2	6.0	6.7	3.8	4.9	7.0	5.4
Mass per unit length, m (t/m)	4.76	5.2[b]	5.8	3.83	8.5[c]	22.74[c]	≥3.3
Freq. first bending, n_B (Hz)	0.6	0.86[b]	0.516	0.329	0.13	0.099	0.17
Remarks[d]	V (B), modify configuration [5.26]	V (B) during erection only	V (B), install TMD, [5.21]	V (B), install TMD [5.22]	V (B and T), torsional flutter [5.19]	V (B), add vane [5.24]	V (B), add vane [5.70]
Parameter, P_b	0.84	0.38	1.86	7.22	3.04	2.69	≤7.52

Source: Courtesy of the Honshu–Shikoku Bridge Authority.

[a] P_b calculated for $V = 40$ m/s in all cases.

[b] Asphalt (10 cm thick) not included.

[c] Suspension bridge, including mass of cables.

[d] V, vortex excitation; B, vertical bending; T, torsion.

and included specific provisions for the Akashi Kaikyo Bridge (Fig. 5.3.1) in 1991 and the Tatara Bridge in 1994 [5.36]. Provisions for flutter instability in long-span bridges require combinations of wind tunnel modeling in both smooth and simulated boundary layer flow, including topographic modeling. The purpose of the wind tunnel tests is to assess competing bridge deck configuration proposals and to refine the proposal being selected. Determining the best deck configuration is achieved in two-dimensional section model tests. Once this task is completed—a task that is not trivial, owing to possible difficulties in the interpretation of the test results—testing of a three-dimensional aeroelastic full model in the best achievable simulated turbulent flow is the most desirable option. However, the cost of such testing is very high.

It is essential to specify an appropriate safety margin for the design wind speed, that is, an appropriate ratio larger than unity between the wind speed that induces aeroelastic instability and the design wind speed. For the critical flutter speed V_{cr} for the Akashi Kaikyo Bridge, the mean wind speed is specified, in terms of a 10-min average, as follows:

$$V_{cr} \geq 1.2\mu_F V_z = 78 \text{ m/s} \qquad -3^\circ \leq \alpha \leq 3^\circ \tag{5.3.2}$$

Figure 5.3.1. Akashi Kaikyo Bridge. (Courtesy of the Honshu–Shikoku Bridge Authority.)

In Eq. 5.3.2, $V_z = (z/10)^{1/8}V_{\text{ref}}$, where $z = 80$ m is the mean elevation of the stiffening girder and $V_{\text{ref}} = 46$ m/s is the estimated 10-min speed with a 150-year mean recurrence interval at a 10-m height over the water level, so $V_z = 60$ m/s. The factor 1.2 is a principal safety factor that accounts for the uncertainties inherent in the section model tests and in the design and construction of the prototype structure, as well as for the social importance of the structure and the fact that flutter has catastrophic consequences. In earlier discussions, a factor of 1.3 was adopted, but it was subsequently agreed that it was appropriate to reduce it to 1.2 in view of the reduced uncertainties believed to have been achieved by wind tunnel and other studies. The coefficient μ_F is a conditional safety factor. Its value, taken to be 1.08, accounts for the imperfect reproduction in the wind tunnel of low-frequency velocity fluctuations and their role in the assumed buildup time for flutter response, estimated to be about 30 s.

Studies showed that flutter response depends strongly on angle of attack α. It was therefore specified that section model wind tunnels be performed for angles of attack $-3° \leq \alpha \leq 3°$. This provision accounts for the effects of low-frequency vertical and longitudinal wind speed fluctuations, expected torsional angles of the deck due to wind loads, and uncertainties with respect to the accuracy of the wind tunnel tests. Originally, angles of attack $-7° \leq \alpha \leq 7°$ were considered, but on the strength of subsequent studies and data, that tentative criterion was relaxed.

Mechanical (viscous) damping is also a key factor, to which the onset of flutter instability is strongly susceptible. Some field observations for long-span bridges recently became available (Tables 5.3.2*a* and *b*). On the basis of those data, and in view of the size of the Akashi Kaikyo Bridge, relatively small structural damping values were selected for the wind tunnel tests as well as for the other verification procedures being used. Mechanical damping was also specified for freestanding towers and for the towers of the completed bridge. Requirements specified for the mechanical damping are as follows (in terms of logarithmic decrement δ):

- *Truss stiffening girder:* 0.03 (bending), 0.02 (torsion)
- *Box stiffening girder:* 0.02 (bending), 0.02 (torsion)
- *Tower under construction:* 0.01; after completion, 0.02
- *Stay or hanger cable:* 0.003

Nearly all the structural members of the truss stiffening girder are controlled by wind loading, primarily through the horizontal drag component normal to the longitudinal bridge axis. The static wind load for the deck girder is specified as

$$P_D = \tfrac{1}{2}\mu_2\rho V_z^2 C_D A_n \tag{5.3.3}$$

TABLE 5.3.2a. Natural Frequencies and Mechanical Damping for Various Bridges

Bridge Name:	Minami Bisan–Seto		Ohnaruto		Innoshima		Rainbow		Ohshima		Hakucho		Akinada	
Spans (m):	274 + 1100 + 274		93 + 330 + 876 + 330		250 + 770 + 250		114 + 570 + 114		140 + 560 + 140		330 + 720 + 330		255 + 750 + 170	
Vibration Mode	Freq. (Hz)	Log. Dec.	Freq. (Hz)	Log. Dec.	Freq. (Hz)	Log. Dec.	Freq. (Hz)	Log Dec.	Freq. (Hz)	Log. Dec.	Freq. (Hz)	Log. Dec.	Freq. (Hz)	Log. Dec.
Vertical														
First asym.	0.151	0.088	0.167	0.108	—	0.024	0.255	0.15	0.189	0.018	0.149	0.016	—	—
First sym.	0.168	0.037	0.166	0.108	0.178	0.035	0.260	0.06	0.232	0.017	0.129	0.034	—	—
Second Sym.	0.228	0.096	0.284	0.085	—	—	0.411	0.16	0.323	0.020	0.218	0.027	—	—
Torsion														
First sym.	0.329	0.026	0.328	0.028	0.335	0.020	0.485	0.06	0.553	0.018	0.469	0.036	0.536	0.040
First asym.	0.457	0.038	0.506	0.062	0.583	—	0.681	0.13	0.759	0.041	0.760	0.033	0.845	0.074
Horizontal														
First sym.	—	—	—	—	—	—	—	—	—	—	0.099	0.290	0.078	0.087
First asym.	—	—	—	—	—	—	—	—	—	—	—	—	0.224	0.157
Second sym.	—	—	—	—	—	—	—	—	—	—	0.560	0.039	0.466	0.066

Source: Courtesy of the Honshu–Shikoku Bridge Authority.

TABLE 5.3.2b. Natural Frequencies and Structural Damping of Cable-Stayed Bridges

Bridge Name:	Tatara		Tsurumi–Tsubasa		Ikuchi		Yokohama Bay		Hitsuishijima	
Spans (m):	270 + 890 + 320		255 + 510 + 255		150 + 490 + 150		200 + 460 + 200		185 + 420 + 185	
Vibration Mode	Freq. (Hz)	Log. Dec.	Freq. (Hz)	Log. Dec.	Freq. (Hz)	Log. Dec.	Freq. (Hz)	Log. Dec.	Freq. (Hz)	Log. Dec.
Vertical										
First sym.	0.226	0.024	0.213	0.07	0.340	0.018	0.351	0.069	0.430	0.080
First asym.	0.263	0.018	0.293	0.04	0.443	0.020	0.565	0.134	0.729	0.088
Second sym.	0.348	0.007	0.518	0.10	0.605	0.031	0.828	0.088	1.030	0.066
Torsion										
First sym.	0.497	0.017	0.545	0.07	0.735	0.038	0.876	0.041	1.058	0.046
First asym.	0.831	0.051	—	—	1.603	0.019	—	—	1.910	0.071

Source: Courtesy of the Honshu–Shikoku Bridge Authority.

where the correction factor μ_2 is 1.55 for the deck and main cables normal to the bridge axis and is 1.25 for the deck girder parallel to the bridge axis; A_n is the frontal area. The drag coefficient C_D of the deck must be verified by wind tunnel testing. The correction factor is used to multiply the static loading by factors accounting for flow turbulence effects. For the Akashi Kaikyo bridge, only the along-wind drag component was taken into consideration for preliminary design purposes; other effects of buffeting were considered in the final design.

5.3.2.2 Analysis and Design for Flutter of the Akashi Kaikyo Bridge. As mentioned earlier, section model tests have been conducted for a large variety of proposed stiffening girders [5.71, 5.72], from which two typical deck configurations have been selected to satisfy the requirement against flutter instability: a compound system of shallow and deep closed box decks, and a truss girder [5.37]. The latter was adopted for the final design. Construction considerations and the need to keep navigation routes open to the maximum extent possible during construction played a role in the selection. Additional section model wind tunnel tests were conducted to study the effects of such factors as the width of the stiffening girder, grating, parapets, curbs and handrails, rails for inspection and maintenance work, pipes for public power lines and information networks, water supply pipes, and so on.

It was inferred from section model tests that [5.37, 5.38]:

1. The critical flutter velocity depends strongly on the size and location of attached appurtenances.
2. Torsional flutter occurs at a wind speed of 40 to 50 m/s. Coupled flutter with vertical bending and torsion occurs at higher wind speeds.
3. The original design for the deck was not acceptable from the point of view of flutter behavior. To raise the critical wind speed, a vertical plate that stabilized the deck's behavior was installed below the center of the floor deck, and a solid central barrier was provided at the floor deck (Fig. 5.3.2).
4. Aerodynamic interference effects between the truss girders and the main cables caused the critical wind speed to decrease.

On the basis of section model test results it was concluded that the coupled flutter and the buffeting response needed to be investigated in as detailed a manner as possible. Also of concern were the very large horizontal static deflection of the deck and the torsion, resulting in a negative angle of attack at the center. For these reasons it was decided to perform tests on an aeroelastic full suspension bridge model with a total length of about 40 m (1/100 scale). The model was tested in a newly built large boundary-layer wind tunnel 41 m wide and 4 m high with a 30-m-long test section (Fig. 5.3.3).

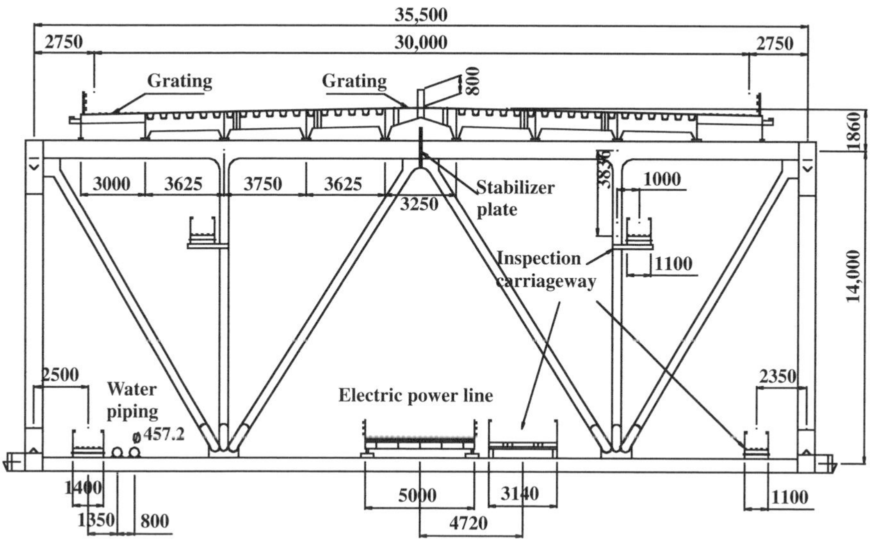

Figure 5.3.2. Truss stiffening girder, Akashi Kaikyo Bridge. (Courtesy of the Honshu–Shikoku Bridge Authority.)

Figure 5.3.3. Aeroelastic 1/100-scale full model. (Courtesy of the Honshu–Shikoku Bridge Authority.)

Test results obtained from the aeroelastic full model in turbulent boundary-layer flow, and for purposes of comparison, in uniform smooth flow as well, are summarized as follows:

1. The critical flutter velocity for the original deck design was higher in smooth flow for the full bridge model than for the sectional model; however, in turbulent flow it was slightly lower. To stabilize the bridge, a vertical plate was installed along the central span.
2. Flutter under the three-dimensionally coupled motion, illustrated in Fig. 5.3.4 for the original deck, indicated dominant torsion and vertical bending accompanied by lateral bending. The flutter motion was accompanied by buffeting caused by the turbulent flow, and at the center of the span, a large lateral deflection, as well as torsion with the windward edge down. Figure 5.3.5*a* shows a set of three displacement components in the flutter mode recorded at every one-eighth interval of one period. The phase lag between the components can be seen in Fig. 5.3.5*b*.

Figure 5.3.4. Wind-induced model deformation. (Courtesy of the Honshu–Shikoku Bridge Authority.)

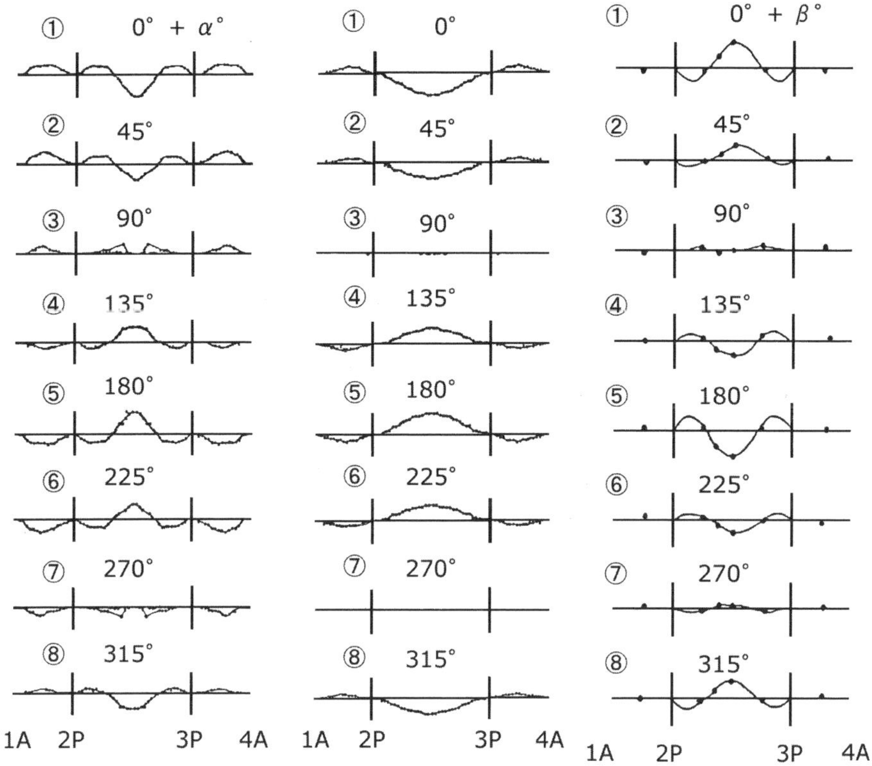

Vertical comp. Torsional comp. Lateral comp.

(*a*)

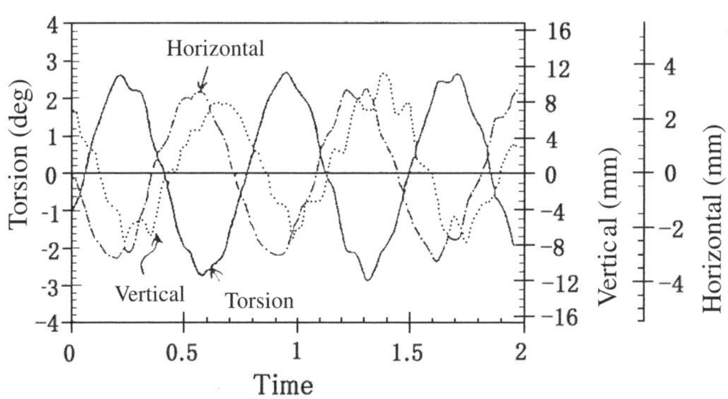

(*b*)

Figure 5.3.5. Typical coupling in full bridge model: (*a*) coupled flutter mode shapes (α and β are mean phase lag angles); (*b*) coupled vertical bending, torsion, and lateral bending. (Courtesy of the Honshu–Shikoku Bridge Authority.)

3. Figures 5.3.6 and 5.3.7 show, respectively, the dependence of aerodynamic damping and response frequency on wind speed measured in the laboratory, on the one hand, and the dependence calculated using analytical models discussed earlier (see Appendix 5B), on the other. Figure 5.3.6 shows that the positive aerodynamic damping occurring at lower wind speeds turns negative at about 8 m/s and remains negative at higher speeds. Results obtained analytically by accounting for coupling of bending and torsional motions, but not accounting for along-wind horizontal motions, are in good agreement with observations for wind speeds of 7 m/s or less. They are inadequate, however, for higher wind speeds. Analyses in which coupling with the horizontal deck motion was taken into account significantly improved the agreement between calculated and measured aerodynamic damping and response frequencies. In particular, they yielded a change of sign of the aerodynamic damping at an 8-m/s wind speed, in agreement with the observations.
4. Analytical results concerning modal shapes associated with flutter were in good agreement with observations on the aeroelastic full-bridge wind tunnel model.

Further analysis confirmed that the critical flutter velocity was reduced by the aerodynamic interference between stiffening the girder deck and the main cables, and was increased by the drag force acting on the cables.

5.3.2.3 Requirements for the Akashi Kaikyo Bridge Main Tower Dampers. The model's main towers experienced reduced but still significant vortex-induced motions after their configuration was improved through the

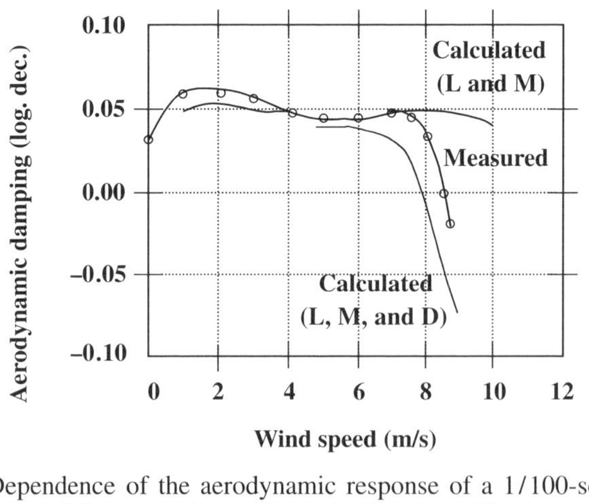

Figure 5.3.6. Dependence of the aerodynamic response of a 1/100-scale full bridge model on wind tunnel flow speed: aerodynamic damping. (Courtesy of the Honshu–Shikoku Bridge Authority.)

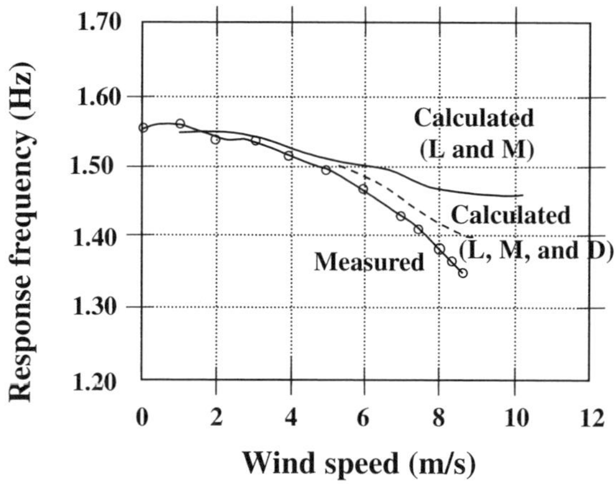

Figure 5.3.7. Dependence of the aerodynamic response of a 1/100-scale full bridge model on wind tunnel flow speed: response frequency. (Courtesy of the Honshu–Shikoku Bridge Authority.)

use of corner-cut cross sections. Figure 5.2.14 shows wind tunnel test results for the towers of the completed bridge. Severe motions are noted in the first in-plane bending mode and for the first torsional mode of the tower, consisting of twin shafts, each shaft being partially constrained at the tip by the main cable connections. Addition of mechanical damping at the connections was quite effective, so no increase in the tower stiffness by changing the tower cross section was judged to be necessary [5.29].

The main towers of the bridge prototype exhibited vortex-induced oscillations whose reduction to allowable levels required the installation of tuned mass dampers and mechanical dampers attached between the tower and the deck's truss girder. The damping to be added was determined as follows. Allowable amplitudes were determined to meet strength and serviceability criteria under various load combinations, with safety margins dependent on the wind speed causing the onset of vortex excitation in various modes of vibration. For each construction stage, from freestanding single towers to the completion of bridge, it was required that the largest stresses in the tower not exceed the allowable stress, that wind-induced uplift forces at the tower base not exceed acceptable levels, and that the number of exceedances of various specified stress levels be sufficiently small for fatigue not to occur.

Allowable amplitudes of the tower and safety margins for the stresses were determined for a number of load combinations and member locations where checking of the amplitudes and/or stresses was necessary. The amplitudes and safety margins depended on whether lower speeds (speeds with frequent occurrences) or higher speeds (speeds with rare occurrences) were considered. The margins were larger for the higher speeds, which are associated with greater uncertainties, and may result in failure of the bridge. The locations being specified differed according to type of vibration (in-plane bending or torsion of the twin-shaft assembly). Both the shafts and the diagonals connecting them were checked. The allowable amplitude of vibrations in the first bending mode in the plane normal to the bridge axis was specified to be 0.3 m. This limit was adopted because the 30-m/s wind speed required to excite such vibrations has a relatively high probability of being exceeded, and at this speed the road may still be open to traffic. Even though they would not cause dangerous stresses, higher amplitudes would be objectionable from a serviceability viewpoint. The allowable amplitude for first-mode torsional motion of the twin-shaft assembly was specified as 0.15 m, because the speed causing the onset of vibrations is close to the design wind speed for the towers; higher amplitudes would correspond to unacceptably high stresses. The acceleration level for the erection stage was specified to be 50 gals (50/980 times the acceleration of gravity), consistent with International Standards Organization requirements.

When calculating required values of the damping to be supplied by dampers, a margin of about 20% is added to account for uncertainties. The requirements are based on the dependence of the peak amplitude of the vortex excitation upon Scruton number Sc for each construction stage of interest, as indicated by wind tunnel test data. The required and design values of the damping are presented in Table 5.3.3 for the completed structure. Additional dampers installed to meet requirements for other stages of the construction process are discussed in Sect. 5.3.3.2.

Information on the devices used to supply the additional mechanical damping required is presented in the next section.

5.3.3 Reducing Wind-Induced Vibrations in Decks and Towers: Aerodynamic and Mechanical Approaches

5.3.3.1 Improvement of Deck Aerodynamics. Functional criteria can result in designs of decks with bluff shapes. Aerodynamic improvements and passive aerodynamic devices that help to reduce or suppress vortex-excited vibrations, galloping, and flutter are discussed in Sects. 5.3.3.1.1 to 5.3.3.1.4 and are illustrated in Figure 5.3.8.

TABLE 5.3.3. Allowable Amplitudes, Required Damping, and Damping to Be Supplied by Damping Devices

Mode of Vibration	Allowable Amplitude (m)	Required Damping (Log. Dec.)	Damping to Be Supplied (Log. Dec.)
Bending, first mode	0.30	0.051	0.061
Torsion, first mode	0.15	0.071	0.085

Source: Courtesy of the Honshu–Shikoku Bridge Authority.

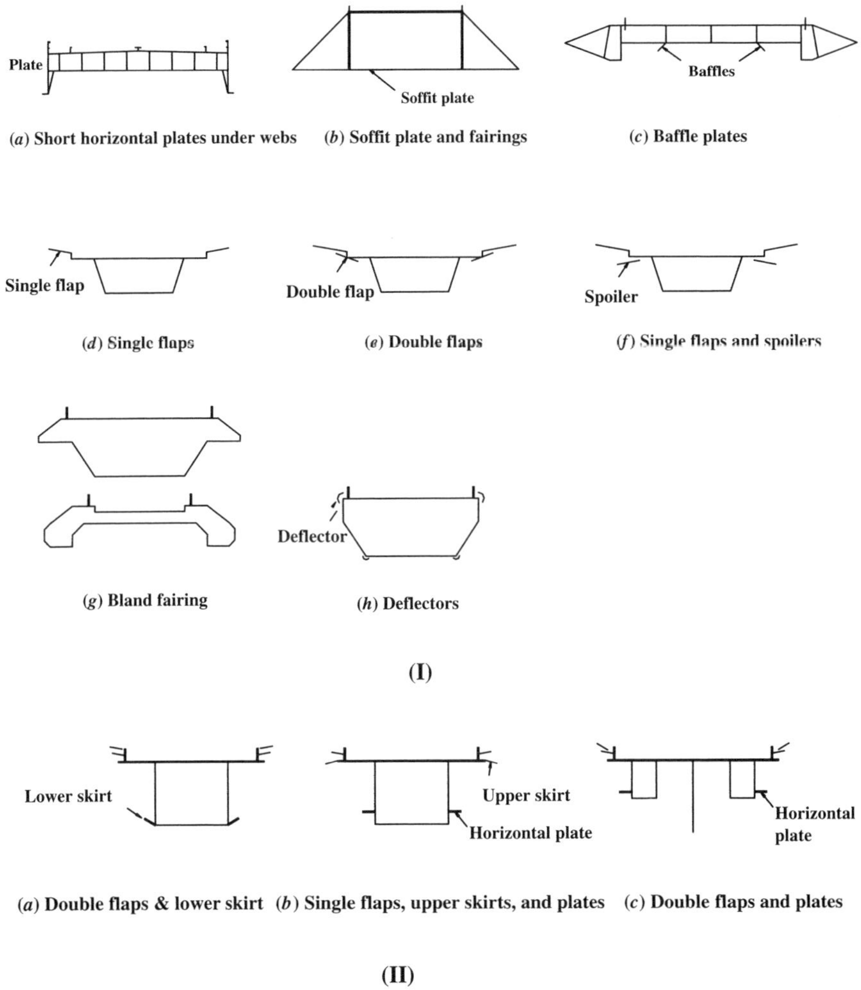

Figure 5.3.8. Aerodynamic improvements to reduce wind-induced vibrations for various types of bridge deck: [I] reduction of vortex excitation; [II] suppression of galloping; [III] suppression of coupled flutter; [IV] aerodynamic improvements, bridge towers and pylons. (From [5.1, 5.2, 5.73].)

5.3.3.1.1 Shallow Bridge Decks. Peak vortex-induced vertical or torsional deck motions in shallow bridge decks are due to flow separation and steady (as opposed to intermittent) flow reattachment. Aerodynamic devices are designed to help render the flow around the deck closer to being streamlined.

1. *For decks with open sections:* (a) *horizontal plates* attached to the lower chord of each web help to produce a less strongly separated flow; (b)

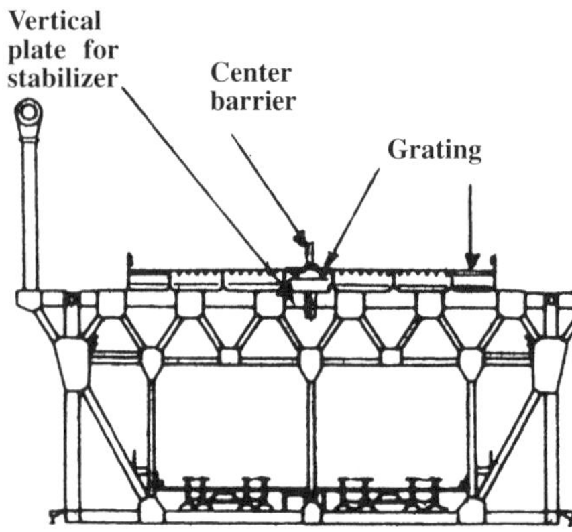

(*a*) Combination for truss girder

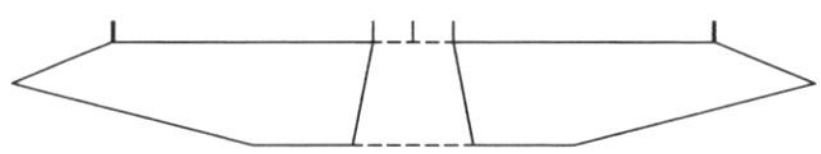

(*b*) Two boxes with open gap

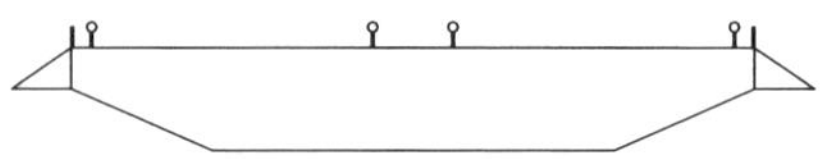

(*c*) Wind nose at both ends

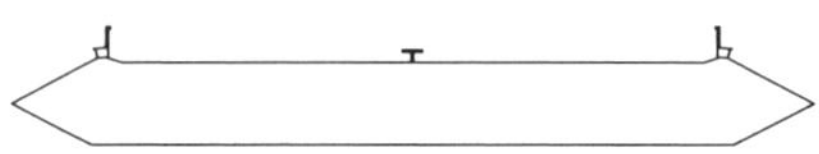

(*d*) Deflectors at ends of floor

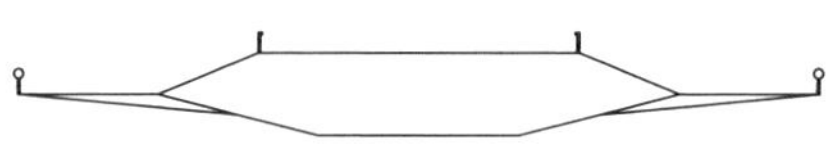

(*e*) Spoilers at both ends

(III)

Figure 5.3.8. (*Continued*)

soffit plates and fairings help to streamline the flow and promote reattachment; (c) *baffle plates* help to break the whirlpool of the separated flow underneath the deck (Fig. 5.3.8I*a–c*).

2. *For box decks with closed sections:* (d) *flaps,* (e) *double flaps,* (f) *single flaps* and *spoilers* enhance the effect of the flaps; (g) *fairings,* and (h) *deflectors* promote streamlining of the flow (Fig. 5.3.8I*d–h*).

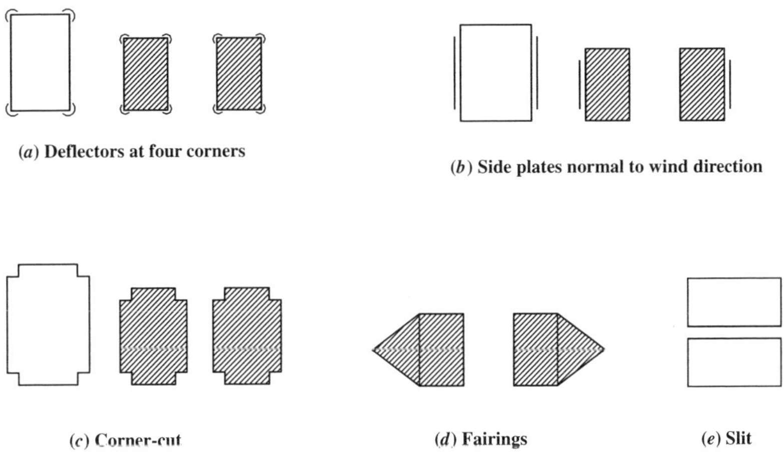

(IV)

Figure 5.3.8. (*Continued*)

5.3.3.1.2 Deep Bridge Decks. Aerodynamic devices are designed to affect (1) vortex excitations, which are due to intermittent (as opposed to steady) reattachment (see Sect. 5.2.1); and (2) galloping, associated with flows in which reattachment does not occur. The use of such devices entails trade-offs, as noted in Sect. 5.2.5.1. Typical devices include (a) *inclined skirt-plates* at lower corner points of a deep box deck; (b) and (c) *short horizontal plates* attached to the webs, which promote reattachment to the lower side of the bridge deck; (a), (b), and (c) *flaps* placed at the ends of the deck floor, which tend to promote reattachment to the upper surface of the deck and to inhibit galloping, owing to the large uplift force that occurs during downward motions (Fig. 5.3.8II*a–c*). Figure 5.3.9 shows a typical example of a box deck, the Tozaki Bridge, with double flaps and lower skirt plates that were confirmed to be effective by results of full-model wind tunnel tests.

5.3.3.1.3 Shallow Box and Truss Stiffening Girders in Long-Span Bridges. Coupled or torsional flutter is a primary concern. The following countermeasures can be helpful: (a) *gratings* installed at both ends and/or at the center of floor deck to accelerate vertical flow; (b) an *open gap* provided within the box deck; (c) *wind-nose devices;* (d) *deflectors* (curved devices similar to those of Fig. 5.3.8I*h*), and (e) *spoilers,* which reduce the flow intensity in the separation zone. In addition, a *vertical plate* installed at the center below the floor deck, or a *closed center barrier* above the floor surface, appears to have effects similar to those of spoilers (Fig. 5.3.8III*a–c*). For the Akashi Kaikyo Bridge [5.37, 5.38], a *vertical plate* placed on the entire center span substantially increased the critical speed for the coupled flutter.

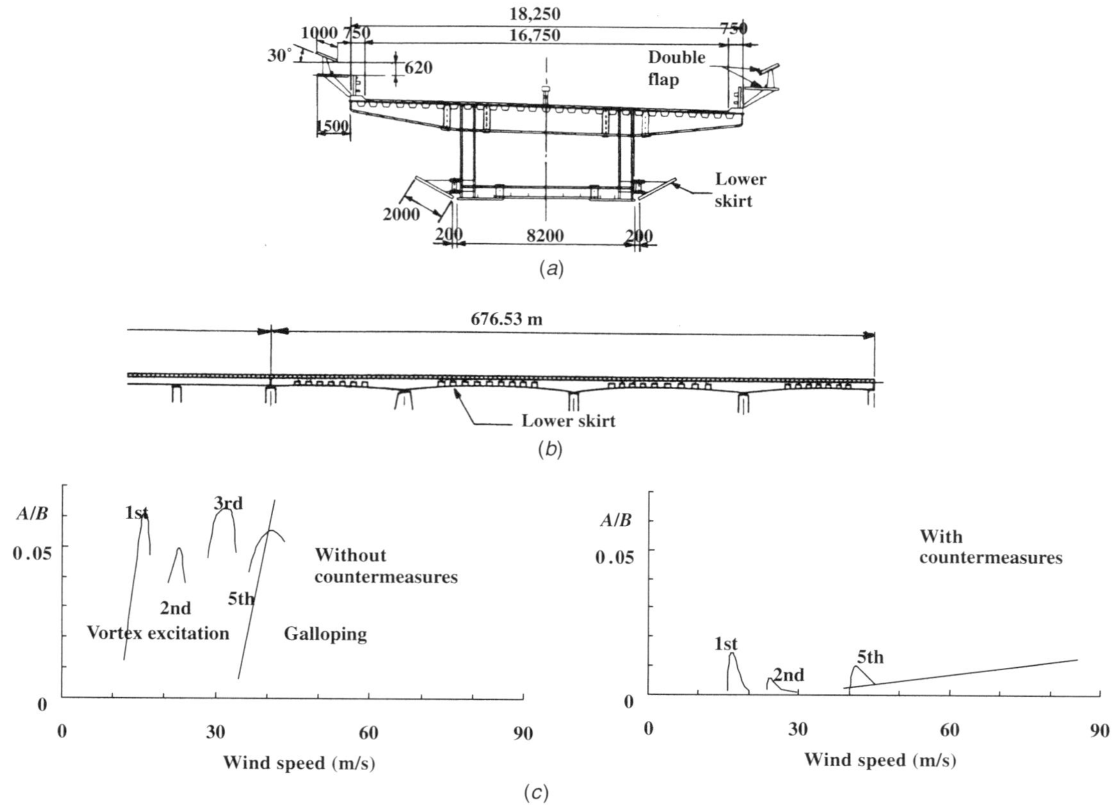

Figure 5.3.9. Reduction of vortex excitation and galloping by double flaps and lower skirts for a box deck bridge: (*a*) prototype section with countermeasures; (*b*) location of lower skirts for a four-span continuous box bridge; (*c*) comparison of wind-induced vibrations in a full-model wind tunnel test with surrounding topography (smooth flow). (From [5.74]; courtesy of the Honshu–Shikoku Bridge Authority.)

5.3.3.1.4 Main Suspension Bridge Towers and Cable-Stayed Bridge Pylons. To mitigate unavoidable bending vibrations due to vortex excitation or galloping, the following devices are used: (a) *deflectors;* (b) *side plates;* (c) *corner cuts* [which produce interferences among separated vortices, as indicated schematically in Fig. 5.2.2, and were used to good effect for the Akashi Kaikyo Bridge towers (see Fig. 5.2.14), although dampers were still necessary]; (d) *fairings,* whose effects are similar to those occurring in decks; and (e) *slits,* which shift separation points windward and decrease the along-wind spatial extent of the separated flow (Fig. 5.3.8IV).

Splitter plates, punched sections, or trip and helical wires attached to the surface of circular pipes are used to mitigate vortex excitation. They tend to increase drag, however, and even their effectiveness as vortex suppressors has been challenged [5.67, 5.75]. As mentioned previously, the effects of turbulence need to be taken into consideration. Turbulence intensifies the flow in the separation region and shifts the reattachment point windward. Those effects differ for various types of deck, hence the need to perform tests in turbulent flow. The presence of topographical obstacles around the bridge site can also affect aerodynamic behavior.

Active devices are currently being investigated. They include flaps or fairings controlled mechanically, and fast-rotating rotors attached at the corners. Such devices are commonly used in mechanical and aeronautical engineering applications. However, their adoption for civil engineering purposes raises operation, maintenance, and cost problems that do not appear to have been solved satisfactorily.

5.3.3.2 Mechanical Dampers. As mentioned earlier, increasing the Scruton number can result in a reduction of the peak vortex-induced response and an increase in the wind speed corresponding to the onset of galloping and torsional flutter. Increasing the Scruton number amounts to increasing the body mass and/or the mechanical damping. Procedures that have been proposed for increasing the damping are:

1. Hanging weights from the bridge deck into the water to increase viscous damping. This procedure is largely impractical.
2. Installing sliding blocks that generate friction. Backstays from the tower top are attached to sleds loaded with ballast and free to slide on inclined steel ramps built at ground level. This approach is used to control vibration during tower construction.
3. Installation of ordinary oil dampers. These devices have the disadvantage of necessitating large installations as well as being of uncertain reliability. Currently, more compact devices are used, which take advantage of modern technological developments. They include the tuned mass damper (TMD) and the tuned liquid damper (TLD), which are passive; the active tuned mass damper (ATMD); and others. ATMDs

have relatively simple mechanisms assuring a favorable dynamic balance of additional mass, spring, and damping, and are more versatile, but they require more maintenance and are more costly. The TMD and TLD have limitations insofar as they are devised for a fixed frequency of vibration.

Compact systems such as TMDs have been installed within bridge decks with a view to controlling peak vortex-induced vertical oscillations. For the Kessock Bridge (Table 5.3.1), eight units of a TMD system were used. The weight of each unit was 2 tons [5.20, 5.21]. As shown in Fig. 5.3.10, in the

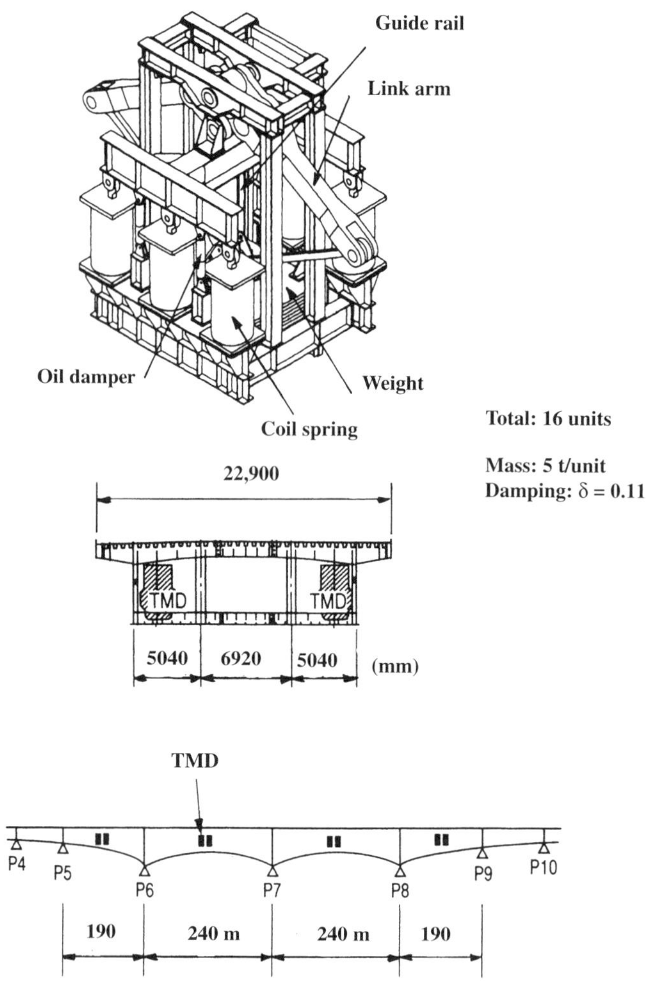

Figure 5.3.10. Tuned mass damper installation inside a box deck for the Trans-Tokyo Bay Bridge. (From [5.22]; courtesy of the Japan Highway Public Corporation.)

Trans-Tokyo Bay Bridge 16 TMD units, each with a weight of 5 tons, were installed over the bridge's four long spans [5.22]. They were designed to provide mechanical damping with logarithmic decrement $\delta = 0.22$ for the first mode and $\delta = 0.14$ for the second mode of vibration. Following their installation, no perceptible vortex-induced response occurred, whereas in their absence the logarithmic decrements for the damping inherent in the bridge structure itself were $\delta = 0.028$ to 0.044 and 0.031 to 0.047 for the first and second modes, respectively, resulting in an amplitude of the peak vortex-induced oscillation of 0.58 m at the bridge center observed in a 16-m/s wind speed. TMD devices for controlling torsional vibrations are more elaborate than those for vertical vibrations, and necessitate large installations (Fig. 5.3.11).

As explained in Sect. 5.3.2.3, it was predicted that large vortex-induced oscillations of various types would occur in the main towers of the Akashi Kaikyo Bridge. To control different types of vibration during construction as well as after completion of the bridge, relatively complex TMD and ATMD systems were installed, which resulted in satisfactory tower performance [5.29].

The main towers are erected using lifting cranes. Higher towers, in particular single freestanding towers, have a high degree of flexibility. After placement of the catwalks for main suspension-bridge cable erection over water, or step-by-step placement of stay cables in a cable-stayed bridge, the flexi-

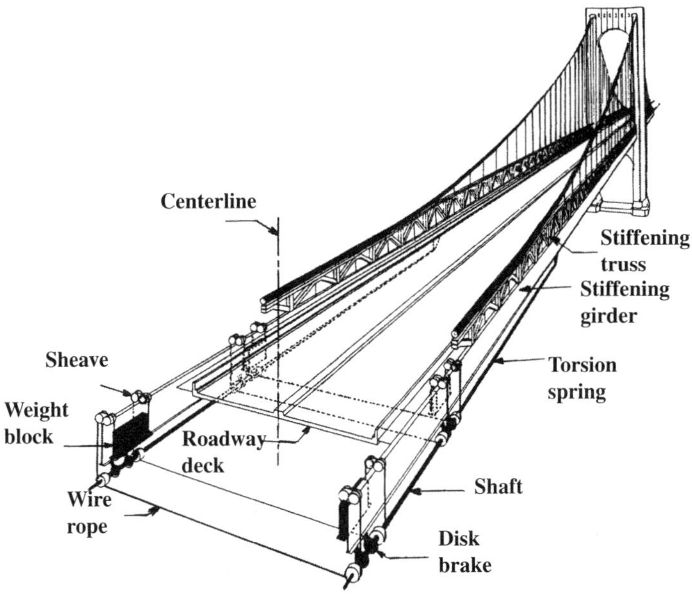

Figure 5.3.11. Large-scale installation of TMD for torsional vibration in a suspension bridge. (By permission of MTA Bridges and Tunnels.)

bility of the towers decreases gradually. Structural damping also increases. In many instances, vortex excitation is a problem during construction of the main towers. Therefore, a method for providing the requisite damping that is sufficiently versatile to accommodate the needs inherent in each construction phase is desirable. The attractiveness of an active control system with variable parameters should be viewed in this light. For this reason, ATMDs are added to systems based on the use of TMDs.

We describe briefly the sequence of steps followed in the investigation of vortex-excitation control for the main towers of the Akashi Kaikyo Bridge. First, basic data on vortex excitation were obtained in wind tunnel tests. An examination of whether the motions were acceptable was based on a parametric study of the effects of mechanical damping. Criteria were developed to assure safety with respect to (1) the strength of the tower structure, (2) the erection equipment, and (3) the work environment during erection of the towers. Criteria are based on requirements on allowable acceleration and displacement amplitudes (see Sect. 5.3.2.3). As noted earlier, the required damping value δ_{req}, based on wind tunnel test data, was affected by safety margins of about 1.2 to account for uncertainties, resulting in the design damping value δ_{des} to be provided by the damping devices, the requisite properties of the TMD and ATMD systems being determined by taking into account the towers' vibration characteristics at the various stages of construction, from single freestanding shafts to the fully completed bridge.

We now discuss briefly the active control method. ATMDs were first used in 1991 on the towers of a short suspension bridge. They have since been applied to other bridges, including the Akashi Kaikyo Bridge and the Kurushima Triple Bridges. The required TMD mass is usually roughly 1% of the tower mass. For ATMDs the mass requirement is as low as about 0.15%. Almost all applications were targeted only to the first mode of out-of-plane bending. Multimode control devices have also been used, targeting several modes in out-of-plane bending, torsion, and in-plane bending. Robust control methods, including optimum feedback control, and fuzzy control methods, with time lags calculated online, are being developed as well.

5.3.4 Special Aerodynamic Effects

5.3.4.1 Twin-Deck Configurations and Other Proximity Effects. Special consideration must be given bridges parallel and in close proximity to each other and to other situations in which parallel decks cause aerodynamic interferences. It is not possible to develop general rules; the only way to acquire the requisite aerodynamic information is to conduct well-designed wind tunnel tests.

Figure 5.3.12 shows wind tunnel test results for the case in which the buffeting response of an existing short suspension bridge is induced by the turbulent wake of a new bridge with a deep box deck. Vortex excitation and galloping were of concern (Fig. 5.3.12*a*). A short horizontal plate was there-

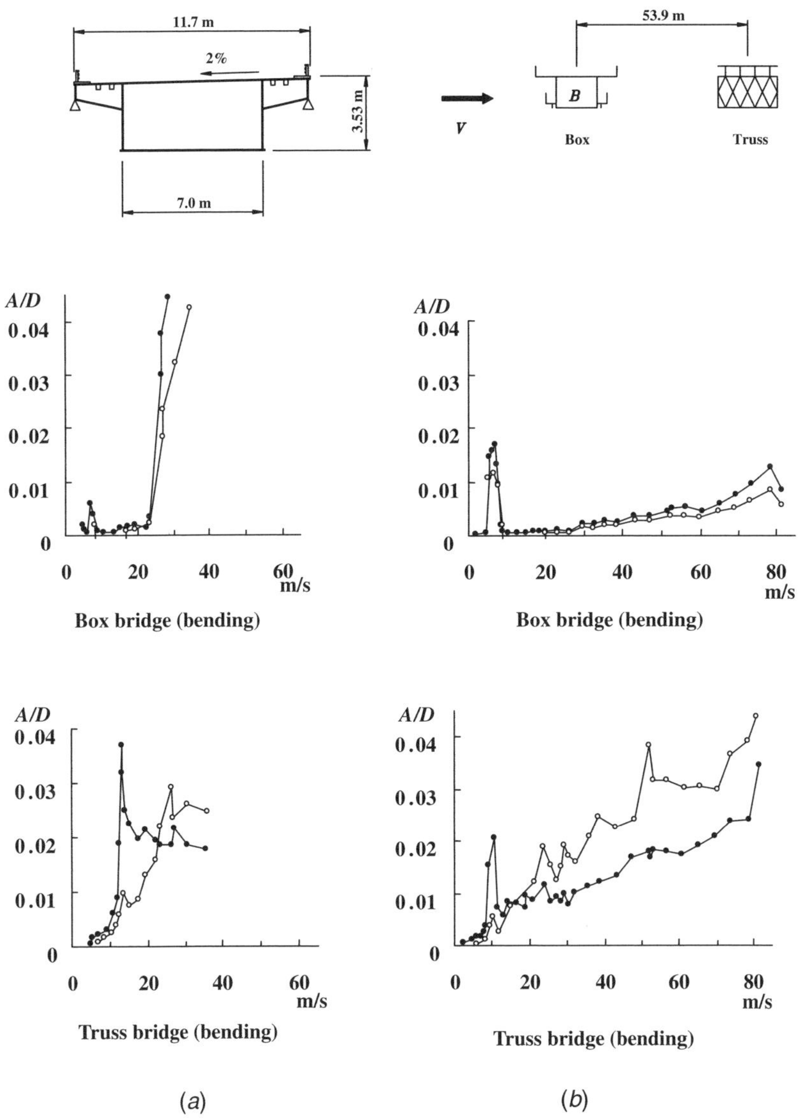

Figure 5.3.12. Interference effects for galloping and random responses in new and existing bridges: (*a*) original box deck (upper left); (*b*) box deck with short horizontal plates. Black circles, point at center of bridge; open circles, point at one-fourth span distance from support. (From [5.74], courtesy of the Honshu–Shikoku Bridge Authority.)

fore attached on each side at the bottom of the box deck model. This led to the suppression of galloping in the box bridge, as shown in Fig. 5.3.12*b*. Wind tunnel tests showed that the buffeting vibration in the existing bridge depended on whether plates were attached to the deck of the new bridge. If no plates were attached, the turbulent wake produced by the new bridge induced in the existing bridge buffeting vibrations larger than those due to atmospheric turbulence in the absence of the new bridge. The effects of attaching the plates were: (1) The new bridge's galloping motion was suppressed, owing to the reattachment of the flow on the lower surface of the box girder, (2) the new bridge's bending vibrations due to vortex excitation were stronger, owing to the vorticity present in the separated flow near the lower surface of the box girder; (3) the leeward bridge's bending vibrations were reduced, at least for speeds of up to about 50 m/s.

Figure 5.3.13 is a plot of mean pressure distributions measured in two parallel shallow box decks with nearly streamlined configuration, for three angles of attack: $\alpha = 3°$, $0°$, and $-3°$. The large differences between the pressures on the two decks are evident.

Guidelines for evaluating proximity effects on twin-deck configurations with a horizontal gap G are provided in the BD 49/01 rules [5.68]. If $G > G_2$, where G_2 is equal to six times the deck width, each bridge deck may be

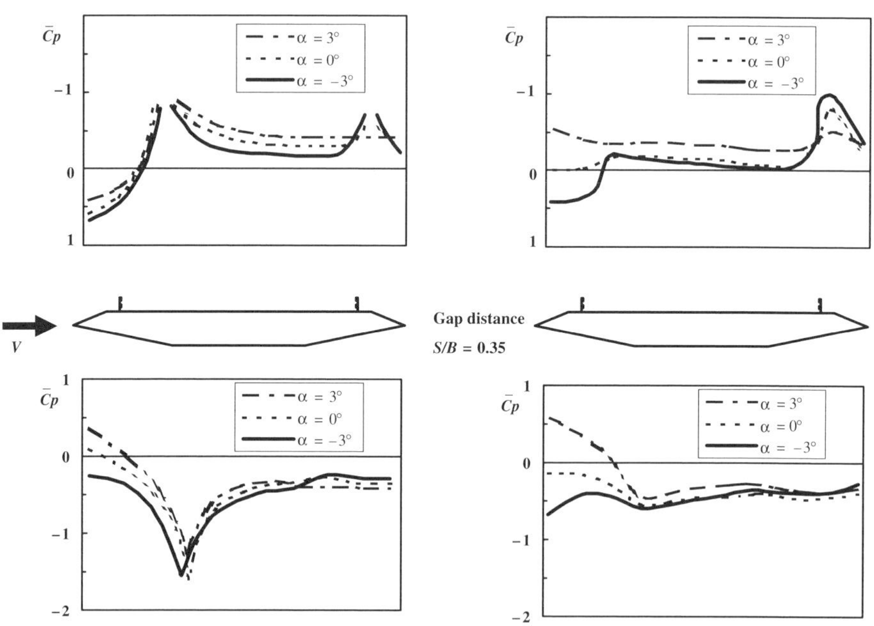

Figure 5.3.13. Proximity effects on mean pressure distribution of parallel streamlined bridge decks. Ratio of gap between decks and deck width: $S/B = 0.35$. (Courtesy of Yokohama National University.)

treated separately with respect to vortex excitation and buffeting due to atmospheric turbulence. For $G_1 < G < G_2$, where G_1 is equal to one-third of the deck height or width, the estimate of the peak vortex-induced response amplitude to vortex excitation should be doubled for each bridge. For 1 m $< G < G_1$ for twin deck configurations, and for $G < G_1$ for all other configurations, special investigations should be made to investigate the interactive vortex response of the dual system. Buffeting by atmospheric turbulence should be halved if there is a parallel structure with $G_1 < G < G_2$, and the parallel structures may be considered as a single structure if $G < G_1$. These provisions are based on numerous wind tunnel tests results [5.69].

Proximity effects caused by parallel obstacles can also be significant in the following cases:

1. Two parallel shafts in main towers (see Fig. 5.2.14)
2. Two nearly-parallel shafts in a variety of cable-stayed bridge pylons with H, A, and reversed-Y shapes
3. Parallel hanger ropes
4. Main cables and stiffening girders

5.3.4.2 Wind Shielding. Large-scale bridge structures are often exposed to high winds which may affect not only the bridge itself, but also the comfort of pedestrians and the safety of vehicles on the bridge floor. Strong winds may require traffic restrictions and closures, limits on the size of some large vehicles, and speed limitations. Wind shielding would help to protect the traffic from such winds. The addition of bluff barriers at the edges of a bridge deck would greatly increase the wind loads on the deck while increasing the bridge's susceptibility to aeroelastic instability. Porous barriers are therefore far preferable, although it is necessary to consider the possible effect of snow or ice buildup.

Figure 5.3.14*a* and *b* show shielding devices provided along the entire bridge span to reduce the wind speed over the floor. Figure 5.3.14*c* and *d* show local wind shielding devices aimed at preventing effects on the main towers of sudden changes of side winds in an area affected by hilly terrain. Sound-shielding walls may also be required, and the wind loads acting on them need to be estimated.

For the Second Severn Bridge [5.77], gust speeds at the deck level are approximately 1.4 times those near the ground. As wind forces are proportional to the square of the wind speed, in the absence of shielding, the forces acting on vehicles would be approximately twice those experienced by vehicles on adjacent ground-level roads. Wind shielding strongly reduces wind forces and therefore the need for traffic restrictions. However, wind tunnel test results showed that the drag forces on the decks were about 2.5 times higher with the barriers in place than in the absence of barriers. To reduce the overall drag forces, the bridge designers adopted a streamlined deck shape.

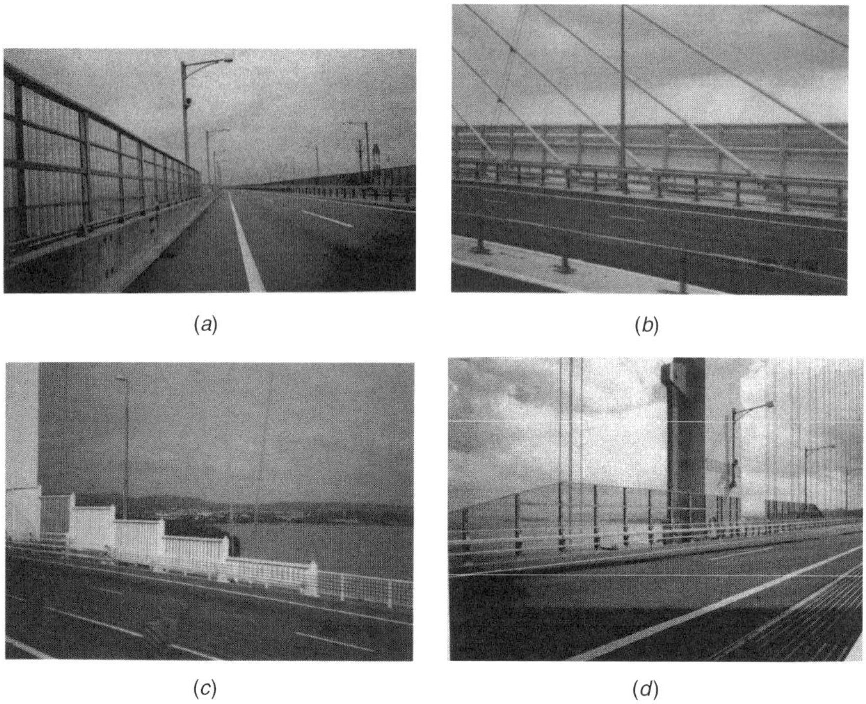

Figure 5.3.14. Examples of wind shielding for highway bridges: (*a*) Honshu–Shikoku–Bisanseto Bridge approaching a road; (*b*) second Severn cable-stayed bridge; (*c*) Severn suspension bridge; (*d*) Honshu–Shikoku–Onaruto Bridge.

For a nonstreamlined deck without barriers, British Standard BS5400: Part 2 provides drag coefficients of 3.52 for a wide deck and 4.81 for a narrow deck. Hence, for bridges where a nonstreamlined deck would be adequate in the absence of barriers, wind shielding can be used almost without penalty by using a streamlined deck instead. Streamlining can be provided by nonstructural components. Its beneficial effects include increasing the width of the deck, reducing the drag, and increasing the wind speed at which aeroelastic instability occurs.

Vortex-induced excitations are sensitive to bridge deck configurations with edge barriers such as hand railings, with or without wind shielding. Tests for the Second Severn Bridge showed that the Strouhal number $\mathrm{St} = nD/V$ (Fig. 5.2.3) varied between 0.13 and 0.15, where D is the overall depth, including the wind shielding. For a wide deck, an outward-leaning barrier produced the highest amplitude of vibration; an inward-leaning barrier produced the smallest. For a narrow deck, a vertical barrier produced the lowest amplitude of vibration. For torsional vibrations, the Strouhal number varied from 0.12 to 0.18, and the vertical wind shielding produced marginally larger amplitudes of vibration. The response was larger when the oncoming wind was inclined upward. The design of the railings took into account the results of snow or ice accretion tests.

The wind shielding provided on the Second Severn Bridge reduced delays due to restrictions to single-lane traffic and the prohibition of traffic by vulnerable vehicles, and thus reduced the associated additional transportation costs. The benefit over a 30-year period of providing a second, wind-shielded crossing was calculated to be greater than the cost of providing and maintaining wind shielding for the First Severn Bridge. The Second Severn Bridge remained open to all traffic at all times following its opening to traffic in June 1996, as the police, which has operational responsibility for traffic safety, did not consider it necessary to impose any traffic restrictions.

5.3.4.3 Suppression of Wind-Induced Vibrations in Cables. Mechanical countermeasures include the installation of dampers or connecting cables to each other. Aerodynamic countermeasures include the improvement of the cable surface coating to reduce the excitation forces causing rain- and wind-induced vibrations.

5.3.4.3.1 Dampers. Dampers currently in use include viscous oil dampers, viscous-shear dampers (Fig. 5.3.15*a*), and dampers with high-performance rubber (Fig. 5.3.15*b*). The performance of dampers deteriorates in time. For details on dampers for stay cables, see [5.78].

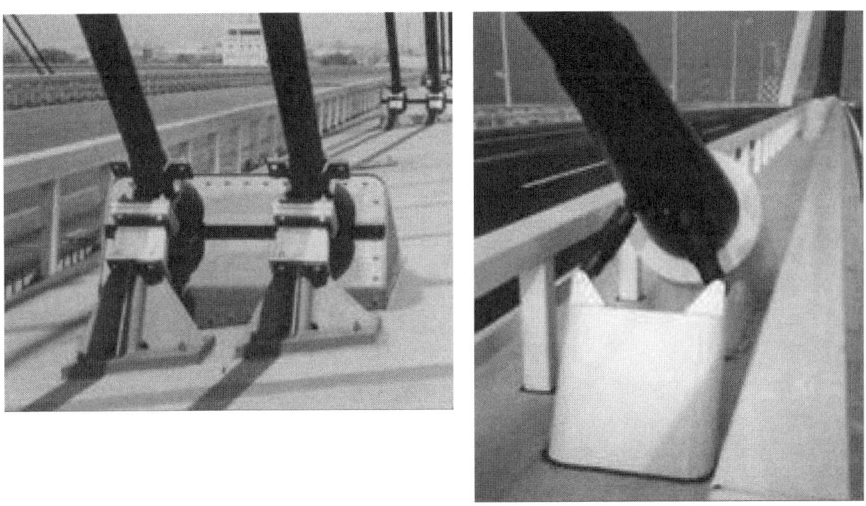

(*a*) (*b*)

Figure 5.3.15. (*a*) Viscous-shear damper; (*b*) oil damper and high-performance damping rubber. [(*a*) Courtesy of M. Yoneda; (*b*) Courtesy of H. Yamada.]

5.3.4.3.2 Cable Connections. A simple method for reducing cable oscillations is to connect stay cables to each other in their plane by using connecting wires (Figs. 5.3.14*b* and 5.3.16). This results in the dissipation of mechanical energy through the transfer of the vibration energy of a cable to adjacent cables with different natural frequencies, and in an increase in the vibrating mass. To our knowledge, no reliable procedure for quantitative evaluation of the effectiveness of this method is currently available. Maintenance may also be an issue, as cases have been reported where connecting wires have been damaged to the point of becoming ineffective.

5.3.4.3.3 Aerodynamic Improvement on Cable Surfaces. To inhibit rivulet formation on the surface of stay cables, cables with parallel protuberances have been used in the East Kobe Bridge (Fig. 5.3.17*a*), cables with parallel U-shaped stripes have been used in the Yuge Bridge (Fig. 5.3.17*b*), and cables with randomly patterned indentations have been used in the Tatara Bridge (Fig. 5.3.17*c*). The protuberances, stripes, and indentations cause larger drag forces than those acting on ordinary cables covered by smooth polyethylene (PE) coating. Drag coefficients C_D for cables with protuberances can be as high as 1.2. For long cables this increase in drag loading can affect the design of the entire structure. On the other hand, the indented cable used in the Tatara Bridge, which effectively inhibits the formation of rivulets and promotes the stabilization of separated flows over the surface of the cable, had a drag coefficient $C_D = 0.7$ (Fig. 5.3.18).

The helical strake method, where wires are affixed onto the cable surface in spiral patterns, has been adapted from its use on stacks. It has been used to inhibit wake galloping in hanger cables of the Akashi Kaikyo Bridge.

Figure 5.3.16. Connecting wires. (Courtesy of H. Katsuchi.)

(*a*)

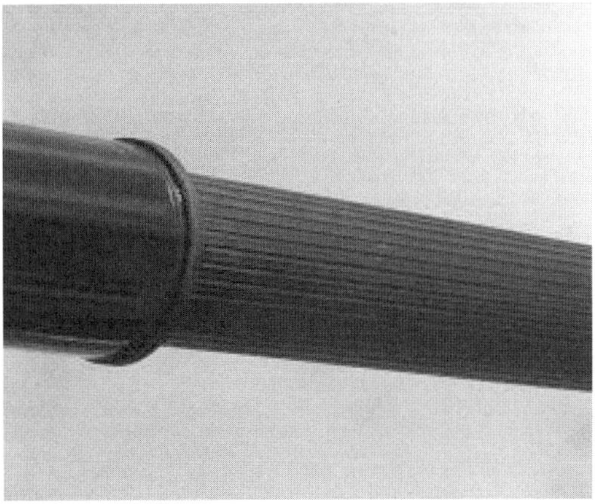

(*b*)

Figure 5.3.17. (*a*) Cable with parallel protuberances; (*b*) cable with U-stripe shaving; (*c*) pattern-indented surface. [(*a*) From [5.65]; (*b*) from [5.65, 5.66]; (*c*) from [5.79, 5.80].]

Usually, only one of the three approaches—use of dampers, connections, or surface processing—is used to mitigate cable vibrations. In some instances, however, two or even three of these methods are applied for the same cable.

5.3.5 Wind Tunnel Testing of Bridges

Wind speeds in commonly used wind tunnels are about 10 to 40 m/s and have test sections that cannot accommodate a full bridge model. It is therefore

(*c*)

Figure 5.3.17. (*Continued*)

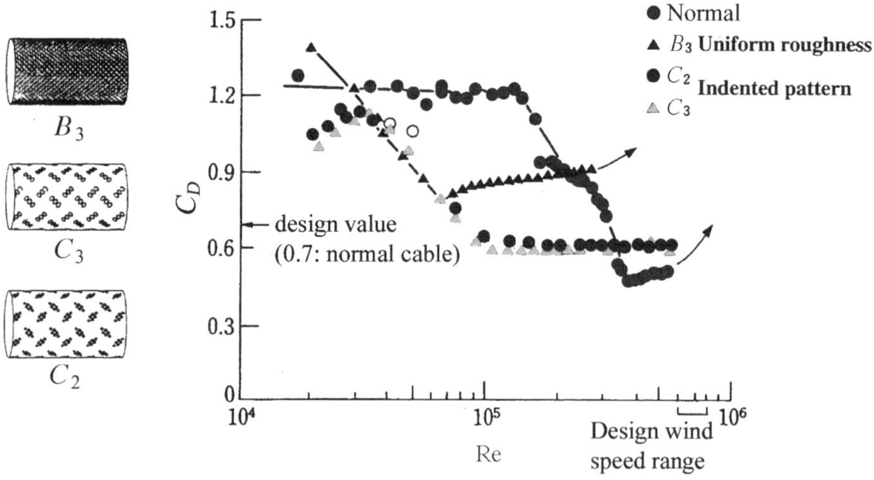

Figure 5.3.18. Drag coefficient vs. Reynolds number for a PE stay cable with a pattern-indented surface. (From [5.79, 5.80].)

necessary to relax similarity requirements. The ability to interpret the information obtained from wind tunnel tests in light of their limitations is essential for evaluating actual bridge behavior.

There are three basic reasons for undertaking wind tunnel tests:

1. To obtain static (steady) force coefficients for basic static design for wind, as input for vortex-excited oscillations, and for buffeting and galloping analyses. Mean force coefficients are also used for flutter analyses. The models used for these purposes are fixed, rigid models.
2. To observe and obtain relevant information on aeroelastic wind-induced vibrations such as vortex excitation and flutter. The models used for this purpose are rigid section models (Fig. 5.3.19) provided with elastic support devices, or aeroelastic models of the full bridge that simulate its geometric and dynamic characteristics. The section model tests yield direct estimates of frequency-dependent forces and aerodynamic derivatives which make it possible to conduct analyses of flutter and buffeting response.
3. To examine the influence of surrounding topography or other perturbations of the incident wind flow due to large structures or other obstacles. A potentially important effect is the inclination of the mean wind with respect to the horizontal.

Accurate measurements of the requisite aerodynamic quantities would require an accurate simulation of turbulence characteristics of natural winds. This is done to the extent practicable.

5.3.5.1 Rigid Section Model Tests. Rigid model tests, whether fixed or elastically mounted, are commonly performed at the largest scale practicable so that relevant details of the structure, appurtenances, and railway or automobile traffic be simulated as accurately as possible. The geometrical scale of the model is usually 1:30 to 1:100. Given these relatively large scales, which do not allow proper simulations of atmospheric flow, it is common to test the section models in uniform, smooth flow. The scale should be selected so that the range of relevant reduced velocities V_r is covered adequately, given the range of wind speeds available in the wind tunnel and the characteristics of the elastic supports.

When measuring vortex-excited response, if large amplitudes are predicted, care should be exercised to ensure that the vibrations do not become so violent as to destroy the model. Elements with circular cross section or with other types of curved surfaces, or rectangular cylinders with cutoff corners, are Reynolds number sensitive, and adjustments of the results based on full-scale data may be necessary even if the model surfaces are roughened or otherwise treated to help overcome Reynolds number dissimilarity. Modeling adjustments are also commonly needed for small elements such as handrails; testing for Reynolds numbers below 500 should be avoided.

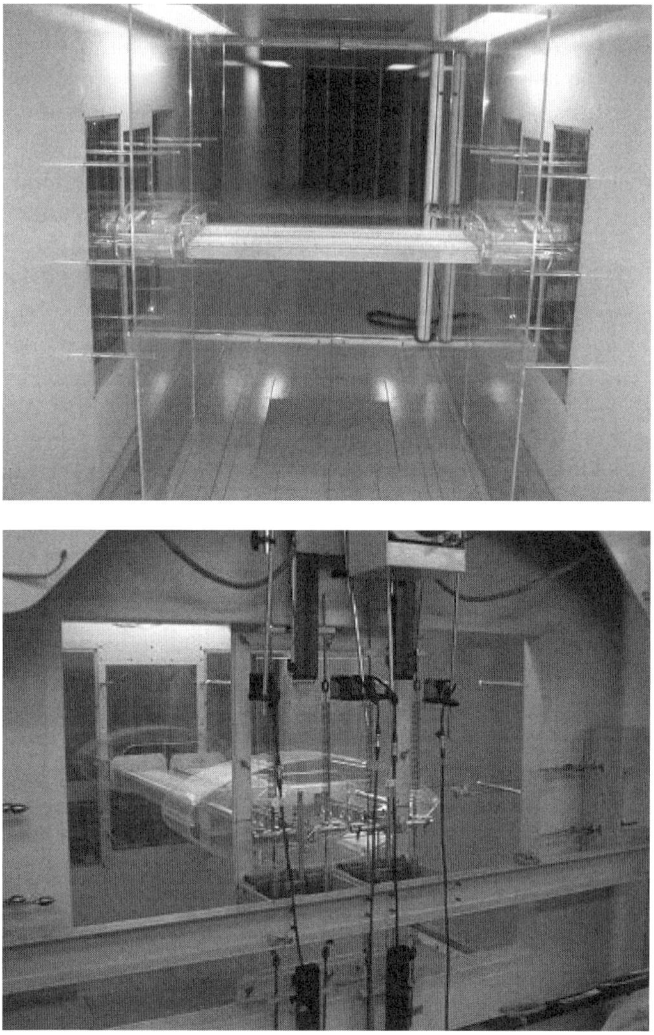

Figure 5.3.19. Views of a rigid section model. (Courtesy of Yokohama National University.)

Section model tests in smooth flow are useful for providing preliminary assessments of the adequacy of deck configurations. In more advanced stages, section models are tested in turbulent flow. As mentioned previously, owing to scaling problems in the wind tunnel, it is generally not possible to simulate the atmospheric turbulence correctly (the more so as its characteristics are highly variable), but a fairly acceptable reproduction of those characteristics is possible. In particular, the turbulence intensity can have an important effect on flow separation, reattachment, and vortex shedding. Given the practical impossibility of reproducing in wind tunnels the prototype small-scale turbulent eddies (this is due to the suppression by internal viscosity of those eddies' wind tunnel counterparts), the turbulence intensity may need to be increased with respect to its full-scale counterpart so that it compensates to some extent for the effects of the imperfect turbulence simulation. Doing so in a consistent, scientifically rigorous fashion is not feasible, so some guesswork is involved in practice. If the tests suggest that differences between results obtained in smooth flow and in imperfect turbulent flow are relatively small, it may be assumed that results obtained in smooth flow are reasonably adequate. If the differences are significant, engineering judgment is required to account for them in analysis and design.

Attention also needs to be paid to the blockage ratio. If the model scale and dimensions are too large in relation to the wind tunnel test section, the data being obtained may be incorrect, owing to increased wind speeds caused by channeling effects and inadequate simulation of the separated flow. Procedures for keeping the flow about a section model as close to two-dimensional as possible require the use of end plates and the placing of auxiliary portions of the model outside the main wind tunnel flow (Fig. 5.3.19).

5.3.5.2 Aeroelastic Full-Model Tests. To obtain direct experimental information on a bridge's response to wind, it is desirable to test in the wind tunnel a dynamic model of the full bridge. This is especially valuable for slender, flexible, and dynamically sensitive structures under wind loads. Such tests must model as faithfully as possible the important characteristics of the natural wind at the site and the significant aerodynamic features of the bridge configuration. It is also necessary to simulate correctly the stiffness, mass, and damping properties of the bridge structure.

It is usually possible to model the full spectrum of atmospheric turbulence in a wind tunnel at a relatively small scale, except for fluctuations with very low frequencies in atmospheric turbulence with large integral scales, and for fluctuations with high frequency (small-scale turbulence), which are suppressed by internal friction in the lower Reynolds number wind tunnel flow. The basic constraint is fitting a full bridge model within the tunnel facility while achieving the modeling scale required to ensure behavior reasonably similar to that of the prototype. This problem is not trivial when, as is commonly the case, test results are sensitive even to small changes in the dimensions of bridge parts and elements.

A closed box deck can be simulated by using a longitudinal steel bar installed inside the deck, whose stiffness models the box deck stiffness. The deck cross section can then be reproduced by a light wooden skin. The requisite mass and its distribution are also reproduced. Following a number of trials, the mechanical damping can be adjusted as well [5.74].

For the truss stiffening girder of the 40-m-long 1:100-scale Akashi Kaikyo Bridge, the size of the model cross section was comparable to that of the section model that had been tested previously, so that the structural details of the truss girder and auxiliary members, including connections, could be modeled accurately [5.37–5.39, 5.81]. For the bridge model, the Froude number similarity criterion must be satisfied [5.2, p. 277]. The Froude number is

$$\mathrm{Fr} = \frac{V^2}{Dg} \tag{5.3.4}$$

where g is the acceleration of gravity. Therefore, it is necessary that

$$\frac{V_m}{V_p} = \sqrt{\frac{D_m}{D_p}} \tag{5.3.5}$$

where the subscripts m and p indicate model and prototype, respectively. In addition, the natural frequencies n of the bridge must satisfy the relation

$$\left(\frac{nD}{V}\right)_m = \left(\frac{nD}{V}\right)_p \tag{5.3.6}$$

It follows therefore that

$$n_m = n_p\sqrt{\frac{D_p}{D_n}} \tag{5.3.7}$$

5.3.5.2.1 Modeling of the Akashi Kaikyo Bridge Stiffening Truss Girder. The stiffening girder model is shown in Fig. 5.3.20*a*. The model was composed of rigid blocks of four truss panels, connected by V-shaped springs that simulate the expansion and contraction of the girder chords, thus reproducing the vertical and lateral rigidities. Torsional stiffness was also simulated by V-shaped springs. A constant shear center was achieved by orienting the four V-shaped springs toward the center of the truss section. The requirement that the mass distribution of the model cross section be similar to that of the actual bridge was satisfied by using light, rigid composite material (0.3-mm carbon fiber plate and foaming polystyrene core) for the truss members and an aluminum plate for the road deck floor. The other parts of the girder deck and auxiliary facilities were made of wood with very low specific mass (balsa wood). The structural and aerodynamic performance of the stiffening girder model was checked before assembling the full bridge model. The section model tests, conducted separately with and without V-shaped springs, showed that the effect of the springs on the aerodynamics of the bridge deck was minor.

5.3.5.2.2 Modeling of the Akashi Kaikyo Bridge Main Towers. The stiffness of the main towers has a significant effect on the behavior of a suspension

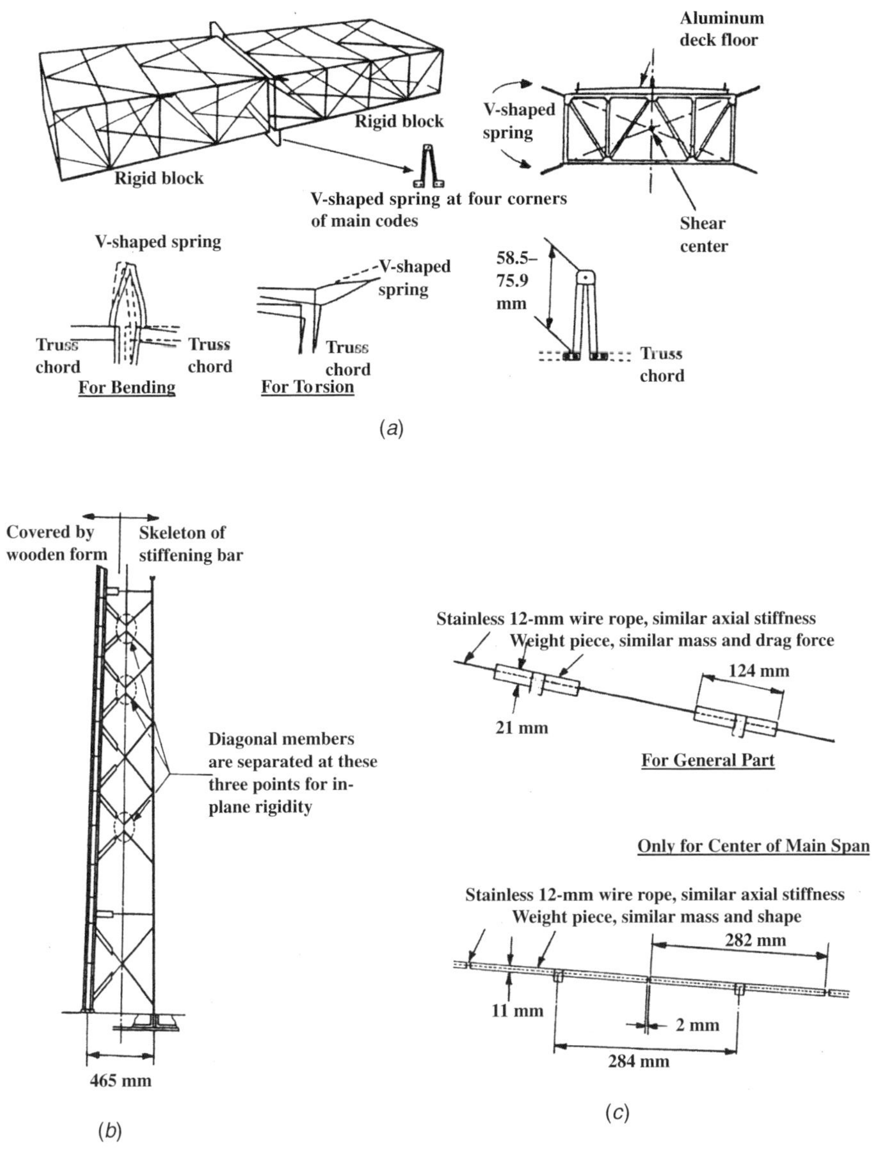

Figure 5.3.20. Aeroelastic full bridge model details, Akashi Kaikyo Bridge: (*a*) stiffening girder model; (*b*) tower model; (*c*) cable model. (From [5.37, 5.81]; courtesy of the Honshu–Shikoku Bridge Authority.)

bridge. Following preliminary studies, the tower model shown in Fig. 5.3.20*b* was used. The model skeleton consisted of a steel frame on which a wood skin was mounted to reproduce the geometric shape. The longitudinal rigidity of the members was significantly greater than required, so that diagonal members had to be disconnected at intersection points to reduce the tower model's rigidity. The static deformation and free vibration of the single freestanding tower showed that the simulation of the prototype properties was satisfactory.

5.3.5.2.3 Modeling of the Akashi Kaikyo Bridge Cables. When modeling the main cable it is necessary to simulate the mass, the longitudinal rigidity, and the steady aerodynamic drag. The solution adopted is shown in Fig. 5.3.20*c*. To obtain the proper longitudinal rigidity as well as sufficient strength, 12-mm-diameter stainless steel woven wire was used. The mass was determined by similarity requirements. The effect of the Reynolds number and of the three-dimensional flow around the cylindrical weights placed to achieve the requisite mass and drag, was studied in detail in wind tunnel tests of various cable models. Results of the tests showed that the model adopted had a satisfactory overall drag.

It was determined through wind tunnel tests on bridge section models with and without modeled cables that the aerodynamic properties were affected by the presence of the cables and that this influence could not be disregarded unless the distance between the girder and the cables was at least as large as the girder height. In view of these aerodynamic interference effects on the behavior of the full bridge, a cable with properly scaled diameter and mass was used for the center portion of the main span.

5.3.5.2.4 Performance of the Akashi Kaikyo Full Model. After assembling all the component parts and elements to create an aeroelastic full bridge model, the calculated static deflections agreed reasonably well with their experimental counterparts. The agreement was also reasonably good between natural modes and frequencies of vibration calculated for the prototype and their measured counterparts for the full bridge model, which were obtained, along with modal damping, by using an electrical excitation device. Table 5.3.4 presents test results for the full model. The frequencies of vertical and horizontal modes generally agreed reasonably well with those calculated for the prototype bridge. Only the frequency of the first symmetric torsional mode was found to be slightly higher than that for the prototype bridge. The logarithmic decrements of the damping for the torsional modes were slightly higher than required in the code, although the vertical and horizontal modes had damping comparable with the required damping. Some discrepancies were judged to be sufficiently small to allow judicious inferences to be made from the test results.

5.3.5.3 Topographic and Local Environmental Effects. Information on the characteristics of the full-scale wind is usually not available if complex topography surrounds a bridge site. Small-scale topographic models can be used to provide estimates for use in subsequent modeling of the wind on a larger scale. The use of computational fluid dynamics techniques, in conjunction with wind tunnel testing, could be helpful in achieving models that account at least approximately for topographic effects.

TABLE 5.3.4. Natural Frequency and Damping of 1:100 Scale Full Model of the Akashi Kaikyo Bridge

Mode Shape	Natural Frequency (Hz)		Logarithmic Decrement	
	Measured	Calculated[a]	Measured	Prescribed by Code
1 Vert., first sym. 	0.647	0.638	0.032	
2 Vert., first asym. (1) 	0.757	0.745	0.031	
3 Vert., first asym. (2) 	0.855	0.835	0.035	0.030
4 Vert., second sym. 	1.245	1.213	0.036	
5 Vert., third sym. 	1.636	1.637	0.055	
6 Tors., first sym. 	1.560	1.497	0.031	0.020
7 Tors. first asym. 	2.075	2.077	0.030	
8 Horiz., first sym. 	0.391	0.387	0.016	
9 Horiz., first asym. 	0.781	0.775	0.032	0.030
10 Horiz., second sym. 	1.270	1.265	0.041	

[a]Based on FEM analysis of a prototype bridge.

Elevated bridges and viaducts in built-up terrain are affected aerodynamically by nearby structures and topographic features. This must be taken into account when conducting wind tunnel tests. Typical geometric scales to be used in various types of wind tunnel studies are recommended in the guidelines of the Annex to the BD 49/01 Design Rules [5.68].

To assure the quality of data obtained in wind tunnel tests, it is desirable to perform full-scale measurements [5.51]. Comparisons between full-scale and wind tunnel measurement results are usually difficult, however. Checking of wind tunnel results against code requirements and available data is strongly recommended.

5.3.6 Computational Fluid Dynamics for Bridges

5.3.6.1 Application to Bridge Deck Configurations. Computational fluid dynamics (CFD) studies for bridges have been reported, for example, in [5.82–5.93]. Analyses by the discrete vortex method [5.85] of unsteady motion-dependent forces on the section of Great Belt East Bridge have yielded useful results, although reliance on this method for application to structural design remains controversial. Motion-dependent forces on square as well as rectangular cylinders with width/height ratio $B/D = 2$ and 4 [5.83, 5.85, 5.88] in across-wind motion have been analyzed by obtaining CFD solutions of the Navier–Stokes equation in which the turbulence is modeled in a simplified manner. Applications of the CFD technique have also been reported for streamlined bridge deck sections [5.87, 5.88, 5.90].

Figure 5.3.21 shows unsteady motion-dependent aerodynamic force coefficients for a rectangular cylinder with width/height ratio $B/D = 5$ and wind acting normal to the height D. The amplitudes of the harmonic motions of the cylinder were $A/D = 0.025$ in nondimensional across-wind motion and $2°$ in torsion. Measured and calculated results agree reasonably well, except for the real part $\mathrm{Re}(C_{Lh})$ of the lift (across-wind) coefficient and the imaginary part of the lift coefficient $\mathrm{Im}(C_{L\alpha})$ (see Sect. 5.2.3.2). The results are consistent with the propensity for torsional flutter inherent in the moment coefficient $C_{M\alpha}$. Note in Fig. 5.3.21 the large discrepancies between computed and experimental results on the one hand, and the Theodorsen thin airfoil values indicated by the solid lines on the other. However, for a shallow rectangular section ($B/D = 20$) the agreement between numerical results, experimental results, and results derived by the Theodorsen thin airfoil was found to be good.

The CFD technique is not yet sufficiently effective for applications to actual bridge deck configurations, including handrails, curbs at floor ends, and various appurtenances. For this reason, current CFD studies tend to limit themselves to neat geometric shapes. An example of a CFD grid used for a streamlined girder is shown in Fig. 5.3.22.

In the BD 49/01 Rules [5.68] and the Eurocode [5.94], the CFD approach is listed, along with wind tunnel testing, as one of the alternative approaches

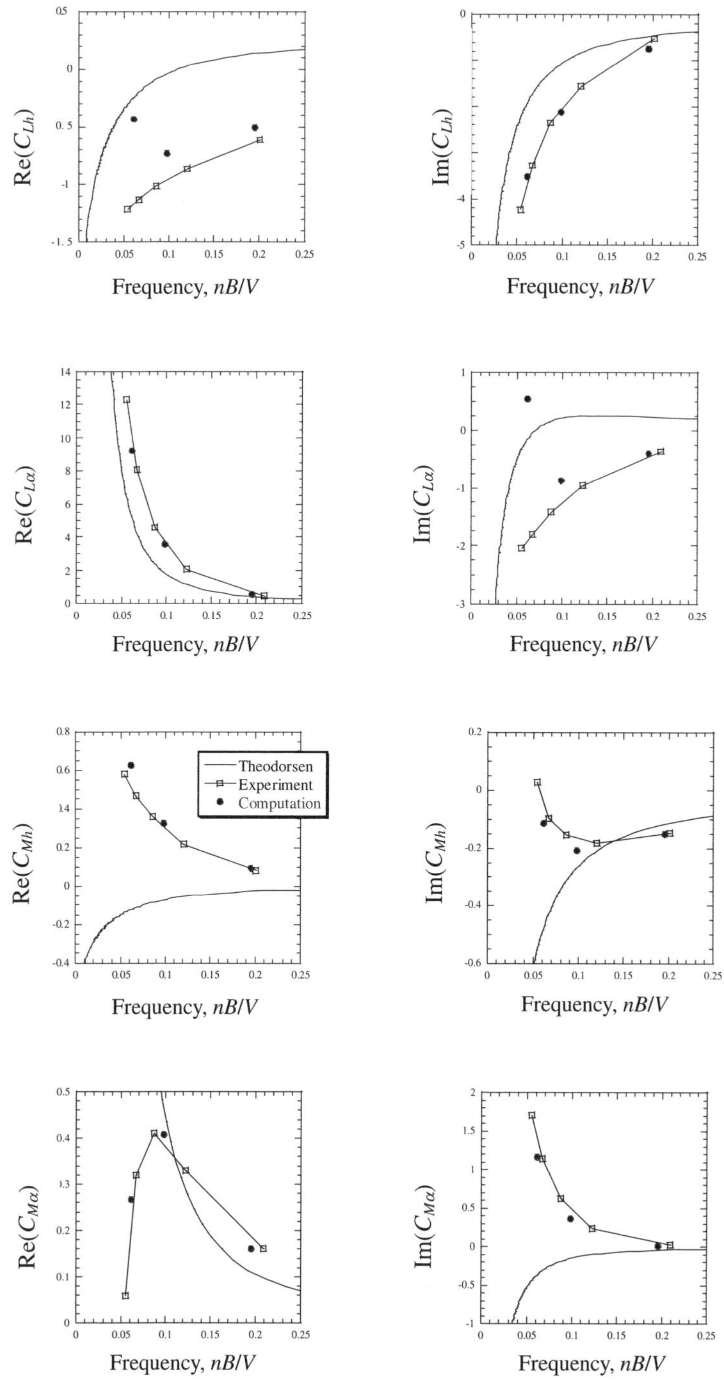

Figure 5.3.21. CFD analysis for unsteady, motion-dependent force coefficients, rectangular cylinder, $B/D = 5$. (From [5.91]; reprinted by permission of the American Institute of Aeronautics and Astronautics, Inc.)

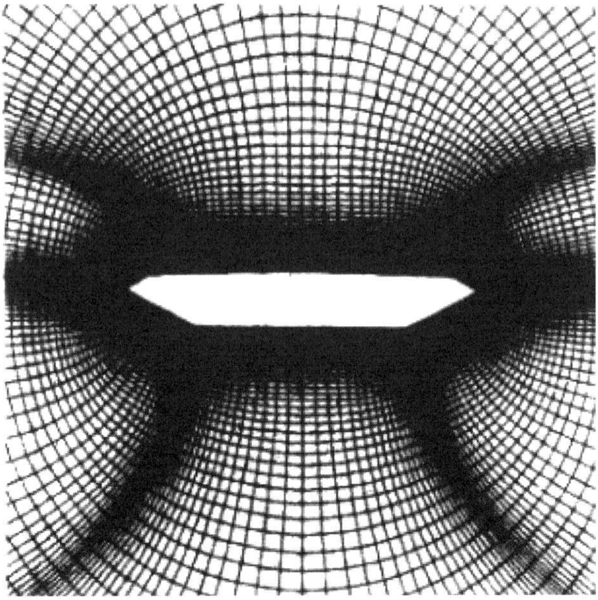

Figure 5.3.22. Grid system for CFD analysis of a streamlined bridge deck without handrails and curbs.

that may be used to verify critical wind speeds or determine aerodynamic forces. The BD 49/01 Rules state that proven theoretical or CFD procedures can provide useful additional guidance. However, no indications are provided as to the adequacy and reliability of various CFD procedures, whereas such indications are given for wind tunnel tests. For this reason, some bridge engineers believe CFD methods are of limited practical value in the present state of the art. This simply means that CFD procedures cannot be relied upon without confirmation through wind tunnel testing, despite uncertainties inherent in such testing due primarily to violation of the Reynolds number in the laboratory, especially for such elements as handrails. However, CFD can in some cases provide useful insights if used with prudence and sound engineering judgment.

APPENDIX 5A: WIND-RESISTANT DESIGN STANDARD FOR THE HONSHU–SHIKOKU BRIDGES (2001) [5.36]

This is a slightly edited and abbreviated English version of the original code that was prepared for information purposes by a study committee on wind-resistant design of bridges [*Code Practice and Recent Developments in Japan, JSCE* (*Japan Society of Civil Engineers*)]. *In case of any conflict between the*

Japanese version of original code and the English version, the Japanese version shall prevail in all contexts.

5A.1 Outline

For the design and construction of the Honshu–Shikoku Bridges (abbreviated as HBS), in which long-span cable-supported bridges were predominant, standards for wind-resistant design have been established due to recognition of the importance of assuring the aerodynamic stability of the bridges. The standards have been subjected to revisions because of accumulated knowledge based on wind tunnel tests, various investigations, related research, and actual bridge construction experience. The JSCE advisory committee was established in 1961. A task committee for wind-resistant design was established in 1963. Guidelines for wind-resistant design were issued in 1964 and revised in 1967. The Honshu–Shikoku Bridge Authority was established in 1970. A wind-resistant design standard was issued in 1972 and revised in 1975. Guidelines for wind-resistant design were issued in 1990 for the Akashi Kaikyo Bridge and in 1994 for the Onomichi–Imabari route. The Honshu–Shikoku Bridges Standard for Wind-Resistant Design was issued in 2001. Construction started in 1975 with the Ohmishima, Ohnaruto, and Ihnoshima Bridges, and proceeded with the Seto–Ohashi, Hakuta Oshima, Ikuchi, Hakata Oshima, Ikuchi, Akeshi Kaikyo, Kurushima–Kaikyo, and Tatara Bridges. For the design of the Akashi Kaikyo, Tatara, and Kurushima–Kaikyo Bridges, tests were conducted in the 1990s in a specially built large boundary-layer wind tunnel facility.

The HBS presented in this appendix is the result of the development of the earlier standards for HBS and is an achievement based on Japanese efforts in bridge aerodynamics over more than 40 years. This standard is established mainly for suspension bridges with spans longer than about 1000 m and for cable-stayed bridges with spans longer than around 500 m. However, the standard is considered to be fully applicable for bridges with shorter spans if its requirements are relaxed appropriately.

5A.2 Scope

This standard shall be applied for the wind-resistant design of the Akashi Kaikyo, Tatara, and Kurushima–Kaikyo Bridges.

5A.3 Definitions and Notations

(1) Definitions. The terminology used in this standard is defined as follows.

(a) *Wind Speed*

1. *Basic wind speed:* the wind speed used for the design of all parts of a bridge, defined as the 10-min average speed at 10 m above the water level at the site.
2. *Design wind speed:* the wind speed used in design and to check the aerodynamic stability of the bridge. It is determined by multiplying the basic wind speed by a modification factor for the elevation of the structure.
3. *Critical wind speed:* the lowest wind speed at which vibrations such as galloping or flutter are determined to occur.
4. *Verification wind speed:* wind speed for which the possible onset of vibrations of the galloping or flutter type must be checked. If the critical wind speed is higher than the verification wind speed, the structure is judged to be safe.

(*b*) *Wind Loads and Aerodynamic Forces*

1. *Design wind load:* wind force (drag component) acting on a bridge or its members in the longitudinal direction or the transverse direction of the bridge, to be calculated by using the design wind speed and modification factors for wind load.
2. *Steady wind force:* time-averaged force.
3. *Fluctuating wind force:* time-varying wind force caused by the turbulence in the oncoming wind.
4. *Unsteady aerodynamic force:* time-varying aerodynamic force acting on an oscillating body, whose fluctuation is accompanied by oscillatory motion of the body.

(*c*) *Wind Characteristics*

1. *Wind direction:* direction in the horizontal plane from which the wind blows.
2. *Azimuth angle:* incoming angle of the wind in the horizontal plane measured from the direction perpendicular to the bridge.
3. *Inclination angle:* incoming angle of the wind in the vertical plane, measured from the horizontal. When the wind blows upward, it is regarded as positive.
4. *Angle of attack:* angle of the oncoming wind measured from the basic axis of the deck in the vertical plane perpendicular to the bridge direction. If the deck is in a horizontal plane, the angle of attack is equal to the inclination angle, but if the deck has a velocity in vertical motion or is statically twisted, the angle of attack differs from the inclination angle. The angle of attack is treated as positive when the wind blows upward.

TABLE 5A.2. Exponent for the Power Law of a Vertical Profile of Wind Speed

Bridge	Exponent α
Akashi Kaikyo	1/8
Kurushima–Kaikyo	1/7
Tatara	

1. *Component along mean flow.* The power spectrum density of $S_u(f)$ the fluctuating wind speed along the direction of the mean flow shall be

$$\frac{fS_u(f)}{\bar{u}^2} = 0.475\,\frac{f}{f'}\left[1+\left(\frac{f}{f'}\right)^2\right]^{-5/6} \tag{5A.3}$$

where f denotes the frequency and $f' = 0.635z^{-0.75}$.

The vertical profile of the turbulent intensity along the mean flow $I_u(z)$ shall be

$$I_u(z)=I_u(10)\left(\frac{z}{10}\right)^{-\alpha} \tag{5A.4}$$

where α is the exponent shown in Table 5A.2, $I_u(10)$ is the turbulent intensity at 10 m height, and $I_u(z) = 0.10$ at the height of suspended deck.

2. *Vertical component.* The power spectrum density $S_w(f)$ of the fluctuating wind speed in the vertical direction shall be

$$\frac{fS_w(f)}{\bar{w}^2} = \frac{2.11f_r}{1 + 11.2f_r^{5/3}} \tag{5A.5}$$

where the nondimensional frequency[3] is $f_r = fz/U_z$.

TABLE 5A.3. Definition of the Basic Reference Height of a Structure

Structural Part	Basic Reference Height, z (m)
Suspended structure	Average height of stiffening or main girder at the center span
Cable and suspender	Mean value of the height for suspended structure and average height of tower top
Tower	65% of tower height

[3] Also known as the Monin coordinate [authors' note].

3. *Spatial correlation of fluctuating wind speeds.* The decay factor to be used for exponential expression of the spatial correlation shall be 8 for both the main flow direction and vertical direction.

(*d*) *Influence of Surrounding Terrain.* If the terrain around the bridge site has complex topography and the wind might be distorted accordingly, it is recommended that special turbulent characteristics be used which are actually observed in wind tunnel tests modeling the surrounding terrain.

(*4*) *Inclination Angle and Azimuth Angle of Wind.* The inclination angle to be considered in wind-resistant design shall basically be in accordance with Table 5A.4. When there is a possibility that a dominant inclination angle exists because of the surrounding terrain, this angle shall be considered in the investigation to be the mean inclination angle. In addition, the angle associated with the torsional deformation of the suspended deck due to wind load shall be properly taken into consideration if this angle is significant.

The azimuth angle considered in the investigation shall be 0° with respect to the normal to the bridge axis. However, an azimuth angle that may be generated by surrounding terrain or by structural characteristics shall also be used when this angle is believed to have a significant influence on the bridge design.

5A.6 Static Design

(*1*) *Static Wind Load.* Only the drag component of the steady aerodynamic force shall be considered for static design. The wind load acting on the main cables and the suspender ropes of a suspension bridge, the stay cables of a cable-stayed bridge, and the suspended deck is

$$P_D = \mu_2 \frac{\rho \overline{U}_z^2}{2} C_D A_n \tag{5A.6}$$

The wind load acting on the tower is

TABLE 5A.4. Wind Inclination Angle

Airflow	Inclination
Uniform	−3° to +3°
Turbulent	0°

$$P_D = \mu_3 \frac{\rho \overline{U}_z^2}{2} C_D A_n \tag{5A.7}$$

where the modification factors μ_2 and μ_3 are as specified in Table 5A.5; the air mass per unit volume ρ shall be 1.18 kg/m^3; the design wind speed $\overline{U}_z$ is determined as in Sect. 5A.5(2); and the drag coefficient C_D as well as the projection area A_n shall be determined by Tables 5A.6 and 5A.7, respectively.

The drag coefficient used in the above-mentioned calculation shall be verified in wind tunnel tests, and redesign is required if the coefficient measured at an angle of attack $\alpha = 0°$ in the flow perpendicular to the bridge axis differs from the value used in the design by 5% or more.

(2) *Design Wind Loading Application and Distribution.* The design wind load shall generally be regarded as a uniformly distributed horizontal load in the directions perpendicular and parallel to the bridge. When necessary, instead of a uniform distributed load, a full load shall be considered over part of the length being considered, while the remaining part is subjected to half

TABLE 5A.5. Modification Factors

Structural Part:	Suspended Structure		Tower	
Direction:	Transverse	Longitudinal	Transverse	Longitudinal
	Akashi Kaikyo Bridge			
Cables, μ_2	1.55	—	1.35	—
Suspenders, Stiffening truss } μ_2	1.55	1.25	1.35	1.25
Towers, μ_3	—	—	1.55	1.50 (top fixed), 1.75 (top free)
	Kurushima–Kaikyo Bridges			
Cables, Suspenders, Stiffening girder } μ_2	0	1.3	1.7	1.3
Towers, μ_3	—	—	1.4	1.55 (top fixed), 1.7 (top free)
	Tatara Bridge			
Cables, Girder } μ_2	1.9	1.35	1.65	1.35
Towers, μ_3	1.9	1.5	1.65	1.5 (top fixed), 1.8 (top free)

TABLE 5A.6. Drag Coefficients

Structural Part	Transverse Direction		Longitudinal Direction	
	C_D	Note	C_D	Note
Suspended structure				
Truss type	Based on results of wind tunnel tests		60% of C_D in transverse direction	Use measured value on similar shape, if available
Box girder type	Based on results of wind tunnel tests		30% of C_D in transverse direction	
Tower	1.8 or based on results of wind tunnel tests	Per one shaft of rectangular shaped tower	1.8 or based on results of wind tunnel tests	For rectangular shaped tower
Suspension cables	0.7			
Suspender ropes	0.7		$0.7/\sqrt{2}$	For truss deck, for box girder deck
Stay cables	0.7		0.7	

the load, so as to yield the most unfavorable effect on a specific member. The loading shall be applied as follows.

(a) *Wind Load in the Direction Perpendicular to the Bridge*

1. *Suspended deck:* the wind load acts on the windward side of the structure.

TABLE 5A.7. Method for Determining the Projection Area

Structural Part	Transverse Direction	Longitudinal Direction
Suspended structure	Windward side of girder, curb, and guard rail	Same as transverse
Tower	Windward and leeward shafts	Both shafts and web members
Suspension cable	Windward and leeward cables	Neglect
Suspender rope	Windward and leeward suspender ropes	Same as transverse
Stay cable	Windward and leeward cables	Same as transverse

2. *Tower:* the wind load acts on each of the axes of the windward and leeward shafts.
3. *Suspension cables:* a uniformly distributed load acts normal to the plane of the cable.
4. *Stay cable:* the wind load acting on the stay cables is distributed equally to the deck and tower(s).

(b) *Wind Load in the Longitudinal Direction of the Bridge*

1. *Suspended deck:* the longitudinal wind load is acting on the structure uniformly.
2. *Tower:* the wind load is based on the entire projected area of shafts and web members and is uniformly distributed.
3. *Suspension cable:* half of wind load acting on the suspender ropes is to be considered, while the load acting on the cable itself shall be neglected.
4. *Suspender ropes of a suspension bridge:* the wind load acting on the suspender ropes is to be distributed equally to the suspended deck and cable.
5. *Stay cables:* the wind load acting on the stay cables is distributed equally to the suspended deck and the tower(s).

(c) *Influence of Skewed Wind.* Simultaneous loading both perpendicular to the bridge axis and in the longitudinal bridge direction shall be applied for designing such structural elements as the deck shoes, expansion joints, stays, and so on, for which the longitudinal displacement is of concern.

(3) Combination of Wind Load and Other Loads, and Increase in Allowable Stress. The combination of wind and other loads, as well as an increase in the allowable stress, shall be in accordance with Table 5A.8.

TABLE 5A.8. Combination of Wind Load and Other Loads, and Increase in Allowable Stress

	Increase in Allowable Stress	
Combination of Loads[a]	Tower	Suspended Structure
D + W + T + SD + E	1.40	1.50

[a]D is the dead load, W is the wind load, T is the influence of temperature change (when it is combined with the design wind load, the temperature shall be 35°C), SD is the influence of movement of the supporting points, and E is the influence of error in the fabrication and erection of superstructure.

5A.7 Verification

(1) Static Verification. The suspended deck shall have enough stiffness to avoid static instability due to the wind load.

(2) Dynamic Verification. This section specifies methods for verification against wind-induced dynamic phenomena such as galloping or flutter, gust response, vortex-induced oscillation, and cable vibration. Mechanical damping, which strongly affects these phenomena, is also specified. When two or more parallel bridges are constructed near each other (this is not the case for the Akashi Kaikyo, Tatara, and Kurushima–Kaikyo Bridges), there is a possibility for the leeward bridge(s) to vibrate due to the wake generated by the windward structure. This needs careful attention.

(a) Mechanical Damping. For the dynamic verification of bridges, the mechanical damping values of Table 5A.9 shall be used as standard values for each vibration mode. When additional structural damping is provided by placing dampers to suppress the vortex-induced oscillation of the towers, the rain vibration of stay cables, the vibration of suspender ropes, and so on, the capability of the damping devices shall be ascertained.

(b) Verification of Galloping and Flutter. For a suspended deck and tower, the critical wind speed uniform flow shall be verified to satisfy

$$U_F \geq 1.2\mu_F\overline{U}_z \tag{5A.8}$$

where μ_F is the fluctuating wind speed modification factor given in Table 5A.10 and $\overline{U}_z$ is the design wind speed.

TABLE 5A.9. Mechanical Damping (Logarithmic Decrement)

Vibration Mode and Structural Part		Logarithmic Decrement	
		Bending Vibration	Torsional Vibration
Deck–vibration dominant mode	Stiffening truss	0.03	0.03 (may be 0.02 for spans of 500–600 m or less)
	Closed-box girder	0.02	0.02
Tower–vibration dominant mode	Tower–cable–girder system	0.02	0.02
	Freestanding tower	0.01	0.01
Stay–cable–vibration dominant mode		0.003	
Suspender–rope–vibration dominant mode		0.003	

TABLE 5A.10. Modification Factor for Galloping or Flutter of a Deck in Fluctuating Wind

Bridge	μ_F
Akashi Kaikyo	1.08
Tatara	1.1
Kurushima–Kaikyo	1.1

(c) Verification of Gust Response. The safety of the suspended deck and tower experiencing gust response shall be verified through a gust response analysis. In addition, the safety shall be checked through wind tunnel tests in boundary-layer flow. If the stress determined in accordance with the gust response analysis or wind tunnel test exceeds the allowable stress by 5% or more, re-design is required. Depending on the scale of the bridge, structural type, wind environment at the site, and so on, verification of the gust response can be omitted if sufficiently large modification factors μ_2 and/or μ_3 are used (see Table 5A.5).

(d) Verification of Vortex-Induced Oscillation. When a bridge or its members have the potential for harmful vortex-induced oscillations, careful checks shall be made for both safety and serviceability.

(e) Verification of Cable Vibration. When there is a possibility for stay cables or suspender ropes to have harmful wind-induced oscillations, countermeasures to secure safety shall be carefully studied.

5A.8 Safety During Construction

(1) Basic Wind Speed During Construction. To investigate safety during construction, the basic wind speed given in Table 5A.11 shall generally be used.

(2) Safety Under Wind During Construction. Bridges under construction, especially freestanding towers before cable erection, and the main cables as

TABLE 5A.11. Basic Wind Speed During Construction

Bridge	Basic Wind Speed During Construction (m/s)
Akashi Kaikyo	37
Kurushima–Kaikyo	31
Tatara	29

well as the suspended deck during erection work, are vulnerable to strong winds. Accordingly, proper countermeasures shall be taken as needed, following verification of static and dynamic safety under winds during construction, through wind tunnel tests and/or other methods.

APPENDIX 5B: THREE-DIMENSIONAL FLUTTER ANALYSIS: EIGENVALUES AND EIGENVECTORS

For a three-dimensional finite element model of a bridge, we consider the governing equation of motion for the model written in the form

$$[M]\{\ddot{u}\} + [C]\{\dot{u}\} + [K]\{u\} = [F_V]\{\dot{u}\} + [F_D]\{u\} \tag{5B.1}$$

where for a structure deformed by static wind forces, $\{u\}$ is the displacement vector, $[M]$ is the mass matrix, $[C]$ is the mechanical damping matrix, $[K]$ is the stiffness matrix, and $[F_V]$ and $[F_D]$ are motion-dependent aerodynamic force coefficient matrices. The aerodynamic coefficients are functions of reduced circular frequency $k = \omega B/V$ and are available from experiments for various angles of attack in increments of 2 or 3°.

Since calculated modal shapes and results of flutter analyses depend on the structural and aerodynamic properties of the bridge, direct FEM analysis can be expected to yield accurate results. An approximate picture of the flutter behavior is required in advance of the detailed FEM analysis.

Assuming a harmonic coupled vibration of circular frequency ω in the form

$$\{u\} = \{\Phi\}e^{i\omega t} \tag{5B.2}$$

the relations

$$\{u\} = -\frac{\{\ddot{u}\}}{\omega^2} \qquad \{\dot{u}\} = -\frac{i\{\ddot{u}\}}{\omega} \tag{5B.3a,b}$$

are obtained. Therefore, the right-hand side of Eq. 5B.1 can be rewritten as

$$[F_V]\{\dot{u}\} + [F_D]\{u\} = \left\{-\frac{i[F_V]}{\omega} - \frac{[F_D]}{\omega^2}\right\}\{\ddot{u}\} = [F]\{\ddot{u}\} \tag{5B.4}$$

To a first approximation, neglecting the effect of mechanical damping (second term on the left-hand side, Eq. 5B.1), the equation of motion is rewritten as

$$[K]^{-1}[F - M]\{\ddot{u}\} = \{u\} \tag{5B.5a}$$

and using Eq. 5B.3a yields

$$[K]^{-1}[M - F]\{\ddot{u}\} = \frac{\{\ddot{u}\}}{\omega^2} \tag{5B.5b}$$

This eigenvalue problem (with eigenvalue $1/\omega^2$ and eigenvector $\{u\}$) is solved by following five steps:

1. Preliminary structural design of bridge subjected to static wind loads.
2. Formulation of equation of motion with appropriately selected force coefficients.
3. Eigenvalue analysis for discrete values of the reduced wind speed. In this step, the complex eigenvalue (ω) and complex mode vector $\{\Phi\}$ are

$$\omega_j = \omega_{jR} + i\omega_{jI} \qquad \{\Phi\}_j = \{\Phi_R\}_j + i\{\Phi_I\}_j \tag{5B.6a,b}$$

The wind speed and the aerodynamic damping for the jth flutter mode can be obtained as

$$V_j = \frac{\omega_{jR}B}{k} \qquad \delta_j = \frac{\omega_{jI}}{\sqrt{\omega_{jR}^2 + \omega_{jI}^2}} \tag{5B.7a,b}$$

where k is the reduced circular frequency.

4. Plot the real and imaginary parts (or the aerodynamic damping) of the eigenvalues versus the wind speed. In each of these plots, the line connecting the respective values corresponding to various wind speeds for mode shape j is referred to as the jth root trace (Fig. 5B.1).
5. Estimate the lowest wind speed causing the onset of flutter (the critical wind speed V_{cr}), that is, the lowest speed for which the aerodynamic damping in Fig. 5B.1*a* vanishes. At this speed the flutter oscillation is harmonic rather than decaying or divergent as would be the case if the imaginary part of the complex exponent ω did not vanish. In Fig. 5B.1, $V_{cr} \approx 57$ m/s.

The static angle of attack of the bridge deck as deformed by wind loads must be accounted for when considering the action of unsteady aeroelastic forces and the deformation of the deck, and may be instrumental in inducing modal coupling in the bridge. The critical wind speed $V_{cr\,j}$ is typically not equal to the wind speed assumed for the preliminary design. An iteration procedure is therefore required for the final design.

Example

We consider two bridges that differ solely in the respective positions of their mass centers (Fig. 5B.2). The spans are assumed to be hinged at the supports

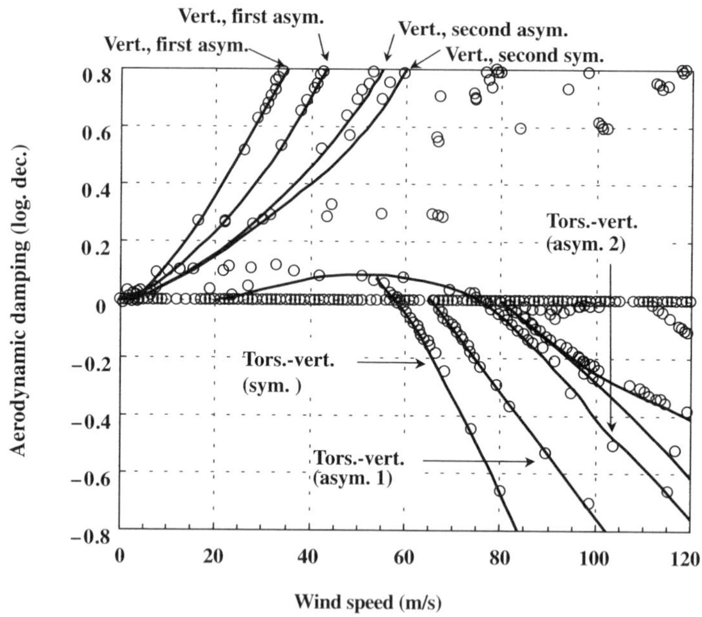

(*a*)

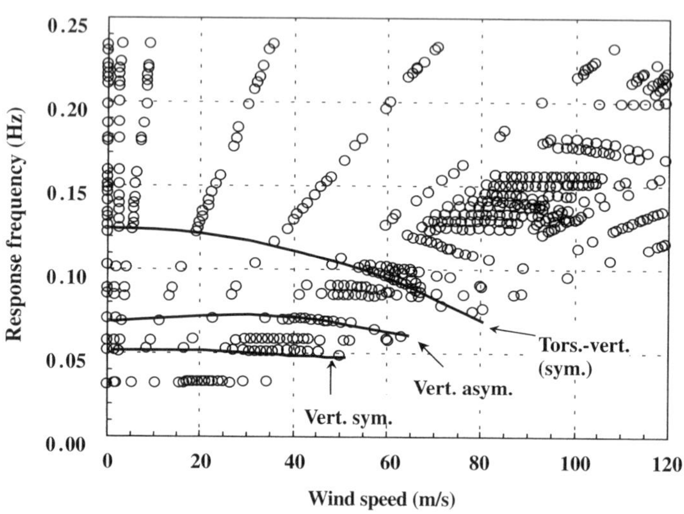

(*b*)

Figure 5B.1. Root traces in three-dimensional FEM flutter analysis: (*a*) aerodynamic damping; (*b*) response frequency.

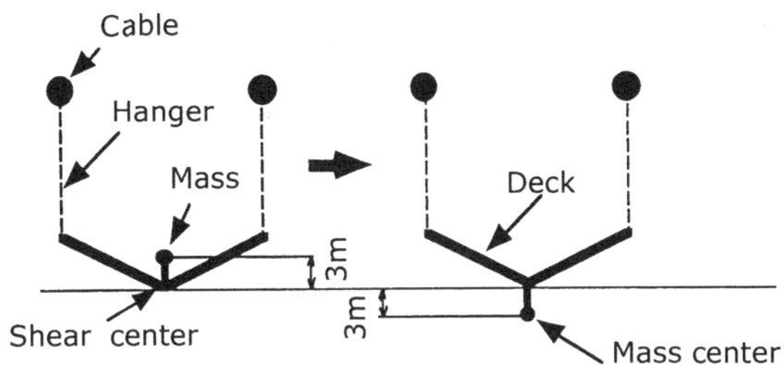

Figure 5B.2. Elastic and mass center configuration: upper and lower deck models.

and have lengths 1750 m + 3500 m + 1750 m. The fundamental natural frequency in torsion (the first symmetric mode) is slightly lower for the bridge with the higher center of mass, all other natural frequencies being the same. No deformation induced by mean wind loads is considered in this calculation. Motion-dependent forces are assumed to be those applicable to a flat plate, as would be appropriate for a very shallow box deck. It is also assumed that eight forces and moments associated with vertical (heaving) motion and torsion are sufficient for the description of aeroelastic effects. The critical wind speed calculated for the model with a high center of mass is lower by about 10% than for the model with a low center of mass. The response modal shapes differ for the two bridge models, as shown in Fig. 5B.3.

APPENDIX 5C: FLUTTER VELOCITY AND BUFFETING RESPONSE[4]

Next, we consider estimation of the flutter velocity and buffeting response in the presence of aeroelastic forces. Unlike in Appendix 5B, this is done within the framework of the Scanlan formulation of the aeroelastic problem.

Let $h(x,t)$, $p(x,t)$, and $\alpha(x,t)$ denote, respectively, the vertical, horizontal, and torsional displacements of the spanwise station defined by the coordinate x; $h_i(x)$, $p_i(x)$, and $\alpha_i(x)$ the ith modal displacements at x; and $\xi_i(t)$, I_i, ω_i, and ζ_i the generalized coordinate, generalized inertia, natural frequency, and damping ratio in the ith mode. We assume that

$$h(x,t) = \sum_{i=1}^{N} h_i(x) B \xi_i(t) \tag{5C.1a}$$

$$p(x,t) = \sum_{i=1}^{N} p_i(x) B \xi_i(t) \tag{5C.1b}$$

$$\alpha(x,t) = \sum_{i=1}^{N} \alpha_i(x) \xi_i(t) \tag{5C.1c}$$

The equation of motion in the ith mode is

[4]The material in this appendix is from [5.40, 5.76].

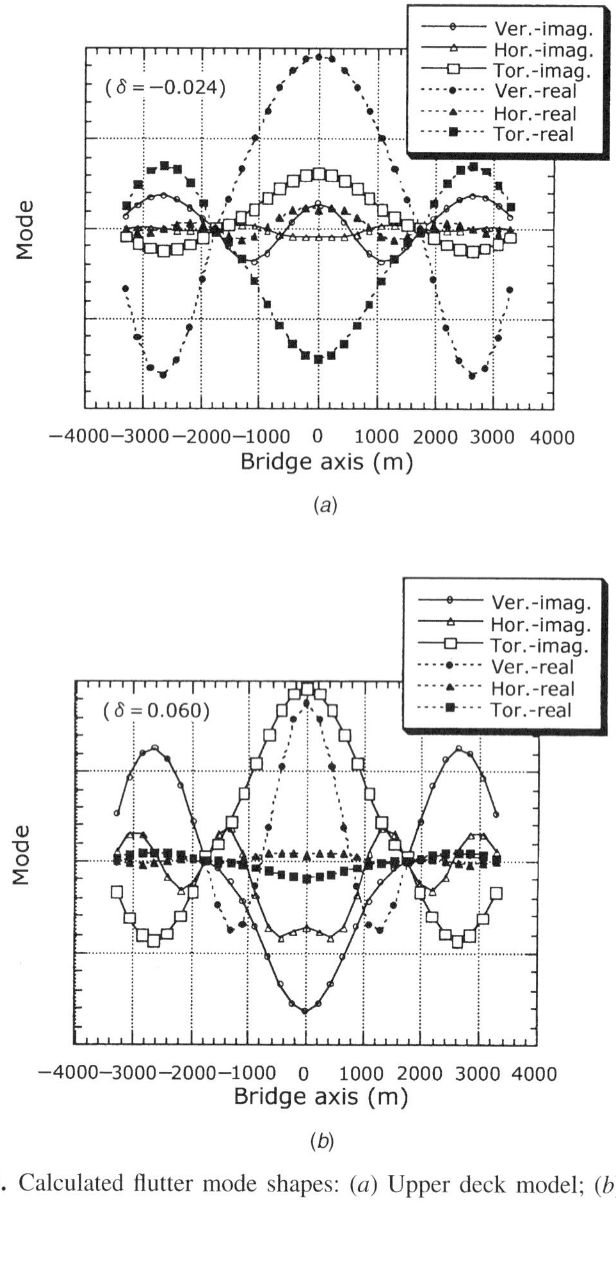

Figure 5B.3. Calculated flutter mode shapes: (*a*) Upper deck model; (*b*) lower deck model.

$$I_i(\ddot{\xi}_i + 2\zeta_i\omega_i\dot{\xi}_i + \omega_i^2\xi_i) = Q_i \tag{5C.2}$$

where the generalized force Q_i is defined by

$$Q_i = \int_0^l [(L_{ae} + L_b)h_iB + (D_{ae} + D_b)p_iB + (M_{ae} + M_b)\alpha_i]\, dx \tag{5C.3}$$

the aeroelastic lift, drag, and moment are given by Eqs. 2.7.5; and the buffeting forces are

$$L_b = \frac{1}{2}\rho U^2 B\left[2C_L\frac{u(x,t)}{V} + \left(\frac{dC_L}{d\alpha} + C_D\right)\frac{w(x,t)}{V}\right] = \frac{1}{2}\rho V^2 B\breve{L}_b(x,t) \tag{5C.4a}$$

$$D_b = \frac{1}{2}\rho U^2 B\left[2C_D\frac{u(x,t)}{V}\right] = \frac{1}{2}\rho V^2 B\breve{D}_b(x,t) \tag{5C.4b}$$

$$M_b = \frac{1}{2}\rho U^2 B^2\left[2C_M\frac{u(x,t)}{V} + \left(\frac{dC_M}{d\alpha}\right)\frac{w(x,t)}{V}\right] = \frac{1}{2}\rho V^2 B^2\breve{M}_b(x,t) \tag{5C.4c}$$

(Eqs. 2.7.9*a*–*c*).

The aerodynamic coefficients and derivatives in Eqs. 5C.3 and 5C.4 should be based on measurements performed in turbulent flow, unless the flow is assumed to be smooth, in which case there is no buffeting by oncoming flow turbulence. Effects of turbulence created by flow–structure interaction (sometimes referred to as *signature turbulence*) are assumed to be negligible. This assumption needs to be verified, however; although effects of vortices shed in the wake of the deck may be assumed to be negligible, this may not be the case if the flow speed being considered is close to the speed associated with lock-in.

The multimodal system of equations of motion of the bridge can be written in matrix notation:

$$\mathbf{I}\xi'' + \mathbf{A}\xi' + \mathbf{B}\xi = \mathbf{Q}_b(s) \tag{5C.5}$$

where ξ is the generalized coordinate vector, $s = Vt/B$, and **I** is an identity matrix. **A** and **B** are the damping and stiffness matrices of the system, respectively, and include terms associated with both structural and aerodynamic damping and stiffness. $\mathbf{Q}_b$ is the generalized buffeting force vector. The terms of matrices **A, B,** and $\mathbf{Q}_b$ are

$$A_{ij}(K) = 2\zeta_i K_i \delta_{ij} - \frac{\rho B^4 l K}{2I_i}\left(G_{h_i h_j}^{H_1^*} + G_{h_i a_j}^{H_2^*} + G_{h_i p_j}^{H_5^*} + G_{p_i p_j}^{P_1^*}\right.$$
$$\left. + G_{p_i a_j}^{P_2^*} + G_{p_i h_j}^{P_5^*} + G_{\alpha_i h_j}^{A_1^*} + G_{\alpha_i \alpha_j}^{A_2^*} + G_{\alpha_i p_j}^{A_5^*}\right) \tag{5C.6a}$$

$$B_{ij}(K) = K_i^2 \delta_{ij} - \frac{\rho B^4 l K^2}{2I_i}\left(G_{h_i \alpha_j}^{H_3^*} + G_{h_i h_j}^{H_4^*} + G_{h_i p_j}^{H_6^*} + G_{p_i p_j}^{P_3^*}\right.$$
$$\left. + G_{p_i p_j}^{P_4^*} + G_{p_i h_j}^{P_6^*} + G_{\alpha_i \alpha_j}^{A_3^*} + G_{\alpha_i h_j}^{A_4^*} + G_{\alpha_i p_j}^{A_6^*}\right) \tag{5C.6b}$$

$$Q_{bi}(K) = \frac{\rho B^4 l}{2I_i}\int_0^l \left[\breve{L}_b(x, s)h_i + \breve{D}_b(x, s)p_i + \breve{M}_b(x, s)\alpha_i\right]\frac{dx}{l} \tag{5C.6c}$$

where δ_{ij} is the Kronecker delta symbol and the modal integrals are given by expressions of the type

$$G_{h_i \alpha_j}^{H_3^*} = \int_0^l H_3^*(K, x)h_i(x)\alpha_j(x)\frac{dx}{l} \tag{5C.7}$$

The coordinate x indexes information that varies along the span. For each vibration mode the angle of attack varies along the bridge. The flutter derivatives are therefore included under the integral sign in Eq. 5C.7, even if the cross section of the deck is uniform along the span. Expressions of the type Eq. 5C.7 imply that the aeroelastic forces are perfectly correlated along the span. This is not truly the case, but it is reasonable to assume that the error inherent in this assumption is small.

The equations of motion obtained by taking into account only diagonal terms ($i = j$) in the matrices **A** and **B** represent single-degree-of-freedom uncoupled equations. The off-diagonal terms (corresponding to $i \neq j$) introduce the aeroelastic coupling.

5C.1 Flutter Analysis

Denote the Fourier transform of ξ_i by

$$\bar{\xi}_i(K) = \lim_{T\to\infty}\int_0^T \xi_i(s)e^{-jKs}\, ds \tag{5C.8}$$

where $j = \sqrt{-1}$. The Fourier transform of Eqs. 5C.5 is

$$\mathbf{E}\bar{\boldsymbol{\xi}} = \overline{\mathbf{Q}}_b \tag{5C.9}$$

where the terms of the matrix **E** are

$$E_{ij} = -\omega^2(B/V)^2\delta_{ij} + j\omega(B/V)KA_{ij}(K) + B_{ij}(K) \tag{5C.10}$$

The flutter condition is identified by solving the eigenvalue problem

$$\mathbf{E}\bar{\boldsymbol{\xi}} = 0 \tag{5C.11}$$

Equation 5C.11 has nontrivial solutions if the real and imaginary parts of the determinant of **E** vanish. The solutions of interest are sought as follows. For a fixed value of K, seek the value of ω in the frequency range of interest until the real part of the determinant of **E** vanishes. Repeat the process for successive values of K until both the real and imaginary parts of the determinant of **E** vanish for the same value of ω. That value of ω is the *flutter circular frequency*. The flutter speed $V_{cr} = \omega B/K$, where K has the value that yielded the flutter circular frequency ω. For a multimodal problem the same procedure must be carried out a number of times equal to the number of modes. The largest solution of K corresponds to the critical flutter condition. The mode corresponding to that solution is the dominant mode in the flutter condition. The eigenvector $\boldsymbol{\xi}$ yields the shape of the flutter oscillatory motion, that is, the relative participation in the flutter motion of each vibration mode. Note that in the linear formulation just described the flutter condition is independent of the buffeting excitation.

The flutter derivative P_1^* may be obtained by equating the following two expressions for the drag:

$$D = \frac{1}{2}\rho(V - \dot{p})^2 BC_D \tag{5C.12a}$$

$$D = \frac{1}{2}\frac{\rho V^2 BKP_1^* \dot{p}}{V} \tag{5C.12b}$$

If the term $\dot{p}^2$ in Eqs. 5C.12 is neglected,

$$P_1^* = \frac{-2C_D}{K} \tag{5C.13}$$

For a streamlined box and an H-shaped section, Eq. 5C.13 was consistent with measurements up to reduced frequencies of $Vn/B = 15$. However, for higher frequencies, Eq. 5C13 was found to underestimate the measured values of P_1^*.

5C.2 Buffeting Analysis

We develop the buffeting analysis with a view to obtaining expressions for the spectral density of the bridge deck response. The Fourier transform of the generalized buffeting force may be written as

$$
\overline{\mathbf{Q}}_b = \frac{1}{2}\rho B^4 l \begin{Bmatrix} \dfrac{1}{I_1}\displaystyle\int_0^l \overline{F}_{b_1} \frac{dx}{l} \\ \dfrac{1}{I_2}\displaystyle\int_0^l \overline{F}_{b_2} \frac{dx}{l} \\ \vdots \\ \dfrac{1}{I_n}\displaystyle\int_0^l \overline{F}_{b_n} \frac{dx}{l} \end{Bmatrix} \tag{5C.14}
$$

$$
\overline{F}_{b_i}(x,K) = \overline{\breve{L}}_b\,(x,K)h_i(x) + \overline{\breve{D}}_b(x,K)p_i(x) + \overline{\breve{M}}_b(x,K)\alpha_i(x,K) \tag{5C.15}
$$

or, in view of Eqs. 5C.3 and 5C.5,

$$
\begin{aligned}
\overline{F}_{b_i}(x,\,K) = \frac{1}{U}\{&[2C_L h_i(x_A) + 2C_D p_i(x_A) + 2C_M \alpha_i(x_A)]\overline{u}(K) \\
&+ [(C_L' + C_D)h_i(x_A) + C_D' p_i(x_A) + C_M' \alpha_i(x_A)]\overline{w}(K)\}
\end{aligned} \tag{5C.16}
$$

Denoting the transpose of the complex conjugate of the vector $\overline{\mathbf{Q}}_b$ by $\overline{\mathbf{Q}}_b^{*\mathrm{T}}$, we have

$$
\overline{\mathbf{Q}}_b\,\overline{\mathbf{Q}}_b^{*\mathrm{T}} = \left(\frac{\rho B^4 l}{2}\right)^2 \begin{bmatrix} \dfrac{1}{I_1 I_1}\displaystyle\int_0^l\int_0^l \overline{F}_{b_1}\overline{F}_{b_1}^{*} \frac{dx_A}{l}\frac{dx_B}{l} & \cdots & \dfrac{1}{I_1 I_n}\displaystyle\int_0^l\int_0^l \overline{F}_{b_1}\overline{F}_{b_n}^{*} \frac{dx_A}{l}\frac{dx_B}{l} \\ & \vdots & \\ \dfrac{1}{I_n I_1}\displaystyle\int_0^l\int_0^{vl} \overline{F}_{b_1}\overline{F}_{b_1}^{*} \frac{dx_A}{l}\frac{dx_B}{l} & \cdots & \dfrac{1}{I_n I_n}\displaystyle\int_0^l\int_0^l \overline{F}_{b_n}\overline{F}_{b_n}^{*} \frac{dx_A}{l}\frac{dx_B}{l} \end{bmatrix} \tag{5C.17}
$$

By definition, the power spectral density of a function φ is

$$
S_{\varphi\varphi}(\omega) = \lim_{T\to\infty} \frac{2}{T}\,\overline{\varphi}\,\overline{\varphi}^{*} \tag{5C.18}
$$

where $\varphi \equiv \varphi(x_A)$ is a stationary stochastic process at a point with coordinate x_A. A similar expression holds for the cross-spectral density of $\varphi(x_A)\varphi(x_B)$. From Eqs. 5C.17 and 5C.18 a spectral density matrix with the following terms can be developed:

$$S_{Qb_iQb_j}(K) = \left(\frac{\rho B^4 l}{2U}\right)^2 \frac{1}{I_i I_j} \int_0^l \int_0^l \{\bar{q}_i(x_A)\bar{q}_j(x_B)S_{uu}(x_A,x_B,K)$$

$$+ \bar{r}_i(x_A)\bar{r}_j(x_B)S_{ww}(x_A,x_B,K)$$

$$+ [\bar{q}_i(x_A)\bar{r}_j(x_B)+\bar{r}_i(x_A)\bar{q}_j(x_B)]C_{uw}(x_A,x_B,K)\}\ dx_A\ dx_B \quad (5C.19)$$

$$\bar{q}_i(x) = 2[C_L h_i(x)+C_D p_i(x)+C_M \alpha_i(x)] \quad (5C.20)$$

$$\bar{r}_j(x = (C_L'+C_D)h_j(x)+C_D' p_j(x)+C_M' \alpha_j(x) \quad (5C.21)$$

$S_{uu}(x_A,\ x_B,\ K)$ is the cross-spectral density of the horizontal wind speed fluctuations u at the spanwise coordinates x_A and x_B, $S_{ww}(x_A,\ x_B,\ K)$ is the cross-spectral density of the vertical wind speed fluctuations w at x_A and x_B, and $C_{uw}(x_{A,}\ x_{B,}\ K)$ is the co-spectrum of u at x_A and w at x_B. It is assumed in Eq. 5C.19 that the quadrature spectrum of u and w is negligible. Aerodynamic admittances are assumed to be equal to unity; in other words, the dependence on frequency of the relation between wind speed fluctuations and the fluctuating forces they induce is neglected. This leads to a slight but probably negligible overestimation of the response.

The power spectral density matrix $\mathbf{S}_{\xi\xi}$ of the generalized coordinate vector $\boldsymbol{\xi}$ is obtained in terms of the matrix $\mathbf{E}$ and the matrix $\mathbf{S}_{Q_bQ_b}$ of the spectral density of the generalized force vector $\mathbf{Q}_b$, that is,

$$\mathbf{S}_{\xi\xi}(K) = \mathbf{E}^{-1}\mathbf{S}_{Q_bQ_b}\ \{[\mathbf{E}^*]^{\mathrm{T}}\}^{-1} \quad (5C.22)$$

The coupling between modes is due to the off-diagonal terms in the matrix **E.** The power spectral density of the displacements, obtained from Eqs. 5C.1, is

$$S_{hh}(x,\ K) = \sum_{i,j} \sum_{i,j} B^2 h_i(x) h_j(x) S_{\xi_i\xi_j}\ (K) \quad (5C.23a)$$

$$S_{pp}(x,\ K) = \sum_{i,j} \sum_{i,j} B^2 p_i(x) p_j(x) S_{\xi_i\xi_j}\ (K) \quad (5C.23b)$$

$$S_{\alpha\alpha}(x,\ K) = \sum_{i,j} \sum_{i,j} \alpha_i(x) \alpha_j(x) S_{\xi_i\xi_j}\ (K) \quad (5C.23c)$$

where the summations are carried out over the number of modes being considered. The mean square values of the displacements are

$$\sigma_h^2(x) = \int_0^\infty S_{hh}(x,\ n)\ dn \quad (5C.24a)$$

$$\sigma_p^2(x) = \int_0^\infty S_{pp}(x,\ n)\ dn \quad (5C.24b)$$

$$\sigma_\alpha^2(x) = \int_0^\infty S_{\alpha\alpha}(x,\ n)\ dn \quad (5C.24c)$$

where $n = \omega/2\pi$ is the frequency.

The cross-spectral density of the horizontal velocity fluctuations u, the cross-spectral density of the vertical fluctuations w, and the co-spectrum of u acting at x_1 and w acting at x_2 may be described empirically by the expressions

$$S_{uu}(x_1, x_2, n) \cong S_{uu}(\omega) \exp\left[-\frac{c_u n|x_1 - x_2|}{U}\right] \tag{5C.25a}$$

$$S_{ww}(x_1, x_2, n) \cong S_{ww}(\omega) \exp\left[-\frac{c_w n|x_1 - x_2|}{U}\right] \tag{5C.25b}$$

$$C_{uw}(x_1, x_2, n) \cong C_{uw}(\omega) \exp\left[-\frac{c_{uw} n|x_1 - x_2|}{U}\right] \tag{5C.25c}$$

respectively, where tentative values of the exponential decay coefficients in Eqs. 5C.25 are to be specified. (According to [5.2], $c_u \cong 15$, $c_w \cong 8$, $c_{uw} \cong 8$. However, the choice $c_u = 8$, which is more conservative, was made in the Wind-Resistant Design Code for the Honshu–Shikoku Bridges (2001) [5.36].) For Eqs. 5C.25*a*–*c*, the following equations were proposed [5.2]:

$$S_{uu}(\omega) \equiv S_{uu}(x, x, \omega) = \frac{200 z u_*^2}{U(1 + 50nz/U)^{5/3}} \tag{5C.26a}$$

$$S_{ww}(\omega) \equiv S_{ww}(x, x, \omega) = \frac{3.36 z u_*^2}{U(1 + 10nz/U)^{5/3}} \tag{5C.26b}$$

$$C_{uw}(\omega) \equiv C_{uw}(x, x, \omega) = \frac{14 z u_*^2}{U(1 + 9.6nz/U)^{2.4}} \tag{5C.26c}$$

where

$$u^* = \frac{kU(z)}{\ln(z/z_0)} \tag{5C.27}$$

is the friction velocity, $k = 0.4$ is von Kármán's constant, z is the height of the bridge deck above water, and it may be assumed that for flow over water the roughness length is $z_0 = 0.003$ m $-$ 0.01 m, say. If the bridge span is over ground rather than over water, z_0 depends on the terrain roughness (see Sect. 2.3.2.1). If topographic effects are significant, micrometeorological measurements should be performed [5.51]. However it should be kept in mind that measurements conducted at low wind speeds, at which stable or unstable stratification effects could be significant, may not be relevant if aeroelastic and buffeting responses occur at high wind speeds. It is emphasized that expressions other than those employed in this section may be used, depending on the choice of the specifying authority.

TABLE 5C.1. Frequencies, Types of Modal Forms, and Modal Integrals for the Golden Gate Bridge

Mode	Frequency (Hz)	Type[a]	$G_{h_i h_i}$	$G_{\alpha_i \alpha_i}$	$G_{p_i p_i}$
1	0.0490	LS	3×10^{-16}	8×10^{-5}	0.33
2	0.0870	VAS	0.3	1.9×10^{-15}	7.7×10^{-15}
3	0.1124	LAS	1.7×10^{-14}	1.24×10^{-2}	0.32
4	0.1285	VS	0.19	1.4×10^{-14}	8.3×10^{-14}
5	0.1340	VAS	0.34	2.7×10^{-14}	6.0×10^{-14}
6	0.1638	VS	0.34	1.8×10^{-14}	4.0×10^{-14}
7	0.1916	TAS	6.7×10^{13}	0.32	0.033
8	0.1972	TS	2.5×10^{-12}	0.18	0.25
10	0.1988	VAS	0.18	9.6×10^{-12}	4.6×10^{-13}
11	0.2021	VS	0.26	8.0×10^{-15}	1.5×10^{-15}

[a] L, lateral; V, vertical; T, torsional; S, symmetric; AS, antisymmetric.

Example: The Golden Gate Bridge[5]

The Golden Gate bridge has a 1280-m center span and two 343-m side spans. The elevation of the deck above water is 67 m. The calculated vibration modes, modal frequencies, and modal integrals for the first 11 modes are given in Table 5C.1, ranked by order of increasing frequency. Static force coefficients are listed in Table 5C.2. A surface roughness length $z_0 = 0.009$ m was used in the calculations. The flutter derivatives were assumed to be independent of position along the span. The flutter velocities were calculated for five distinct sets of flutter derivatives, corresponding to $-5°$, $-2.5°$, $0°$, $2.5°$, and $5°$ mean angles of attack. Figure 5.2.16*a* shows flutter derivatives A_i^* ($i = 1, 2, 3, 4$) obtained in smooth flow from a 1:50 section model of the Golden Gate bridge for a 0° mean angle of attack. Flutter derivatives are shown in [5.2] for a 5° mean angle of attack. Only the flutter derivatives H_i^*, A_i^* ($i = 1, 2, 3, 4$), and P_1^* were accounted for in the calculations.

Buffeting calculations were carried out using flutter derivatives for the

TABLE 5C.2. Static Force Coefficients at 0° Angle of Attack for the Golden Gate Bridge

C_D	C_L	C_M	C'_D	C'_L	C'_M
0.304	0.211	0.0044	0	3.25	−0.177

[5] This example is from [5.40, 5.41].

TABLE 5C.3. Calculated Flutter Velocities for the Golden Gate Bridge (0.5% Modal Damping Ratios)

Angle	−5°	−5°	−2.5°	−2.5°	0°	0°	2.5°	2.5°	5°	5°
Number of Modes	7	8	7	8	7	8	7	8	7	8
K	1.27	1.27	1.28	1.28	1.34	1.34	1.43	1.43	1.49	1.49
ω (rad)	1.18	1.18	1.17	1.17	1.18	1.18	1.19	1.19	1.19	1.19
V_{cr} (m/s)	25.5	25.5	25.2	25.2	24.1	24.1	22.7	22.7	22.0	22.0

bridge at 0° mean angle of attack under a 22.2-m/s mean velocity. The largest values along the deck calculated for 2% modal damping ratios and a mean wind speed of 34 m/s at the deck elevation were σ_{hh} = 0.31 m and $\sigma_{\alpha\alpha}$ = 0.77×10^{-3} rad. Smaller damping ratios would result in larger buffeting response.

For flutter calculations it was assumed that the damping ratios were 0.5%. Calculations showed that under the assumptions used, no flutter occurred for combinations including only the first six modes. Inclusion of the seventh mode, and of the seventh and eighth modes, yielded the reduced frequencies K, the circular frequencies ω, and the critical flutter velocities V_{cr} listed in Table 5C.3 for each of the five mean angles of attack. If a 1% modal damping was assumed in the calculations, the multimodal flutter analysis yielded a flutter velocity of 44.6 m/s rather than 22.0 m/s as was the case for the assumption of 0.5% modal damping.

In addition to a multimodal analysis, a single-mode analysis was performed. Mode 7 yielded the lowest flutter velocity in all cases. The single-mode flutter velocities obtained were almost identical to those obtained by a multimodal analysis based on a combination of seven modes. This result, as well as the shape calculated for the oscillatory flutter motion, suggests that the flutter is dominated by mode 7, which is the first torsional asymmetric mode (i.e., the type of mode in which the flutter of the Tacoma Narrows Bridge occurred).

In addition to section model tests in smooth flow, tests in flow with turbulence intensities 7.4 and 11% were conducted for a 1:150 scale model of half the central span. Depending on the turbulence intensity, the mean flutter velocities were about 10 to 20% higher than the flutter velocity for smooth flow conditions [5.40, 5.41].

REFERENCES

5.1. Japanese Society of Steel Construction, *Wind Effects on Structures,* T. Miyata, et al., eds., Tokyo Denki University Press, Tokyo, 1997 (in Japanese).

5.2. E. Simiu and R. H. Scanlan, *Wind Effects on Structures,* 3rd ed., Wiley, New York, 1996, Sects. 6.4 and 13.1.2.

5.3. R. H. Scanlan and E. Simiu, "Aeroelasticity in civil engineering," in *A Modern Course in Aeroelasticity,* E. A. Dowell, ed., Kluwer Academic, Dordrecht, The Netherlands, 2004.

5.4. A. Okajima, "Strouhal number of rectangular cylinders," *J. Fluid Mech.,* **123,** 379–398, 1982.

5.5. A. Okajima, K. Sugitani, and T. Mizota, "Strouhal number and back pressure coefficients of rectangular cylinders (height to width ratio 1:9)," *J. Mech. Eng. Jpn. B,* **49,** No. 447, 2551–2558, Nov. 1983 (in Japanese).

5.6. A. Okajima et al., "Influence of Reynolds number on aerodynamic characteristics of rectangular cylinders with corner-cut," *J. Wind Eng.* (*Jpn.*), **49,** 1–13, Nov. 1991 (in Japanese).

5.7. T. Miyata and M. Miyazaki, "Turbulence effects on aerodynamic responses of rectangular bluff cylinders," *Proceedings of the 5th International Conference on Wind Engineering,* Fort Collins, CO, J. E. Cermak, ed., July 1979, Vol. 1, pp. 631–642.

5.8. M. Miyazaki, "Suppression of wind oscillations of long span box girder bridge decks," *Int. J. Wind Eng. Ind. Aerodynam.,* **14,** 187–194, 1983.

5.9. T. Miyata, M. Miyazaki, and H. Yamada, "Pressure distribution measurements for wind induced vibrations of box girder bridges," *Int. J. Wind Eng. Ind. Aerodynam.,* **14,** 223–234, 1983.

5.10. G. Schewe, "On the force fluctuations acting on a circular cylinder in cross flow from subcritical up to transcritical Reynolds numbers," *J. Fluid Mech.,* **133,** 265–285, 1983.

5.11. Y. Nakamura, "Bluff body aerodynamics and turbulence," *Int. J. Wind Eng. Ind. Aerodynam.,* **49,** 65–78, 1993.

5.12. A. Okajima and Y. Nakamura, "Flow around circular cylinder with surface roughness at high Reynolds number," *Rep. Res. Inst. Appl. Mech.,* Kyushu Univ., **40,** No. 378, 1973 (in Japanese).

5.13. Y. Nakamura, T. Mizota, and K., Yoshimura, "Aerodynamic three force components on rectangular and H-shape cylinders," *Rep. Res. Inst. Appl. Mech.,* Kyushu Univ., **40,** 245–255, Nov. 1973 (in Japanese).

5.14. Y. Nakamura and Y. Tomonari, "The effects of turbulence on the drag of rectangular prisms," *J. Soc. Aeronaut. Space Sci.,* **19,** No. 44, 1976.

5.15. T. Theodorsen, "General theory of aerodynamic instability and the mechanism of flutter," *NACA Rep.,* **496,** 1935.

5.16. Y. C. Fung, *An Introduction to the Theory of Aeroelasticity,* Wiley, New York, 1955.

5.17. R. L. Bisplinghoff et al., *Aeroelasticity,* Addison-Wesley, Reading, MA, 1955.

5.18. T. Miyata, "Historical view of long-span bridge aerodynamics," *Proceedings of the Engineering Symposium to Honor A. G. Davenport for His 40 Years of Contributions,* London, Ontario, June 2002 (available on CD).

5.19. F. B. Farquharson et al., *Aerodynamic Stability of Suspension Bridges with Special Reference to the Tacoma Narrows Bridge,* Bulletin, University of Washington Engineering Experiment Station, Seattle, WA, 1949–1954.

5.20. A. A. A. C. Wallace, "Wind influence on Kessock Bridge," *Eng. Struct.,* **7,** 18–22, Jan. 1985.

5.21. L. Clements et al., "Design and construction of Kessock Bridge," *Proc. Inst. Civ. Eng.,* Pt. 1, 179–214, Feb. 1986.

5.22. H. Katsura et al., "Vortex-induced oscillation of the Trans-Tokyo Bay Highway Bridge, *Proceedings of the 8th U.S. National Conference on Wind Engineering,* Baltimore, N. Jones, ed., May 1997.

5.23. A. Larsen, "Aerodynamic aspects of the final design of the 1624 m suspension bridge across the Great Belt," *Int. J. Wind Eng. Ind. Aerodynam.,* **48,** 261–285, 1993.

5.24. A. Larsen, S. Esdahl, J. E. Andersen, and T. Vejrum, "Storebaelt suspension bridge: vortex shedding excitation and mitigation by guide vanes," *Int. J. Wind Eng. Ind. Aerodynam.,* **88,** 283–296, 2000.

5.25. J. S. Hay, "The wind induced response of the Erskine Bridge," *Proceedings of the Conference at the Institute of Civil Engineers,* Mar. 1981, No. 6, pp. 81–90, Thomas Telford, London, 1981.

5.26. R. L. Wardlaw, *A Wind Tunnel Study of Modifications to Improve the Aerodynamic Stability of the Long Creek Bridge,* NAE LTR-LA-8, National Research Council of Canada, Toronto, Ontario, 1968.

5.27. A. G. Davenport, N. Isyumov, and T. Miyata, "The experimental determination of the response of suspension bridges to turbulent wind," *Proceedings of the 3rd International Conference on Wind Effects on Buildings and Structures,* Tokyo, 1971, No. IV.54-1, pp. 1207–1219.

5.28. C. Scruton and A. R. Flint, "Wind-induced oscillations of structures," *Proc. Inst. Civ. Eng.,* **27,** 673–702, 1964.

5.29. M. Tatsumi et al., "Vibration control of the main towers of the Akashi Kaikyo Bridge," *Proceedings of the 8th U.S.–Japan Bridge Engineering Workshop,* Chicago, 1992.

5.30. W. J. Duncan, "Flutter of systems with many freedoms," *Aeronaut. Q.,* **1,** 59–76, May 1949.

5.31. Y. Nakamura and T. Yoshimura, "Binary flutter of suspension bridge deck sections," *J. ASCE,* **102,** No. EM4, 685–700, 1976.

5.32. O. Mahrenholtz et al., "Wind-induced oscillations of some steel structures," *Proceedings: Practical Experiences with Flow-Induced Vibrations, IAHR/IUTAM Symposium,* Karlsruhe, Germany, E. Naudascher et al., eds., 1979, pp. 643–649.

5.33. H. Tanaka, "Aerodynamic self-excited force considerations of bridge decks," *Trans. Jpn. Soc. Civ. Eng.,* **1,** Pt. 2, 1969.

5.34. F. Bleich et al., *The Mathematical Theory of Vibration in Suspension Bridges,* Bureau of Public Roads, U.S. Government Printing Office, Washington, DC, 1950.

5.35. A. Selberg, "Oscillation and aerodynamic stability of suspension bridge," *Acta Polytech. Scand.,* Ci 13, 1961.

5.36. Honshu–Shikoku Bridge Authority, *Wind Resistant Design Standard for Honshu–Shikoku Bridges,* Bridge and Offshore Engineering Association, Tokyo, 2001 (in Japanese); English version for Japan Society of Civil Engineers, 2005.

5.37. T. Miyata and K. Yamaguchi, "Aerodynamics of wind effects on the Akashi Kaikyo Bridge," *Int. J. Wind Eng. Ind. Aerodynam.,* **48,** 287–315, 1993.

5.38. T. Miyata, K. Tada, H. Sato, et al., "New findings of coupled-flutter in full model wind tunnel tests on the Akashi Kaikyo Bridge," *Proceedings of the International Symposium on Cable-Stayed and Suspension Bridges,* Deauville, France, Oct. 1994, Vol. 2, pp. 163–170.

5.39. T. Miyata, M. Kitagawa, H. Yamada, et al., "Design considerations on 3-dimensional gust response in the Akashi Kaikyo Bridge," *Proceedings of the International Symposium on Cable-Stayed and Suspension Bridges,* Deauville, France, Oct. 1994, Vol. 2, pp. 171–178.

5.40. A. Jain, N. P. Jones, and R. H. Scanlan, "Fully-coupled buffeting analysis of long-span bridges," *Proceedings of the 9th International Conference on Wind Engineering,* New Delhi, 1995, pp. 962–971.

5.41. P. A. Irwin, J. Xie, and G. Dunn, *Wind Tunnel Studies for the Golden Gate Bridge,* Rowan Williams Davies & Irwin, Guelph, Ontario, June 1995.

5.42. S. H. Tiffany, and W. M. Adams, *Nonlinear Programming Extensions to Rational Function Approximation Methods for Unsteady Aerodynamic Forces,* NASA Technical Paper 2776, 1988.

5.43. N. N. Minh, T. Miyata, and H. Yamada, "Complex-mode buffeting analysis of long-span bridges," *JSCE, J. Struct. Eng.,* **44A-II,** 929–936, 1998.

5.44. N. N. Minh, H. Yamada, and T. Miyata, "Complex-mode buffeting analysis of long-span bridges, II: Extensive formulation and further results," *JSCE, J. Struct. Eng.,* **45A-II,** 1075–1086, 1999.

5.45. H. Katsuchi, N. P. Jones, and R. H. Scanlan, "Multi-mode coupled flutter and buffeting analysis of the Akashi Kaikyo Bridge," *J. Struct. Eng.,* **125,** 60–70, 1999.

5.46. G. Diana, M. Falco, S. Bruni, A. Cigada, G. L. Larose, A. Damsgaard, and A. Collina, "Comparisons between wind tunnel tests on a full aeroelastic model of the proposed bridge over Stretto di Messina and numerical results," *Int. J. Wind Eng. Ind. Aerodynam.,* **54–55,** 101–113, 1995.

5.47. Y. L. Xu, "Buffeting of the Tsing Ma suspension bridge: analysis and comparison," *Wind Engineering into the 21st Century: 10th International Conference on Wind Engineering,* A. Larsen et al., eds., 1999, Vol. 2, pp. 1075–1080.

5.48. T. Miyata, "Significance of aeroelastic relationships in wind-resistant design of long-span bridges," *Int. J. Wind Eng. Ind. Aerodynam.,* **77–78,** 1479–1492, 2002.

5.49. T. Miyata and H. Yamada, "New ideas for detailed investigation on aerodynamic behaviors of long span suspension bridges," *15th Congress Report,* International Association of Bridge and Structural Engineers, Copenhagen, June 16–20, 1996.

5.50. T. Miyata, "Comprehensive discussions on aeroelastic-coupled flutter control for very long span suspension bridge," in *Long-Span Bridge and Aerodynamics,* T. Miyata et al., eds., Springer-Verlag, Tokyo, 1999, pp. 181–200.

5.51. C. Barré, O. Flamand, and G. Grillaud, "The Millau viaduct: special wind studies for an exceptional structure," in *Wind Engineering into the 21st Century, 10th International Conference on Wind Engineering,* A. Larsen et al., eds., A. A. Balkema, Rotterdam, The Netherlands, 1999, Vol. 2, pp. 827–832.

5.52. Y. Fujino, K. Wilde, et al., Rational function approximation of aerodynamic forces on bridge deck and its application to active control of flutter," *Proceedings of the 9th International Conference on Wind Engineering,* New Delhi, 1995, pp. 994–1005.

5.53. Y. Hikami and N. Shiraishi, "Rain induced vibrations of cables in cable stayed bridges," *Int. J. Wind Eng. Ind. Aerodynam.,* **29,** 409–418, 1988.

5.54. M. Matsumoto, N. Shiraishi, et al., "Aerodynamic behavior of inclined circular cylinder: cable aerodynamics," *Proceedings: Bluff Body Aerodynamics and Applications,* Kyoto, N. Shiraishi et al., eds., 1988, pp. 103–112.

5.55. H. Yamada, "Control of wind-induced cable vibrations from the viewpoint of the wind resistant design of cable-stayed bridges," *Proceedings of the International Symposium on Cable Dynamics,* Tokyo, AIM, Belgium, 1997, pp. 129–138.

5.56. H. Yamaguchi and Y. Fujino, "Stayed cable dynamics and its vibration control," in *Bridge Aerodynamics,* A. Larsen et al., eds., A. A. Balkema, Rotterdam, The Netherlands, 1998, pp. 235–253.

5.57. H. Langsø and O. Larsen, "Generating mechanisms for cable-stay oscillations at the Farø bridge," *Proceedings of the International Conference on Cable-Stayed Bridges,* Bangkok, 1987, pp. 1023–1033.

5.58. O. Flamand, "Rain–wind induced vibration of cables," *Int. J. Wind. Eng. Ind. Aerodynam.,* **57,** 353–362, 1995.

5.59. T. Yoshimura, A. Inoue, K. Kaji, and M. G. Savage, "A study on the aerodynamic stability of the Aratsu Bridge," *Proceedings of the Japan–Canada Workshop on Bridge Aerodynamics,* 1989, pp. 41–50.

5.60. T. Yoshimura, M. Savage, and H. Tanaka, "Wind-induced vibrations of bridge stay-cables," *Proceedings of the International Symposium on Cable Dynamics,* AIM, Liege, Belgium, 1995, pp. 437–444.

5.61. J. A. Main and N. P. Jones, "Full-scale measurements of stay cable vibration," *Wind Engineering into the 21st Century: 10th International Conference on Wind Engineering,* A. Larsen et al., eds., 1999, Vol. 2, pp. 1075–1080.

5.62. M. Matsumoto, T., Yagi, H. Shirato, et al., "Field observations of wind-induced cable vibrations using a large-scale inclined cable model," *Proceedings of the 11th International Conference on Wind Engineeing,* Lubbock, TX, D. A. Smith and C. W. Letchford, eds., 2003, pp. 2149–2156.

5.63. S. Cheng, P. A. Irwin, J. B. Jakobsen, J. Lankin, G. L. Larose, M. G. Savage, H. Tanaka, and C. Zurell, "Divergent motion of cables exposed to skewed wind," *Proceedings of the 5th International Symposium on Cable Dynamics,* Santa Margherita, Italy, AIM, Belgium, 2003, pp. 271–278.

5.64. M. Miyazaki, "Aerodynamic control method for vibration of bridge cables," *Proceedings of the First World Conference on Structural Control,* Los Angeles, U.S. Panel on Structural Control Research, July 13–22, 1995, pp. WA 4.

5.65. *Technical Report on Aerodynamics of Stay Cables of Cable-Stayed Bridges,* Honshu–Shikoku Bridge Authority and Japan Institute of Construction Engineering, 1988 (in Japanese).

5.66. *Technical Report on Stay Cable Vibration Control of a Long-Span Cable-Stayed Bridge,* Part II, Appendix, Honshu–Shikoku Bridge Authority and Public Works Research Center, 1993 (in Japanese).

5.67. T. Miyata, H. Yamada, and H. Katsuchi, "Wake-induced flutter of twin cables and its countermeasure," *Proceedings: Bluff Body Aerodynamics and Applications V,* Ottawa, Ontario, 2004, pp. 381–384.

5.68. The Highways Agency, UK, *Design Manual for Roads and Bridges,* Vol. 1, *Highways Structures: Approval Procedures and General Design,* Sect. 3, *General Design,* Part 3 BD 49/01 *Design Rules for Aerodynamic Effects on Bridges,* May 2001.

5.69. B. W. Smith, T. A. Wyatt, et al., *Background to the Development of BD 49/01: Design Rules for Aerodynamic Effects on Bridges,* TRL Report 528, TRL Ltd., Workingham, Berkshire, UK, 2002.

5.70. B. Isaksen, E. Strommen, and K. Gierdin-Smith, "Suppression of vortex shedding vibrations at Osteroey Suspension Bridge," *Strait Crossings 2001: Proceedings of the 4th Symposium on Strait Crossings,* Bergen, Norway, Sept. 2–5, 2001.

5.71. T. Miyata, I. Okauchi, et al., "Preliminary design considerations for wind effects on a very long-span suspension bridge," *Proceedings of the 7th International Conference on Wind Engineering,* Aachen, Germany, July 1987; *Int. J. Wind Eng. Ind. Aerodynam.,* **29,** 379–388, 1988.

5.72. M. Ohashi, T. Miyata, I. Okauchi, et al., "Considerations for wind effects on a 1,990 m-main span suspension bridge," *Prereport: 13th Congress of the International Association of Bridge and Structural Engineers,* June 1988, pp. 911–916.

5.73. R. L. Wardlaw, "The improvement of aerodynamic performance," in *Aerodynamics of Large Bridges,* A. A. Balkema, Rotterdam, The Netherlands, 1992, pp. 59–70.

5.74. A. Tanaka, N. Nomura, and C. Miyashita, "Aerodynamic stability of a long span box girder," *12th Congress Report, International Association of Bridge and Structural Engineers,* Vancouver, British Columbia, Canada, 1984, pp. 163–174.

5.75. H. Ruscheweyh, "Vortex-induced vibration," in *Wind-Excited Vibrations,* H. Sockel, ed., Springer-Verlag, New York, 1994, pp. 51–84.

5.76. A. Jain, *Multimode Aeroelastic and Aerodynamic Analysis of Long-Span Bridges.* Doctoral thesis. Department of Civil Engineering, Johns Hopkins University, Baltimore, MD, 1996.

5.77. K. R. Wilson, "Wind resistance design in a large scale bridge project," in *Long-Span Bridge and Aerodynamics,* T. Miyata et al., eds., Springer-Verlag, Tokyo, 1999, pp. 231–250.

5.78. B. M. Pacheco, Y. Fujino, and A. Sulekh, "Estimation curve for modal damping in stay cables with viscous damper," ASCE, *Journal of Structural Engineering,* **119,** No. ST6, 1961–1979, 1993.

5.79. T. Miyata, H. Yamada, and T. Hojo, "Aerodynamic response of PE stay cables with a pattern-indented surface," *Proceedings of the International Symposium Cable-Stayed and Suspension Bridges,* Deauville, France, Oct. 1994, Vol. 2, pp. 515–522.

5.80. T. Miyata, H. Yamada, T. Fujiwara, and T. Hojo, "Wind resistant design of cables for the Tatara Bridge," *Symposium Report, International Assocication of Bridge and Structural Engineering,* Kobe, Japan, 1998, pp. 51–56.

5.81. T. Miyata, "Full model testing of large cable-supported bridges: a state of the art in wind engineering," *Proceedings of the 9th International Conference on Wind Engineering,* Wiley Eastern, New Delhi, 1995, pp. 249–280.

5.82. T. Nomura and T. J. R. Hughes, "An arbitrary Lagrangian–Eulerian finite element method for interaction of fluid and a rigid body," *Comput. Methods Appl. Mech. Eng.,* **95,** 115–138, 1992.

5.83. T. Nomura, "A numerical study on vortex-excited oscillations of bluff cylinders," *Int. J. Wind Eng. Ind. Aerodynam.,* **50,** 75–84, 1993.

5.84. T. Tamura and Y. Itoh, "Three-dimensional vortical flows around a rectangular cylinder in unstable oscillations," *Int. J. Wind Eng. Ind. Aerodynam.,* **67–68,** 141–154, 1997.

5.85. A. Larsen and J. H. Walther, "Aeroelastic analysis of bridge girder sections based on discrete vortex simulations," *Int. J. Wind Eng. Ind. Aerodynam.,* **67–68,** 253–265, 1997.

5.86. M. Kato, "Two-dimensional flow analysis around stationary and oscillating square cylinder by using a modified k–ε model," *J. Jpn. Soc. Civ. Eng*., No. 577/I-41, 217–230, 1997.

5.87. S. Kuroda, "Numerical simulation of flow around a box girder of a long span suspension bridge," *Int. J. Wind Eng. Ind. Aerodynam.,* **67–68,** 239–252, 1997.

5.88. R. P. Selvam, M. J. Tarini, and A. Larsen, "Computer modeling of flow around bridges using LES and FEM," *Int. J. Wind Eng. Ind. Aerodynam.,* **77–78,** 643–651, 1998.

5.89. K. Shimada and T. Ishihara, "Prediction of aeroelastic vibration of rectangular cylinders by a k–ε model," *J. Aerosp. Eng.,* **12,** No. 4, 122–135, 1999.

5.90. C. B. Jensen et al., "Computational methods for FSI simulations of slender bridges on high performance computers," *Proceedings of the International Symposium on Computational Methods for Fluid–Structure Interaction,* Tapir Publishers, 1999, pp. 31–40.

5.91. S. Kuroda, "Numerical calculations on unsteady forces on a long span bridge by a two equations turbulence model," *J. Jpn. Soc. Civ. Eng.,* No. 654/I-52, 377–387, 2000.

5.92. S. Kuroda, "Numerical computations of unsteady flows for airfoils and non-airfoil structures," AIAA 2001-2714, *31st AIAA Fluid Dynamics Conference,* Anaheim, CA, June 2001.

5.93. L. Thiesemann, D. Bergmann, and U. Starossek, "Numerical and experimental evaluation of flutter derivatives by means of the forced vibration method," *Proceedings of the 11th International Conference on Wind Engineering,* D. A. Smith and C. W. Letchford, eds., Lubbock, TX, 2003 (available on CD).

5.94. European Committee for Standardization, *Eurocode 1: Basis of Design and Actions on Structures*—Part 2–4: *Actions on Structures—Wind Actions,* DD ENV 1991-2-4, 1997.

5.95. P. Sarkar, *New Identification Techniques Applied to the Response of Flexible Bridges to Wind,* Department of Civil Engineering, Johns Hopkins University, Baltimore, 1992.

INDEX